George H. Stankeyffle

Marjatta Hytönen

9.20

Multiple–use forestry in the Nordic countries

Marjatta Hytönen (ed.)

1995

METLA, The Finnish Forest Research Institute

Cover photo: A view from Sotkamo in eastern Finland by Erkki Oksanen/ METLA.

Distribution: The Finnish Forest Research Institute, Library. Address: P.O.Box 18, FIN–01301 Vantaa, Finland. Phone: +358–0–8570 5580, fax: +358–0–8570 5582, email: library@metla.fi.

ISBN 951–40–1421–9

Paper: Designer's Matt Art 100 gsm/m^2 (the paper is entitled to use the Nordic environmental label).

Printed in Finland by Gummerus Printing, Jyväskylä 1995.

Foreword

This publication is a result of Nordic research cooperation under the Nordic Council of Ministers and its Forest Research Cooperation Committee SNS (Samarbetsnämnden för Nordisk Skogsforskning). Since the 1970s, the SNS has financed Nordic research in many fields of forestry. Because of growing interest in non–timber products and services, the SNS established a working group on multiple–use forestry in 1986.

After several meetings and seminars, the working group decided to produce a book on multiple–use forestry to enhance multifunctional forestry research, education and practice in the Nordic countries, and also to increase international information exchange and contacts. With this in mind, the book has been written in English.

The SNS, the Nordic Council of Ministers, and forest research organizations from each Nordic country accepted the proposal to write the book in 1991. About half of the costs of the project have been financed by the SNS and the Council of Ministers, and the rest is covered by the national research organizations.

The steering group of the project was formed by Dr. Aarne Reunala from the Finnish Forest Research Institute, who also acted as its chairman, Dr. Christina Axelsson Lindgren from the Institute of Landscape Planning of the Swedish University of Agricultural Sciences, Director Sigurdur Blöndal from the Iceland Forestry Service, Professor Lars Helge Frivold from the Department of Forestry of the Norwegian Agricultural University, and Director Niels Elers Koch from the Danish Forest and Landscape Research Institute. Later, Dr. Katarina Eckerberg from the Department of Political Science of the University of Umeå joined the group.

The steering group prepared an outline for the contents of the book and invited researchers from all the countries involved to cover the topics. All authors have made an extra effort to write not only about their own country, but also about the other Nordic countries. Research assistants helped the authors to get information on their fields. The assistants were Christina Axelsson Lindgren in Sweden, Sigurdur Blöndal in Iceland, Jann Fernand in Norway, Marjatta Hytönen in Finland, and Lene Kristiansen in Denmark. At a later stage of the project, Marjatta Hytönen was appointed as the editor of the book.

The following referees have helped the authors with their comments: Oluf Aalde, Martti Aarne, Jørund Aasetre, Jette Baagøe, Monica Bennett–Gårdø, Paul Christensen, Finn–Egil Eckblad, Lars Emmelin, Raija–Riitta Enroth, Olof Eriksson, Jan Falck, Ann–Katrine Geelmuyden, Risto Heikkilä, Jan Heino, Pekka Helle, Finn Helles, Eeva Hellström, Hans Fredrik Hoen, Erik Holmsgaard, Sven–G. Hultman, Hanne Hübertz, Liisa Kajala, Harri Karjalainen, Matti Kärkkäinen, Matti Leikola, Michael Linddal, Harto Linden, Mirja Miettinen, Arto Naskali, Timo Nikunen, Nils–Erik Nilsson, Matti Palo, Börje Pettersson, Päivi Piispa, Jens Nytoft Rasmussen, Arne

Rognmo, Jyrki Salmi, Risto Savolainen, Heikki Seppälä, Yrjö Sevola, Tuija Sievänen, Birger Sohlberg, Håvard Steinsholt, Anne Toppinen, Bo Wallin, Martin Welp, Esa–Jussi Viitala and Lars Östlund.

We are grateful for the contribution of all the people and institutions which have been involved in preparing this book. Compiling a publication by many authors and referees from different countries has been a rewarding experience, but, also, much more time consuming than expected in the planning phase of the project. Finally, warm thanks are due to Arja Suokas and Anna–Kaisu Korhonen for the lay–out of the book, and to Allan More for correcting the language.

Helsinki, March 1995

Aarne Reunala　　　　Marjatta Hytönen

1 Introduction

The character of forestry in northern Europe is changing. The economic significance of timber production and the pulp and paper industry is decreasing in relation to other branches of the economy. New trends affecting traditional wood–based livelihoods include increasing production of wood in southern countries where it grows faster, reduction of wood fibre content in paper, and recycling of paper products. In Finland, Norway and Sweden, the supply of wood has exceeded demand for decades, and the gap has been widening further during recent years. People in Denmark and Iceland have always been less dependent on wood as a source of living. In all these countries, the recreational and nature protection functions of forests are gaining in importance because of growing environmental awareness. These changes allow for the elaboration and promotion of non–material forest benefits. They also call for new economically profitable uses of wood and non–wood forest resources.

From the global point of view, the Nordic countries may look quite similar to each other with a cool and humid climate, and high standard of living. Still, they are very different from each other. The functions of forests in densely populated Denmark, in almost treeless Iceland, in mountainous Norway and in sparsely inhabited conifer–dominated Finland and Sweden are partly the same, but also very specific to their location. Moreover, the northern parts of the area are inhabited by the Sámi people, whose culture crosses the national borders.

The naming of the book was problematic. There is no exact geographical term which would cover the five countries involved. The often used term "Scandinavia" usually refers only to Denmark, Finland, Norway and Sweden. Because of historical reasons, Iceland has close relations to the Scandinavian countries and participates in many forms of cooperation. The forms which include Iceland are referred to by the word "Nordic". Because of this tradition, the term is also used in this book. A few authors use the term Fennoscandia. It refers, most often, to Norway, Sweden, Finland and the northwestern part of Russia.

The chapters of this book have been written during the period 1991–1994. A lot has happened since starting the preparation of the publication. However, all the chapters are well rooted in their historical context and provide valuable information for the future. A few of the details are likely to change soon, mainly due to institutional rearrangements. Furthermore, forestry and other rural livelihoods are being affected by strong external forces. For example, air pollution, climate change, environmental movements, trade regulations and international agreements all have their impacts on forests and utilization of forest resources.

One of the biggest challenges facing the Nordic societies today is the high unemployment rate together with the cutback in the state–financed welfare systems, which have been largely based on public organizations. Other present trends include the privatization and commercialization of state–owned enterprises, which also affects to some degree the organizations responsible for forests, and the spreading of decision–making power to local level. As in many other regions, interest conflicts concerning the utilization and protection of forest resources are common in the Nordic countries and they are widely covered by the media.

Simultaneously with domestic restructuring, the Nordic societies are adapting to European integration. Denmark has been a member of the European Union since 1973. Finland and Sweden became members in 1995. Iceland and Norway have decided to stay outside the Union for the time being. Consequently, the character of Nordic cooperation, which has been carried out through hundreds of more or less formal organizations and arrangements, is changing and new contacts are being established. The new cooperation partners, in addition to increased contacts with central Europe, include the Baltic states of Estonia, Latvia and Lithuania. Cooperation with the northwestern parts of Russia is promoted, among other things, by the Barents treaty, agreed upon in 1993 by the European Union, Russia and the Nordic countries.

Most of the information presented in this book is based on research results. The articles mainly discuss the traditional fields of multiple–use forestry, namely wood and non–wood forest products, nature conservation, recreation and cultural issues. Employment problems and forest–related interest conflicts in the Nordic and neighbouring countries will increase the need for research on the social and livelihood aspects of forest resources. In addition, there is an evident need for more emphasis on research on integrated multiple–use forestry management and associated forest policy measures.

Urbanization has been affecting the Nordic societies for a long time and has led to geographical differentiation of values. The interests of the urban majority are threatening the cultural identity and livelihood opportunities of people living in the countryside. This confrontation was clearly demonstrated by the referendums of Finland, Norway and Sweden in 1994, in which most rural people voted against joining the European Union, while urban people voted for it. Thus, a relevant future research topic will be the role of multiple–use forestry in bridging the gap between the culture in rural and urban areas.

Forest–related livelihoods and hobbies have been and still are an essential part of the Nordic culture. They are not disappearing, but instead being reshaped according to the new values of people and changes in social structures. This book is the most complete overview of Nordic multiple–use forestry in English until now. Still, it cannot give a very detailed description of all the fields of multiple use. For a deeper study, many of the chapters include comprehensive reference lists containing more specific sources. The authors and their organizations can also provide further information. It is to be hoped that this book can serve as a source of inspiration for more multiple–use discussion, research and practical applications.

2 A short history of forest uses

Bo Fritzbøger[1]
Poul Søndergaard[2]

Abstract

The last ice age ended in Scandinavia about 15,000 years ago. The land was gradually occupied by the present tree species. Human settlement followed the vegetation. The first forest uses were hunting and gathering. Animal husbandry and forest grazing came later. Shifting cultivation was a widespread form of agriculture, especially in Sweden and Finland. Wood was first used for domestic purposes and for construction. In the 18th century, wood became a commercial product. Overcutting, grazing and agricultural cultivation caused deforestation and degradation of forests. These problems accelerated the creation of forest policy. The principles of sustained yield wood production were first applied in Scandinavia in the 18th century. In the 19th century, forests became a source of raw material for the forest industry. Recently, increasing urbanization has created the need for renewed contact with nature and led to the development of modern multiple–use forestry management.

Keywords: ice age, forests, grazing, agriculture, deforestation, logging, forest policy.

---

[1] University of Copenhagen
Department of History
Njalsgade 102, DK–2300 Copenhagen S, Denmark

[2] Royal Veterinary and Agricultural University
Department of Botany, Dendrology and Forest Genetics, Arboretum
Kirkegårdsvej 3 A, DK–2970 Hørsholm, Denmark

2.1 Introduction

The extensive woodlands of Scandinavia have for thousands of years provided human society with a wide range of indispensable natural resources. Traditional forest uses have always been many–sided, whereas the one–sidedness of modern wood production is a relatively recent phenomenon. The multiplicity of uses reflected the great variety of woodland types ranging from the vast coniferous plains of inland Fennoscandia (Finland, Norway and Sweden) over the southwestern tracts of deciduous forests to the birch forests of Iceland and southern Greenland. Many forest uses have induced a number of changes in the forest ecosystems in a continuous interaction with the general development of society. These man–made changes became a threat to human society in many areas and eventually gave rise to the introduction of management systems to assure sustained yield from the forests.

However, during recent decades, the public demand for recreation and contact with nature has changed forest uses from mainly wood production to new kinds of multiplicity.

Regional diversities

When the ice started its retreat at the end of the last glaciation, the Weichselian, about 15,000 C–14 years ago, the main forest tree species began to recolonize central and northern Europe. Birch and pine were the predominant species, followed by oak. Later came lime and elm, attracted by a still warmer climate, which reached its postglacial peak about 8,000 calendar–years ago. At that time, the primeval forests were dominated by lime and oak, and they covered most of the soils in southern Scandinavia. In northern Scandinavia and on poorer soils in the south, the most common species were pine and birch.

About 4,000 years ago, spruce entered Finland from Russia, and continued its westward expansion into Sweden and Norway. At approximately the same time, beech entered Scandinavia from the south (*Figure 2.1*). Gradually spruce became a dominant component in the forests of middle and northern Fennoscandia, whereas it never reached Denmark and southwest Norway by natural dispersal. Eventually, beech attained a dominant position in the broadleaved forests of southernmost Scandinavia, while lime and elm lost their importance and disappeared from many areas. Thus, 2,000 years ago, the present forest zones of northern Europe were more or less established.

Denmark, southernmost Sweden, and a fringe of western Norway belong to the central and western European deciduous forest zone (Rubner 1960), in which the most characteristic species are beech (*Fagus sylvatica*), pedunculate oak and sessile oak (*Quercus robur* and *Q. petraea*). Central Sweden, southeast Norway and a fringe of southern Finland are covered by mixed forests of broadleaves and conifers, of which the most important species are Norway spruce (*Picea abies*), Scots pine (*Pinus sylvestris*), pedunculate oak, silver

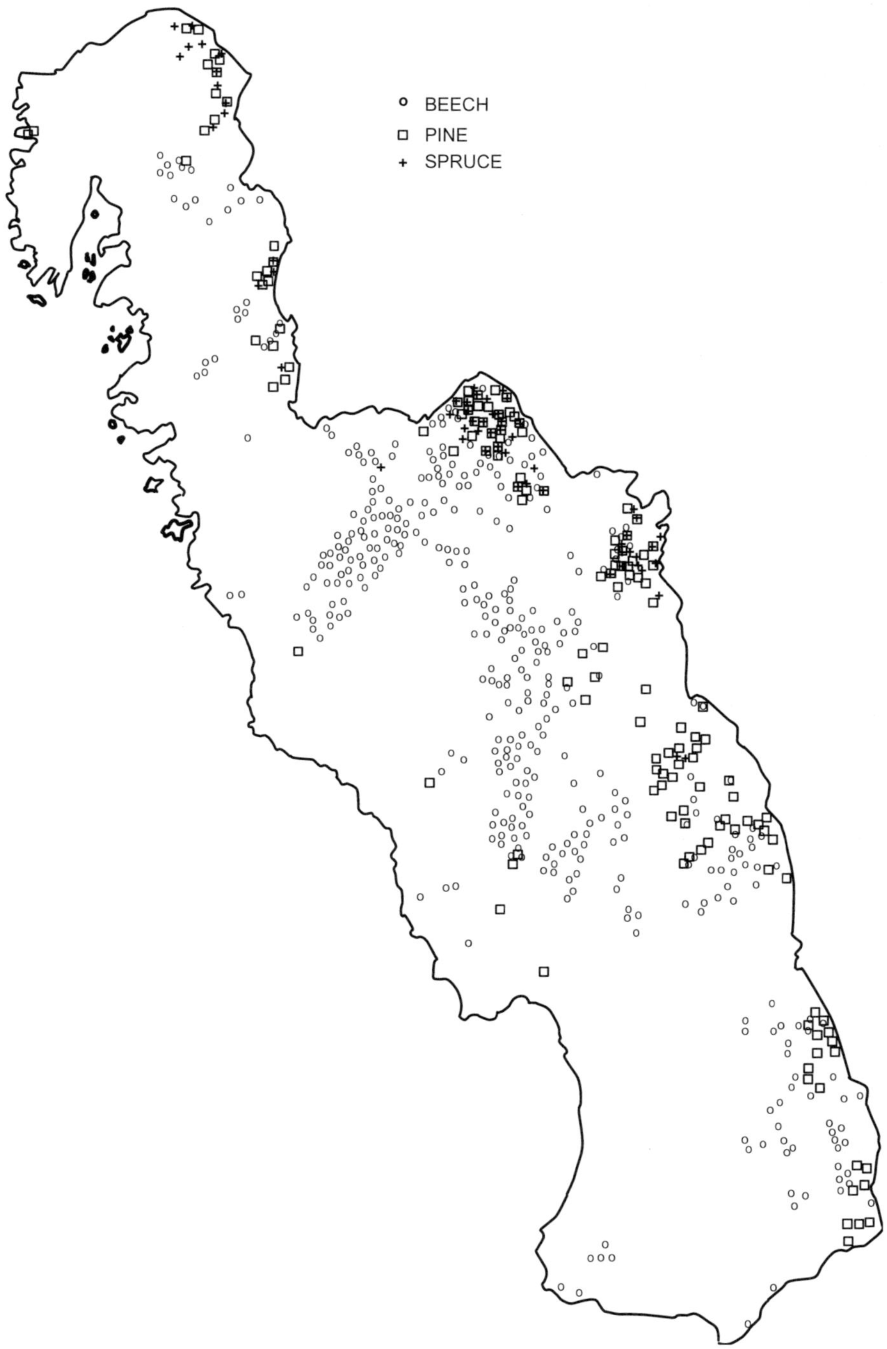

Figure 2.1 Distribution of stands of beech, pine and spruce in the southwestern Swedish province of Halland according to forest inventories in 1729. The area is situated on the northeastern edge of the natural dispersal of *Fagus sylvatica* and on the southern border of *Picea abies*. Spruce seems to have abstained from dispersal into the coastal lowland where saline winds may have impeded its growth (Malmström 1939).

birch (*Betula verrucosa*) and downy birch (*Betula pubescens*). The remaining parts of Finland, Norway and Sweden belong, with minor exceptions, to the North European conifer–birch region, which covers approximately 90 % of the three countries. This zone is the western outreach of the Russian Taiga, or the Boreal Conifer Zone with Scots pine, Norway spruce, silver birch and downy birch as the dominant species.

Iceland, the Faeroe Islands and Greenland belong to the Nordic sphere of cultural and political influence. Consequently, their traditions in forestry and tree planting are closely linked to Scandinavia. Iceland, the Faeroe Islands and southernmost SW–Greenland can be classified as a potential subarctic boreal forest zone. This has so far materialized in birch climaxes (*Betula pubescens s.l.*) in Iceland and SW–Greenland, while the Faeroe Islands have had no natural forests in historical times.

Continuity and change

The immense contrasts in the natural geography of Scandinavia have been reflected in the regional patterns of forest uses. In spite of such differences, the development of the relationship between man and forest seems to have followed the same four fundamental stages (Kardell 1988):

1) As long as population was small and nomadic, and woodland resources correspondingly superfluous, people's elementary requirements for wood were easily satisfied. Access to utilize forest resources was therefore free.

2) As soon as population density increased and settlements became more stable, local overcutting and overgrazing imperceptibly produced cultural landscapes with unequally distributed woodland resources. The limited goods of these woodlands (timber, fuel, etc.) were treated and valued as (potential) commodities, and their continuous production had to be protected by legal regulations.

3) These merely inactive, protective measures were rarely sufficient to maintain a sustained yield of wood. As a result of greater innovative reforms in land ownership and forestry technology, trees were domesticated and bred in active silviculture.

4) The ultimate phase in the development of silviculture was the introduction of new species, selection, improvement and genetic engineering.

Transition through these roughly outlined stages did not take place concurrently in all parts of Scandinavia. Their chronology was, so to speak, displaced from southwest towards northeast. Several forest uses which had been abandoned in Denmark, southern Sweden and western Norway during prehistoric or medieval times still had great importance in 19th century eastern Finland. This delay was not only caused by extremely dissimilar geographical conditions. Man's approach towards the forest, and consequently his influence on its reproduction, can, broadly speaking, be regarded as an interaction between four interdependent factors.

It is evident that density of population is of immediate importance to the impact of man on nature. For example, an increasing population pressure encourages new clearings and settlements, thus reducing the extent of woodland. With unaltered technologies, the consumption of woodland resources will develop in direct proportion to the population. Hence, the most densely populated areas of Scandinavia were the first to experience shortage of wood. Secondly, not only the number of people but also their technological abilities have always determined forest uses and their ecological consequences. Thirdly, demands from abroad can determine the kind and intensity of forestry. Since the Middle Ages, the international economic setting of trade relations and market trends accordingly became a still more perceptible factor in the shaping of forestry. Finally, the socio–political context (e.g. property forms and legislation) has, since the Late Middle Ages, developed into a more and more dominant factor in the shaping of forestry. Political regulation and even the formulation of property rights seem primarily to have followed local scarcity of wood and specific demands of the market.

2.2 Traditional multiple–use forestry

Settlements and clearings

Already by the beginning of this millennium man had for generations manipulated the landscape in which he lived and died. This impact became still more perceptible with the development of a more stable settlement structure in southern Scandinavia following, among other things, the building of Christian village churches during the Early Middle Ages (10th–12th centuries). Around new hamlets or single farms and in the fringes of old village grounds, woodland was cleared for crops and pasture. With these clearings, the fundamental distinctions between woodland and plain were formed.

The core areas of Scandinavian settlement have remained largely unaltered since the Middle Ages, whereas the density of population has varied greatly. In southwest Scandinavia, the demographic boom of the 10th–13th centuries was followed by the setback of the Black Death (1349–50). The repercussions of this Late Medieval decrease were not counterbalanced until the 16th–18th centuries, when a number of extremely destructive wars caused a new temporary decrease. The gross population development AD 1000–1993 was, nevertheless, positive. Whereas the Danish average 13th century population density has been estimated as about 20 persons/km^2, Norwegian, Swedish and Finnish population density did not exceed 2. Around 1850, the average densities still varied from 40 persons/km^2 in Denmark to 8 in Sweden and 5 in Finland and Norway.

Figure 2.2 Distribution of the Norwegian population in the 1660s. Each dot represents 1,000 persons in rural districts; each square a town with the number of inhabitants not given (Tuxen & Hellesen 1988).

Within the four countries, the population was very unevenly distributed. Finland was until 1809 a province of Sweden and during the following 110 years a Russian grand duchy. Early settlement took place in the southern river valleys, where considerable areas were cleared around large stationary villages. In Sweden, the largest concentration of settlements were found in the great plains of the central Svealand and in the southernmost province of Scania, which was until 1658 a part of Denmark. The Norwegian population was primarily situated along the fjords of the west coast and in the fertile river valleys north of Oslo (*Figure 2.2*). In many areas of western and southern coastal Norway, southern Sweden and Jutland, large parts of the indigenous wildwoods had been replaced by *Calluna* heathland already during the neolithic period.

Forests as a resource for hunting and gathering

An abundance of animal and plant resources formed the basic means of prehistoric human subsistence in the deciduous and coniferous wildwoods of Scandinavia. But ever since the Neolithic Revolution, beginning in southern Scandinavia some 6,000 calendar years ago and gradually spreading towards the north and east, the importance of hunting, picking of wild fruits and berries, and gathering of mushrooms and bark declined.

Both hunting and gathering nevertheless continued to form an important supplement to agriculture and animal husbandry in times of famine. For example, flour substitutes made from bark were well known all over Scandinavia. There are numerous local examples of the general economic importance of these ancient means of subsistence, especially in some parts of northern Scandinavia. Until the emergence of Dutch–American trappers during the 16th century, northern Fennoscandia and Russia were the main international suppliers of furs. From the southernmost islands of Denmark, considerable quantities of nuts were exported during the 16th and 17th centuries to the Hanseatic cities south of the Baltic, and so collecting of hazelnuts was a notable factor in the local economy.

Together with the woodland in which they lived, the game population first decreased in the densely populated southern Scandinavia. Mainly for this reason, hunting in Denmark and southern Sweden was increasingly regarded as a royal or noble sport. Since the Late Middle Ages the liberty to hunt, especially large game, was notably restricted. These limitations were partly caused by certain royal and noble ideological pretensions. The chase, so to speak, replaced the tour de force of medieval tournaments. However, the enormous consumption of venison at court could alone explain the royal interest in hunting (*Figure 2.3*).

The legal restraints on common hunting rights produced a number of special royal hunting domains. The English term "forest", known also in the Latin "Foresta", the German "Forst" and the French "forêt", originally referred to a hunting domain belonging to the king or another magnate. Thus, not necessarily to an area of woodland. Such game preserves were established in Denmark during the High Middle Ages and could consist either of rather small territories ("dyrehaver" (anglo–saxon: "derhage") or somewhat more sizeable tracts "vildtbaner"). Both types normally seem to have been placed in regions dominated by woodland, and the special attention paid to the wild forest animals considerably influenced the composition and reproduction of the woodland.

In most of northern Scandinavia, the hunting rights of independent farmers on the other hand remained fairly untouched. In 1729, Swedish and Finnish farmers obtained hunting rights similar to those of the nobility. During the 19th century, Danish hunting monopolies were dissolved parallel with gradual political liberation and the formation of constitutional democracy. In the vacuum between zealously guarded restrictions and modern protective measures,

Figure 2.3 Allegorical representation of winter. In the foreground a noble lady and her children enjoy the heat from open fire. In the background a log is being dragged from the forest by two horses while a noble master and his hounds chase a wild boar. Ceiling painting from the Danish manor Næsbyholm, dated approximately AD 1580. (The National Museum, Copenhagen).

19th century hunting experienced a veritable boom. In Denmark, poaching seems to have become a non–negligible menace to forest owners.

Animal husbandry and pastoralism in the forests

Animal husbandry has not up till now shown any specific preference for the coniferous forests of northern and eastern Scandinavia, but in the deciduous forest zones there is an old affinity between livestock and woodland. The extent of arable land cleared in the forest by neolithic or medieval settlers in general remained modest, and livestock consequently was of great importance. In many places, animal husbandry was even so closely related to the presence of wood pasture that manorial rents from forest settlements normally consisted of butter, meat, cheese or living animals rather than grain or money.

The woodland first of all provided extensive areas of permanent pasture among scattered trees. During the summertime, large herds of domestic animals occupied the forests, equally grazing the forest floor and browsing its trees and bushes. In southern Scandinavia, tree cover protected the herb and grass layer against the desiccating sunlight, so that grass production of wood pastures in general seems to have been more stable than that of other types of pasture.

Originally herdsmen must have followed the herd in order to protect it against predators in the vast prehistoric wildwoods. As the degree of cultivation and the extent of stock raising soon underwent notable regional differentiation, a diversity in pasture forms developed. First of all, the duration of the summer pasture varied significantly with the latitude.

In mountainous regions, the livestock was normally relocated from the homestead to extensive areas of mountain pasture ("sæter", "fæbod") with thickets of stunted birch trees. In the conifer–birch zone south of the tundra, forest glades were used as pasture. The dark coniferous forest was not a very hospitable environment for herbs or grasses, nor to grazing animals. But since the "slash–and–burn" cultivation (see below) practised in these regions included an interval of forest regeneration dominated by broadleaved pioneer species like birch, some livestock could be fed in the tree covered infields near the settlement.

In southern Scandinavia at least three different historical categories of wood pasture can be discerned:

1) Stretches of commons between the enclosed fields of individual villages were normally used for permanent pasture and would often be wooded.

2) Substantial cattle grazing took place in the fields after harvest. Since arable land in large parts of Scandinavia as late as the 19th century was characterized by the intermingling of field strips with moist, infertile depressions, thickets and scattered, old, solitary trees, even this could be typified as "wood pasture".

3) Finally, in woodland regions, by the Middle Ages each settlement had acquired their own woodlots which in turn were used as fenced paddocks. Such "hagar" were a particularly distinct feature of central Scandinavia.

Concerning animal husbandry, broadleaved trees represented a far greater diversity of uses than conifers. Consequently, deciduous woodland was normally located in the infields surrounding the settlements, whereas conifers were confined to the outfields. This feature was, and still is, particularly conspicuous in the birch–conifer dominated zones of northern–central Scandinavia.

"Pannage", i.e. the fattening of domestic swine in the forest, was exclusively related to the presence of deciduous forest. As in most parts of western and central Europe, this royal and noble pannage regulation produced a rich source material which could give the misleading impression that precisely <u>this</u> kind of forest use was the most important. Every summer all woodlots were surveyed according to their prospective yield of acorns and beech mast. Following the survey, the appropriate number of swine was counted and branded before being led to the forest, where they spent the last months of their lives until the traditional great Christmas slaughter (*Figure 2.4*).

The vegetation of the forest floor did not contribute to livestock reproduction only during the summer pasture season. Especially in northern Scandinavia, fenced and wooded meadows corresponding to the grazing paddocks secured production of winter fodder. Even though meadows were equally indispensable throughout Scandinavia, the wooded type had rather different functions in the north and south. In the south, hay and straw seem, at least since the Middle Ages, to have provided sufficient fodder for the wintertime, and trees in the meadows were mainly coppiced to produce fuelwood. In the north, where the winter was longer and harder, and where production of the meadows did not meet the demand for stall feeding, tree leaves were cut down and dried to be used as highly nutritious fodder. Special knifes for this purpose from the Iron Age have been excavated in western Norway. Primarily elm, birch and ash were used as leaf fodder, and in general these species, together with hazel, lime and oak were preferred by browsing livestock. On the other hand, alder, willow and beech did not appeal to most animals.

From the area of present day Denmark, scant evidence of post medieval use of leaf fodder exists, even though its application in southern Sweden is amply recorded. In years with bad hay making or harvest failures, the foliage and twigs of trees formed an appreciated substitute everywhere. 18th and 19th century evidence furthermore indicates that leaves have been used as fertilizer in both vegetable gardens and forest nurseries, whereas the central European traditions for "Streunutzung" (collection of litter) do not seem to have been practised in any part of Scandinavia.

Figure 2.4 Forest swine as depicted in O. Brunfels: Contrafayt Kreüterbuch, 1535.

Agriculture and woodland

The widespread ideological antagonism between woodland and arable fields symbolizes man's struggle against nature. The two represent primeval wilderness vs. civilized society. This explanation is equally promoted by archaic myths and modern alienation from nature. It overlooks the fact that the close,

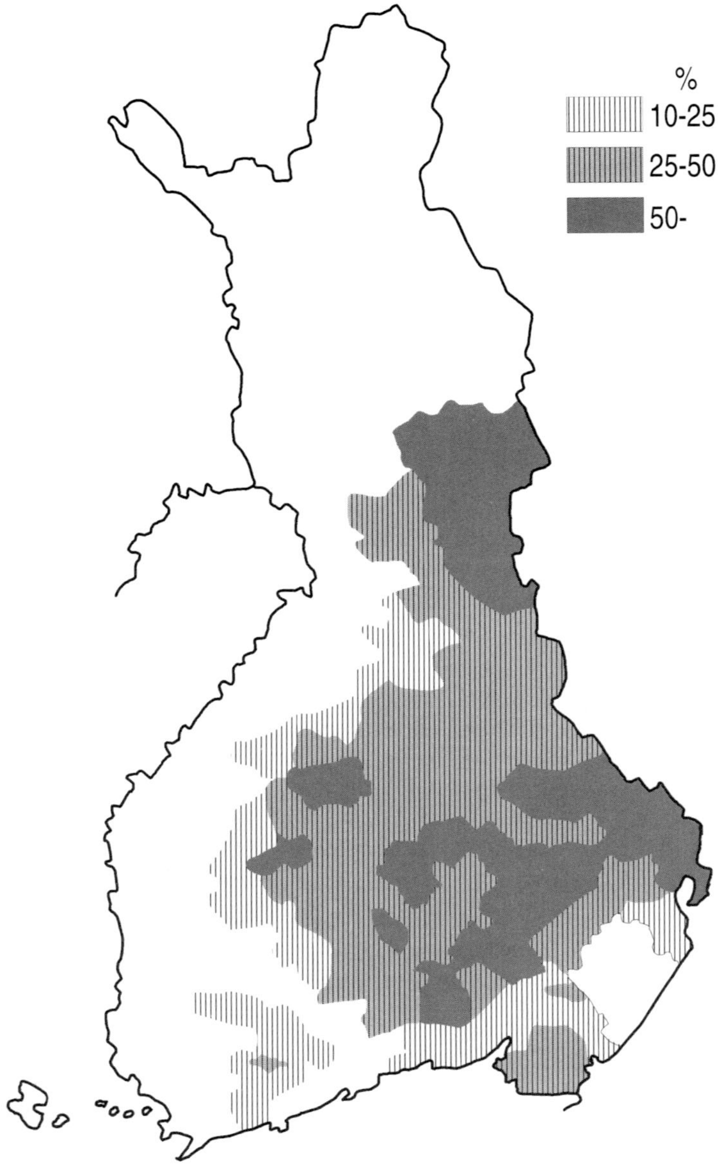

Figure 2.5 Areas of Finland under slash–and–burn cultivation in the 1830s. The rye yield from this type of agroforestry is expressed as a percentage (per parish) of the total rye crop (Linkola 1988).

original connection between deciduous woodland and arable farming was dissolved only a few centuries ago.

In regions dominated by woodland, no clear distinction between farm land and forest was discernible. Throughout large parts of northern Scandinavia, clearing and regrowth alternated in various field systems based on long periods of fallow. Everywhere, the wooded village outskirts served as buffers between cultivated infields and extensively utilized commons. In times of demographic and economic expansion, trees and thickets were cleared to give way to increased cultivation. When recession prevailed, these marginal fields were once more left uncultivated and converted into new woodland.

Fire has been used in forest clearance since the very beginning of farming and animal husbandry, and was one of man's most efficient tools to control and suppress forests. In Finland and Sweden it was still used extensively in a "burnbeating", "slash–and–burn" or "swidden cultivation" system as late as the early 20th century. The system was particularly well developed in Finland, and was at times of utmost importance for the country's provision with grain (*Figure 2.5*).

Both deciduous and coniferous forests seem to have provided the necessary conditions for this type of agriculture. Cultivation was rotational, and, especially in coniferous forests, it demanded many years of hard labour to have merely one to four crops of rye. First, the trees were killed by stripping off their bark. After a period of about ten years, when an understorey of broadleaves was established, the burning took place. The crop, usually rye, was then sown in the ashes, and, after reaping, the area was used as pasture for a number of years. During this period it reverted to forest, and the rotation could be repeated (*Figure 2.6*).

The "slash–and–burn" culture opened up the dark Fennoscandian spruce forests and made room for broadleaved pioneer species such as birch and al-

Figure 2.6 Slash–and–burn rye cultivation photographed in Kihniö in the 1930s. Photo: E. Nikkilä 1934. (Museovirasto, Helsinki).

der. In the vicinity of rural settlements, the landscape was therefore characterized by a far greater diversity than in the primeval coniferous woodland. This improved the living conditions for game, making hunting a feasible supplement to agriculture.

No matter how the alternation between a woodland state and cultivated clearings was organized, it generally seems to have improved soil quality. Even without the vegetation being burnt down. Earthworms are attracted by the litter from trees and bushes and its decomposition leads to an accumulation of nutrients in the topsoil. Furthermore, tunnels made by earthworms seem to contain large quantities of nitrogen–fixing bacteria. In cases where trees were cut down on a regular basis (as coppices or pollards), a subsequent decomposition of roots would liberate additional nutrients.

In densely populated parts of southern Scandinavia, there rarely were sufficient woodland to exercise these various methods of long–term fallow agroforestry. Where access to common pastures made animal husbandry superior to agriculture, it was possible to maintain a system of permanently sown fields. Grazing and browsing livestock were brought back to the farm at milking time and during the night. In this way, nutrients from huge areas outside the village or farm were concentrated through manure from the livestock. The majority of

Figure 2.7 Peasants taking a rest while mowing a meadow with scattered standards and surrounded by dense thickets. The open field landscape dominating southern Scandinavia from the Middle Ages until the great Land Reforms in about 1800 was characterized by the coexistence of several landscape types in the same area. In this case, the standards belonged to the landlord. The remote thickets were normally coppiced by the peasant, with little or no interference by the landlord. Finally, the produce of the meadow beneath and in between the trees would pertain to the members of the village community. A drawing from the mid–18th century, attributed to C. A. Lorenzen. (The Museum of National History, Frederiksborg, Denmark).

these pastures were wooded, at least until the major 16th century deforestation (*Figure 2.7*).

Preindustrial wood production

The production of firewood and timber has always depended on the accessibility, amplitude and quality of woodland resources. Since these resources, as a reflection of the general settlement structure, were rather unevenly distributed, wood production was organized in a number of different ways. And as local demands increased, wood production was further differentiated and, at times, forests were devastated. However, overall development can not simply be described as a transition from thoughtless exploitation to modern sustained yield forestry.

Archaeological findings of large quantities of homogeneous poles and twigs in fences and road surfaces seem to indicate that some kind of underwood management providing a sustained and regular production of wood was well known in prehistoric times. It is thus very likely that the basic principles of sustained yield wood production were developed several thousand years ago. Through the centuries, still more sophisticated varieties of rotational coppice management evolved. The regulated "Niederwaldwirtschaft" based on a rotation among well defined underwood compartments, and widespread in central and western Europe by the Middle Ages, apparently did not penetrate southern Scandinavia until the advent of modern forestry during the late 18th century.

The greater part of wood production was used for domestic purposes by the rural population living in or nearby the forest. Until the 19th century, trees were normally cut down with axes and the further splitting or division into beams and boards was carried out with different types of saws and axes. Where wood production was not influenced by world market demands, there seems to have been little or no technological innovation since the Middle Ages. The diffusion of, for example, new felling techniques, using saws instead of axes was evidently very slow.

Wood is a truly multi–functional material. Apart from meeting domestic demands for timber and fuelwood, forest trees provided raw materials for a great variety of local manufacturers. Coppices furnished the corn producing plains with poles and twigs to fence the fields against livestock and game. In coastal areas near Copenhagen, a quite considerable manufacturing of hoops took place. From the Russian Taiga, birch bark techniques for basketry etc. were imported to Finland, and then Finnish settlers passed them on to Sweden and the East–Norwegian Finnskog.

In some localities early industries made very substantial demands on fuelwood. Most conspicuous were the Swedish and Norwegian mining districts, and the extensive Finnish regions of tar production, all developed during the Middle Ages but expanded immensely during the 16th century. In general, none of these enterprises seem to have caused serious deforestation. They all provoked latent social skirmishes between industrial entrepreneurs and the local peasant population wanting to use the woodland in the same way as the generations before them. It was in connection with the manufacturing of charcoal for the Swedish mining industry that the first government attack on extensive swidden cultivation was launched. The last notable confrontation between agriculture and wood production took place in Norrland less than one hundred years ago and in some parts of eastern and northern Finland just before World War I.

Metal was originally extracted from deposits in bogs all over Scandinavia, but from the 16th century production was concentrated around Norwegian and Swedish ores of iron, silver and copper. Extraction of bog iron made a strong impact upon the forests in West Scandinavia, where the deposits were particularly well developed. The metallurgical industries connected with the mining areas also used large quantities of charcoal when manufacturing the ore. During the second half of 19th century approximately 5,000,000 m^3 of wood were annually converted to charcoal in Sweden. Charcoal was not exclusively burned in mining areas. Coals could be burned in both deciduous and coniferous forests and even in Denmark charcoal burning seems to have been widespread at least until the 17th century. With the extraction of ore (and particularly iron ore), pressure on the forests was transferred from western Scandinavia, which was by then almost woodless, to the mining regions in southern Norway and central Sweden (*Figure 2.8*).

Wood as a commodity

Already in prehistoric times some parts of northern Jutland in Denmark apparently relied on imported Norwegian fuelwood and turf. At least until the

Figure 2.8 The location of mines (squares) and forges (dots) in the Swedish county of Värmland, north of Lake Vänern in 1600, 1700 and 1800. Originally the forges were placed near the iron deposits in the eastern part of the county. As this profitable industry was built up it gradually spread all over the county (Rydbo & von Hofsten 1984).

Middle Ages woodland resources generally seem to have been distributed so that local scarcity could always be met by purely regional trade. As demands from abroad increased, wood production of many coastal areas of western Scandinavia nevertheless went into export. This submission to mercantile demands contributed to the introduction of new technologies which unavoidably affected the subtle balances of traditional peasant forestry.

As mentioned above, southwest Norway served as a timber and fuel exporter even in prehistory. The production process was slow and incapable of creating any shortage of wood. Timber was cut with axes, and normally a full grown tree would be cut into 2 boards. In the very first years of the 16th century, the first water driven sawmills appeared in Norway. They made it possible to divide a tree into 7–8 boards at a much higher speed than the old cutting, and within less than a century this simple technological innovation made timber export to western Europe flourish. In their quest for raw materials, sawmill owners gradually capitalized forest ownership. The escalating demands soon led to deforestation of vast areas along the coast making river floating from inland forests to the coastal ports imperative.

In the centuries when Norwegian timber supplies supported the building and maintenance of the naval powers such as Great Britain, the Netherlands and the Danish–Norwegian monarchy itself, the same maritime demands made tar the most dominant export article of Finland. Much of the artillery of that same bellicose era originated from Swedish iron works using gigantic quantities of local charcoal. Because of incessant international demands, the manufacturing of these three woodland products (timber, tar, charcoal/iron) thus came to determine forest uses in large parts of early modern Scandinavia (*Figure 2.9*).

The production of tar and potash demanded huge quantities of wood, and tar burning from pine during the 18th century became one of Sweden's three major export industries. This trade climaxed during the following century. In Finland, the eastern parts first constituted the main area of tar production, but as later happened with timber, production from about 1720 moved northwards to Ostrobothnia and later to Kainuu and southern Lapland. The burning of potash from birch wood experienced a similar dislocation during the 19th century, and total annual exports of potash from the Västerbotten region amounted to no less than 4,500 tons (*Figure 2.10*).

2.3 Forestry, forest policies and woodland development

Woodland as a cultural landscape

Historical and regional differences among the woodland ecosystems of Scandinavia did not only reflect natural diversities and climatic changes. They

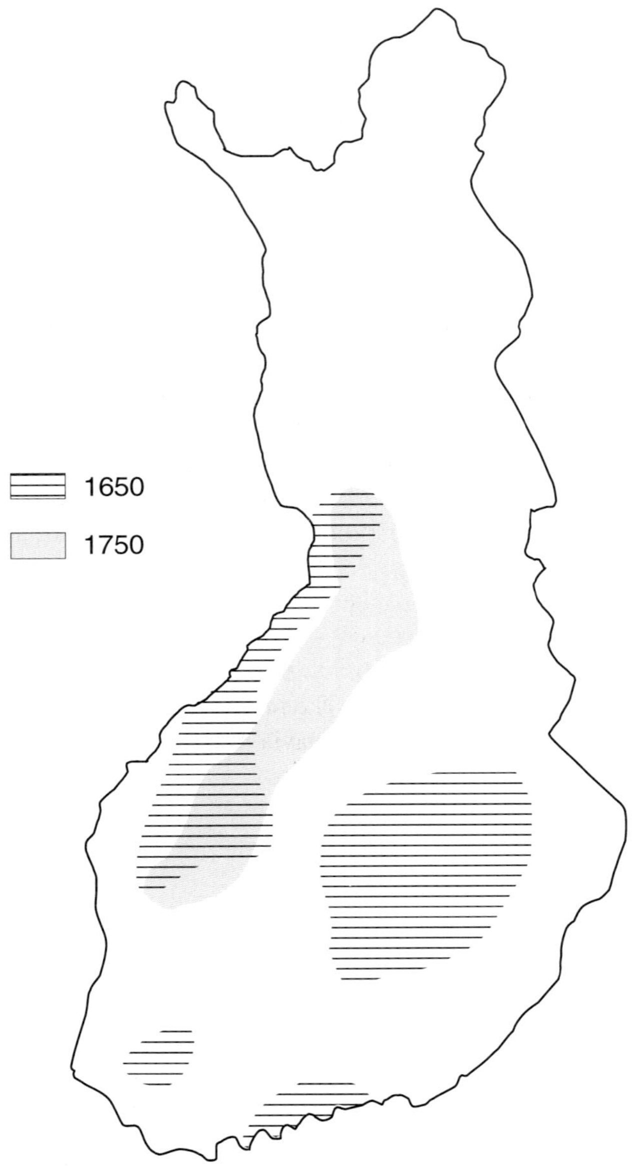

Figure 2.9 Areas in Finland with intensive tar production in 1650 and 1750. As forest resources decreased the main areas of tar production were displaced from central–eastern Finland towards more easily navigable areas in the Northwest (Appelroth 1987).

also mirrored anthropogenuous effects originating from the immense variety of forest uses. The woodland formed a cultural landscape.

Apparently prehistoric hunters had resorted to burning down stretches of woodland in connection with hunting of elk and wild reindeer. During the 17th and 18th century, the establishment of royal and noble deer parks in southern Scandinavia produced open forests consisting of scattered, old trees, the branches of which characteristically ended at a browse–line approximately one and a half meters above the ground.

Many forms of management affected the composition of species. The Finnish swidden cultivation produced a local dominance of broadleaved pioneer species such as birch, aspen and rowan. Only when the rotation period for various reasons was prolonged, the original coniferous state would reoccur. Heathland areas in western Scandinavia were recurrently burned down in order to rejuvenate the heather plants, on which agriculture was based. The fire served to select the tree species, and as a consequence thickets of oak prevailed whereas beech in general disappeared.

Also pannage seems to have affected the distribution of species significantly. Swine would break up the soil and, since they avoided seedlings of beech, they favoured the growth of this species. In Denmark and southern Sweden, the gradual dominance of beech since the Iron Age was likely brought about by forest swine. During the 18th and 19th centuries, forest swine were deliberately used to reproduce stands of beech.

Deforestation

The most radical effect of traditional multiple–use forestry was the disappearance of extensive areas of woodland. The process of deforestation was neither equally intensive in all parts of Scandinavia nor one of incessant destruction.

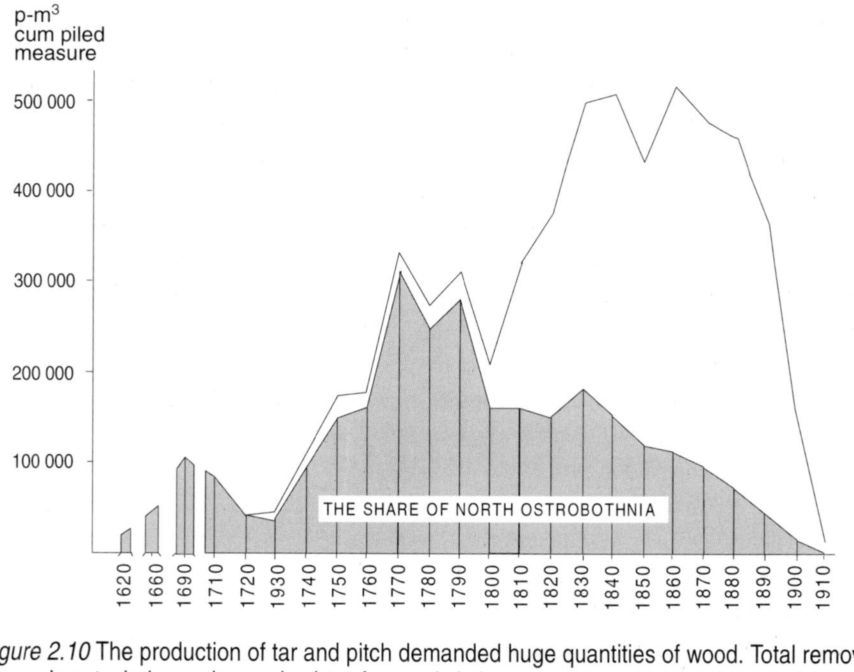

Figure 2.10 The production of tar and pitch demanded huge quantities of wood. Total removal of growing stock due to the production of tar and pitch exported from Oulu and Raahe from the 1620s to 1910, and the share of North Ostrobothnia (Alho 1968).

Its momentum differed, as mentioned above, according to factors such as the density of population, technological abilities, economic setting and the socio–political organization of society. The first sign of serious deforestation was evident in a shortage of timber and fuelwood.

Historical records seemingly establishing a shortage or even deficiency of wood products are not altogether credible. Alarming outcries in 17th century literature most frequently attest fears for the future rather than immediate scarcity. Defective supplies did not necessarily reflect inadequate resources but could also be the result of an exorbitant price level or insufficient means of transportation.

Clearings for settlement and agriculture had the most permanent impact on the distribution of Scandinavian woodland. Consequently, deforestation accelerated in areas and periods of population pressure and wherever market trends favoured corn–growing. In southern Scandinavia, this was the case both in the 16th and the 18th centuries. The conclusive reduction of the Danish forest acreage from 7 % to 4 % during the last decades of 18th century took place in conjunction with radical land reforms.

The most remarkable deforestation took place in Iceland. By the time of the Norwegian settlement ("Landnam") in the 9th and 10th centuries, the vegetation was largely dominated by birch forests. The greater part of this primeval woodland disappeared through the Late Middle Ages as the result of an enormous demand for fuelwood (e.g. for the significant Icelandic salt production and burning of charcoal) combined with burdensome sheep grazing. Today only insignificant relics are preserved.

To talk about regional or even general lack of wood in northern Europe during the Middle Ages is obviously an exaggeration. Examples of local wood shortage generally occurred in Scandinavia one or two hundred years later than in central Europe. Fuel deficiency had become a problem in German mining areas during the late 17th century and legislative and silvicultural measures were taken to protect their surrounding forests. Early Swedish and Norwegian decrees (during the first half of the 17th century) were aimed particularly at the naval shipyards and the mining areas (as in Central Europe), while decrees from the latter part of that century had more general objectives. The Danish Forest Acts, of which no less than 6 were issued between 1665 and 1733, sought to protect the forests in general, but with the royal inclination towards hunting and game keeping as strong underlying motives.

This difference between Norway and Sweden, on the one hand, and Denmark, on the other, can be explained by the fact that Denmark very early on became a timber importing country. Danish export of oak was prohibited for strategic reasons during the 15th century, while the export of wood products such as tar and potash early became important sources of income for both Norway and Sweden. Whereas there was no real risk of timber shortage in the Swedish monarchy, Denmark rested heavily upon Norwegian supplies.

Causes of deforestation

Forests do not disappear simply because the major trees are cut down. Hence the exploitation of forest trees for charcoal, fuelwood or timber cannot by itself explain the temporal or permanent deforestation of large parts of (especially southern and western) Scandinavia. This negative process could be understood only as the concurrent effects of several forest uses. Only the combination of, for example, poor soil, excessive cutting and constantly immoderate grazing or deliberate grubbing and cultivation could lead to a total conversion of woodland to treeless plain.

Woodland is the favourite habitat of many Scandinavian game species, and hunting is the oldest of all forest uses. Partly because of this affinity, the royal Danish administration of hunting and forestry remained closely connected until the late 18th century. Game preservation was one of the most significant duties of the forest rangers. Consequently, during the 18th century many attempts to introduce silviculture in Danish forests were hampered by enormous population of protected deer. Modern forestry thus was unable to make its mark until effective campaigns of deer elimination had been carried through in the 1790s. On the other hand, it cannot be ignored that larger predators such as the wolves, widespread in most of Scandinavia (but eliminated in Denmark in about 1820), could have curtailed local attempts to use wood pasture, thus protecting the forest against harmful domestic animals.

Browsing cattle are normally regarded as the chief accelerator of the incontestable deforestation of southern and western Scandinavia during the 17th and 18th centuries. Even if this explanation is basically sound, it needs modification.

A great number of southern Scandinavian wood pastures <u>did</u> degenerate to stands of scattered old bollings and lifeless trunks or even to treeless plains, simply because cattle hampered regeneration and the establishment of a vigorous understorey. However, this development was not universal. Several examples can be found of wooded commons surviving the continuous grazing pressure.

The detrimental effects to trees and bushes depended first and foremost upon the actual number of grazing animals. Secondly, the grazing period had an influence on the amount of damage, since young sprouts are most vulnerable during the springtime. Thirdly, the animal species significantly determined the extent of tree browsing. For example, the goat, being a renowned tree killer, was banned from royal Danish forests in the 16th century. Finally not all trees were equally damaged by browsing animals, partly because of the animals' likes and dislikes, and partly owing to different species' dissimilar power of resistance. Oak being on the one hand more resistant than beech but on the other much more favoured as a nutrient. Together these factors largely differentiated the effect of wood pastures upon the tree cover.

In most areas of eastern Denmark and southern and east–central Sweden, the arable land rotated annually between states of cultivation and fallow. In years of fallow as well as every autumn after harvest, arable land was used as pasture. Since fields normally consisted of mosaics comprising a multitude of small and large trees and bushes, they also formed a sort of forest pasture. In contrast to the extensive commons, it seems that the cultivated infields generally were better preserved than the outfields. The disappearance of trees in the infields was usually due to escalating agricultural needs rather than to damage caused by browsing animals.

Even though wood production could not by itself cause major deforestation prior to the 19th century, forestry was, in large parts of Scandinavia, dominated by an increasing disparity between wood production and regrowth. This imbalance was most evident in southern Scandinavia, where the rapidly growing population placed ever increasing demands on fuelwood and timber. The timber exporting coastal regions of Norway and the mining districts of central Sweden also experienced shortages. During the 16th and 17th centuries, a series of royal acts decreed forest preservation and silvicultural measures, but generally they were ineffectual. The main reason for this vivid legislative activity was not the alleged, all–embracing wood shortage but rather fear induced by the prospects of failing supplies for an expanding centralistic state apparatus. Thus, forestry soon turned into one of the most favoured domains of absolutist power play.

The political economy of deforestation

First the need for timber, and then the problem of fuel shortage, formed the background for an ideologically tinted forest concern formulated as politics and legal intervention. The first instances appeared in Denmark during the late 15th century. During the following centuries wood shortage became a serious problem in large areas of southern and western Scandinavia.

Since the earliest royal acts of the Middle Ages, the primary purpose of forestry regulation was to maintain the woodland resources by securing their reproduction, and to safeguard an unimpeded fuel and timber supply for the Crown. This objective first became clearly formulated when and where scarcity (initially of timber) threatened. Even in 15th century Norway, the government would incite farmers to practise swidden cultivation in order to increase arable land area, whereas the forest laws of the following century explicitly prohibited this practice, for the benefit of escalating timber exports. Meanwhile this export was zealously regulated for strategic reasons. As can be learned from modern Third World experiences, restrictions in traditional forest uses only function when the average living standard of the population has reached a certain minimum level. So, for centuries protective efforts were mostly futile.

At times, other interests than protection and woodland regeneration influenced forest politics. In large parts of northern Scandinavia, the early modern states encouraged colonization of uninhabited woodland areas for a number of reasons. In Sweden, to secure a supply of charcoal for the iron works and in Finland to establish permanent settlements, that could be taxed as farmland by the monarchy.

Another consequence of the imminent scarcity of timber and fuelwood was the severely limited access to utilize the forests of notably densely populated areas during the 15th and 16th centuries. Whereas all sorts of forest uses during the Prehistorical and Medieval era of expansion had been common goods, the development of property rights, and escalating social and political inequality, induced decisive restrictions on access to woodland resources. In northern Scandinavia, forestry continued to be free for centuries, but in the south ordinary tenants would normally only have the right to specific forest uses. For instance, the use of timber and mast trees (oak and beech) was restricted to the landlords, while coppice continued as a specific "peasant right".

However, none of the early ventures to control traditional Scandinavian multiple–use forestry managed to prevent accelerating deforestation. Attempts were made to protect the forests and to assure their regeneration, but they only succeeded on a very modest scale. This, again, could be due to the fact that the farmers either considered the forest resources inexhaustible, or that they (in the areas of southern Scandinavia where deforestation was most vigorous) did not own their forests. Furthermore, the confusion of pasture, pannage and wood liberties made it virtually impossible to find a common denominator that could assure an overall strategy which would preserve and improve the forest.

Early modern deforestation caused a distinctive differentiation among the Scandinavian countries. The southern and Atlantic parts became importers of forest products, whereas the Norwegian, the North Swedish and the Finnish forests, through the application of sustained yield principles, seized their present dominant place in the world market of timber.

2.4 The advent of sustained yield forestry

Coppicing was common in early Scandinavian woodland management and has in many areas been used to secure a stable yield of specific wood products. Balanced swidden cultivation also was a form of sustained yield forest use.

The first attempts to introduce a modern system of sustained yield wood production in Scandinavia were made in the 1730s. German forestry experts, headed by the von Langen brothers, were engaged by the Danish/Norwegian king. Their main task was to get the Norwegian forests under a regular management system (with a particular view to mining areas). The intentions were good, and a lot of work was done, not least in producing high quality maps.

However, the forest owners and particularly the Norwegian farmers, could not accept the restraints placed upon their forest uses, and the initiative faded during the 1740s.

A similar attempt was made on a smaller scale in Denmark from 1763, when Johann Georg von Langen was appointed by the Danish King to introduce his forest management system in the royal forests of Denmark. Even if this attempt was far from being carried through, it coincided with Danish land reforms, which eventually separated forestry from farming in Denmark. A new era was opened up in Danish forestry, and sustained yield management principles were adopted for most of Denmark's forests during the first fifty years of the 19th century.

Until the middle of the 19th century, the impact of milling was limited to areas with sufficient and exploitable waterpower. With the advent of steam mills around 1850, mills could be located in any area where the transport of timber was most economical, and milling became independent from fluctuating waterflow in the rivers. The strong increase in export of timber and other forest products led to increased pressure on the forests of Fennoscandia, and at the end of the 19th century deforestation was considered a real threat to society. As late as 1914, Johan B. Bull wrote about the Norwegian forests: "Norwegian timber is declining in dimension as rapidly as the amount of exported timber is increasing, and if an immediate change by rational silviculture is not brought about, the Norwegian realm of forests, protector of our climate, breathtakingly beautiful and highly original will soon be only a saga" (freely translated from Tveite 1964). However, when the Norwegian forests were inventoried during the 1920s, the total yearly increment proved to be about 4 % superior to yearly harvesting (*Figures 2.11, 2.12*).

Similar situations could be found in Finland and Sweden. Accordingly, the Scandinavian forests were not disappearing, but they had lost a considerable part of their big and marketable timber trees. The need for sustained yield management was recognized both by state and private forestry. The silvicultural systems, which had been introduced from Central Europe since the beginning of the 19th century, were now developed and adapted to Scandinavian conditions. Selective cuttings by measures were, from the beginning of the 20th century, replaced by natural regeneration cuttings. With the advent of mechanized logging – in the middle of the 20th century – clearcuttings followed by planting became generalized. These clearcuttings often covered large areas, and they have in many cases proved difficult to regenerate, particularly in environments where small changes in microclimate can be crucial to the establishment of seedlings and young plants. For example, there have been problems in forests near the timberline, which in some areas have been severely reduced due to badly regenerated clearcuttings since the middle of the 20th century.

Sustained yield management was introduced much earlier in Denmark, which had faced the threat of complete deforestation during the late 18th century. The decisive turn was made by the Forestry Act of 1805, by which all

Figures 2.11 and 2.12 The appearance of the waterpowered sawmill at the beginning of the 16th century caused a tremendous increase in the export of timber from Finland, Norway and Sweden. An old sawmill above. Floating was the most important way of transporting timber in Scandinavia until the middle of the 20th century. Photos by M. Risdal (1967) from the Arendal River Basin, Telemark, Norway.

high forests in Denmark were preserved and exploitation of the forests was strictly regulated. Simultaneously, a reforestation programme was launched, slowly at first, but gathering momentum during the latter part of the 19th century, so that the forest area of Denmark by the middle of the 20th century had doubled compared to its area in 1805. New forest tree species had been introduced (e.g. by von Langen in 1763), and in terms of acreage Norway spruce was now the most important species in Danish forestry.

In contrast to Denmark, forest legislation was much more liberal in Finland, Norway and Sweden, first of all because forest products were among the main export articles. The Finnish forest ordinance of 1805, for instance, permitted private owners the free use of their forests as long as the forests were not destroyed and the ability to pay taxes was maintained.

Plantation forestry was started in the other Scandinavia countries during the late 19th century. For example, in western Sweden plantations were established to combat drifting sands. From the beginning of the 20th century, plantation forestry gathered momentum in West Norway, where Sitka spruce (*Picea sitchensis*) became important. Later in the century, plantation forestry became widespread in the rest of Scandinavia in connection with rationalized logging procedures, and more exotic species were introduced into forestry, such as Lodgepole pine (*Pinus contorta*), which has been planted on a large scale, particularly in northern Sweden.

Planting of trees and plantation forestry was initiated in Iceland and on the Faeroe Islands around the beginning of the 20th century. The plantings have been successful, particularly during the last 40 years, even if modest in extent. In southwest Greenland, promising results have also been obtained in planting trials during the last 40 years, but so far with limited commercial prospects.

2.5 Forests for leisure and modern multiplicity

Forests have probably always inspired the human imagination and they have served as a place for refuge in times of trouble. Kings and noblemen combined pleasure and necessity by hunting in the forests, outlaws and persecuted minorities found shelter in the wilderness, people inspired by religious ideals of hermitage combined introspection with woodmanship, and to ordinary people the woodlands represented a vividly imaginative environment. The forest was simultaneously captivating and perilous.

When modern methods of forest management in the 19th century began to change the forest picture of Scandinavia, man's role in the forest changed as well. From being a natural and indispensable part of most peoples lives, the forest became a source and producer of raw material for the forest industry and it gradually became a field of action for specialists. But, like in other industrialized regions of the world, increasing urbanization created a need for renewed contact with nature. This has in many areas submitted the forests to such a strong pressure that new ideas had to be developed, and foresters had to revise many of their traditional views on forest management. In the most densely populated areas of Scandinavia, forestry is now a mixture of park– and landscape management and traditional commercial forestry. A lot of different uses must be reconciled.

This modern multiple–use forestry is completely different from the multiple uses of former times, before the forest was conquered by man. At that time people passed through the forests and left their imprints in the form of winding tracks, traces of work and fire, and they came to the forests all year round, and at all times. Nowadays, people are more or less guided, when they visit the forests. They follow roads or tracks which have been conceived and established according to management plans. Most people do not visit the forests as part of their everyday activity. As a result of this development, the forests of Scandinavia have again become common "property" due to improved means of access, and above all due to a more affluent society.

References

Alho, P. 1968. Pohjois–Pohjanmaan metsien käytön kehitys ja sen vaikutus metsien tilaan. Summary: Utilization of forests in North Ostrobothnia and its effect on their condition. Acta Forestalia Fennica 89. 216 pp.

Andersen, S.T., Aaby, B. & Odgaard, B. 1983. Environment and man: current studies in vegetational history at the Geological Survey of Denmark. Journal of Danish Archaeology 2: 184–196.

Appelroth, S.–E. 1987. Nine thousand years of forests in Finland. Paper presented at the IUFRO s1.05–12 Symposium "Northern Forest Silviculture and Management", in Lapland, August 16–22, 1987. 7 pp.

Bjerke, S. 1957. Nogle træk af de sydskandinaviske løvskoves udvikling i de sidste århundreder. (A few features of the development of South–Scandinavian broadleaved forests during the recent centuries.) Dansk Dendrologisk Årsskrift IV: 373–413. (In Danish.)

Björklund, J. 1984. From the Gulf of Bothnia to the White Sea. The Scandinavian Economic History Review No. 1.

Björklund, J. & Östlund, L. (eds.). 1992. Norrländsk skogshistoria: människan, skogen och industrin. (The history of forest in Norrland: people, forest and industry.) Umeå Universitet/Sveriges lantbruksuniversitet, Umeå. 120 pp. (In Swedish.)

Blöndal, S. 1982. Fremmede treslag i Hallormstad skogområde, Öst–Island. (Foreign tree species in Hallormstad forest area in East Iceland.) Tidsskrift for Skogbruk 1. Oslo. Repr. 11 pp. (In Norwegian.)

Blöndal, S. 1988. Skov og skovdyrkning på Island. (Forests and forestry in Iceland.) Dansk Natur Dansk Skole. Årsskrift 1987/88: 75–86. (In Danish.)

Blöndal, S. 1991. Socioeconomic importance of forests in Iceland. In: Alden, J., Mastrantonio, J.L. & Odum, S. (eds.). Forest development in cold climates. Plenum Press, New York. 13 pp.

Buttenschøn, J. & Buttenschøn, R.M. 1978. The effects of browsing by cattle and sheep on trees and bushes. Natura Jutlandica 20: 79–93.

Carbonnier, C. 1978. Skogarnas vård och föryngring. (Silviculture and regeneration.) In: Skogshögskolan 150 år: problem och ideer i svenskt skogbruk 1828–1978. (The faculty of forestry 150 years: problems and ideas in Swedish forestry 1828–1978.) Sveriges lantbruksuniversitet, Uppsala. p. 85–126. (In Swedish.)

Cate, C.L. ten 1972. Wan god mast gift.... Bilder aus der Geschichte der Schweinezucht im Walde. Wageningen. 300 pp.

Emanuelsson, U. 1987. Översikt över det nordiska kulturlandskapet. (Overview of the Nordic cultural landscape.) In: Biotoper i det nordiske kulturlandskapet. (Biotopes in the Nordic cultural landscape.) Nordiska ministerrådet, Miljörapport 6/ Nordrapport 63:13–52. (In Swedish.)

Emanuelsson, U. 1988. The relationship of different agricultural systems to the forest and woodlands of Europe. In: Salbitano, F. (ed.). Human influence on forest ecosystems development in Europe. Proceedings of a workshop held in Trento, Italy, 26–29 September, 1988. Pitagora Editrice, Bologna. p. 169–178.

Erixon, S. 1955. Djurfångst och jakt samt insamling av vegetabilska födoämnen under nyare tid. (Capturing animals, hunting and gathering of plants for food during the newer time.) Nordisk Kultur XI–XIIa: 110–123. (In Swedish.)

Erixon, S. 1956. Lantbruket under historisk tid med särskild hänsyn till bondetraditionen. (Agriculture during the historical time, especially from the point of view of the peasant tradition.) Nordisk Kultur XIII: 43–215. (In Swedish.)

Fritzbøger, B. 1988. Forestry in crisis: society in crisis. Economic and ecological consequences and restrictions in the peasant forestry: Falster, Denmark 1660–1685. In: Salbitano, F. (ed.). Human influence on forest ecosystems development in Europe. Proceedings of a workshop held in Trento, Italy, 26–29 September, 1988. Pitagora Editrice, Bologna. p. 349–352.

Fritzbøger, B. 1989. Skove og skovbrug på Falster 1652–1685. (Forests and forestry in Falster 1652–1685.) Landbohistorisk Selskab, Odense. 311 pp. (In Danish.)

Fritzbøger, B. 1992. Danske skove 1500–1800: en landskabshistorisk undersøgelse. (Danish forests 1500–1800: a landscape history study.) Odense. 345 pp. (In Danish.)

Fritzbøger, B. 1993. Dansk skovhistorie. (Danish forest history.) Kgl. Veterinær– og Landbohøjskole, København. 324 pp. (In Danish.)

Gårdö, M.B. 1991. Bevarande av kulturmiljöer vid skogsbruk. (Preserving cultural milieus in forestry.) Riksantikvarieämbetet, P.M. 26 pp. (In Swedish.)

Gissel, S. et al. 1981. Desertion and land colonization in the Nordic countries c. 1300–1600: comparative report from the Scandinavian research project on deserted farms and villages. Almqvist & Wiksell, Stockholm. 304 pp.

Haglund, B. (ed.). 1955. Den levande skogen. (The living forest.) Stockholm. (In Swedish.)

Hamilton, H. 1988. 175 år: ett skogsbruk vaknar långsamt och reser på sig. (175 years: forestry wakes up slowly and gets up.) Kungl. Skogs– och Lantbruksakademiens Tidskrift, Supplement 20: 7–30. (In Swedish.)

Harstveit, A. 1989. Fangstmannsminne. (Hunting memories.) Aslak Harstveit fortel til Andreas Vevstad. Åmli. 95 pp. (In Norwegian.)

Huse, S. 1990. Litt om Norges skogforhold i fortid og nåtid, med særlig vekt på deres rolle for kulturlandskabet. (Little about Norwegian forest situation in the past and today, with special emphasis on its role in cultural landscape.) Nordisk Bygd 4: 2–3. (In Norwegian.)

Hustich, I. 1979. Ecological concepts and biogeographical zonation in the North: the need for a generally accepted terminology. Holarctic Ecology 2: 208–217.

Iversen, J. 1973. The Development of Denmark's Nature since the last Glacial. D.G.U. Ser. V 7–C. 126 pp.

Jakobsen, B. 1972–73. Skovens betydning for landbrugets udvikling i Danmark indtil ca. 1300. (The importance of forests for the development of agriculture in Denmark until c. 1300.) Det Forstlige Forsøgsvæsen i Danmark, Rapport 33: 345–396. (In Danish.)

Johannesson, T. 1955. Näringsfång och därmed sammenhörande bebyggelse på Island under Medeltiden. (Subsistence hunting and the related settlements in Iceland during the Middle Ages.) Nordisk Kultur XI–XIIa: 103–109. (In Swedish.)

Jørgensen, E.L. & Nielsen, P.C. 1964. Nordsjællands skove gennem 200 år. (The forests of Nordsjælland during 200 years.) København. (In Danish.)

Jutikkala, E. 1963. Bonden i Finland genom tiderna. (Peasants in Finland through the ages.) LT, Helsingfors. 523 pp. (In Swedish.)

Kaland, P.E. 1979. Landskapsutvikling og bosetningshistorie i Nordhordlands lynghei–område. (The development of landscape and settling history in Nordhordland's lynghei region.) In: Fladly, R. & Sandnes, J. (eds.). På leiting etter den eldste garden. Universitetsforlaget, Oslo. p. 41–70. (In Norwegian.)

Kardell, L. 1976. Skogsägandet genom tiderna. (Forest ownership through the ages.) Skogsägaren 54(10): 42–46. (In Swedish.)

Kardell, L. 1976. Svedjebruket. (Shifting cultivation.) Skogsägaren 54(5): 27–30. (In Swedish.)

Kardell, L. 1988. Skogsvårdens uppkomst. (The origin of forestry.) Kungl. Skogs– och Lantbruksakademiens Tidskrift 127(3): 163–181. (In Swedish.)

Kardell, L., Dehlen, R. & Anderssen, B. 1980. Svedjebruk förr och nu. (Shifting cultivation in the past and today.) Sveriges lantbruksuniversitet, Avdelningen för landskapsvård, Rapport 20. 92 pp. (In Swedish.)

Koch, N.E. & Kennedy, J.J. 1991. Multiple–use forestry for social values. Ambio 20(7): 330–333.

Leikola, M. 1987. Metsien hoidon aatehistoria. Summary: Leading ideas in Finnish silviculture. Silva Fennica 21: 332–341.

Linkola, M. 1988. The influence of the slash–and–burn cultivation on forest ecosystems and forest landscapes in Finland. In: Salbitano, F. (ed.). Human influence on forest ecosystems development in Europe. Proceedings of a workshop held in Trento, Italy, 26–29 September, 1988. Pitagora Editrice, Bologna. p. 79–89.

Linkola, M. 1988. Skogen som finländskt kulturlandskap. (Forest as cultural landscape in Finland.) Nord Nytt 33/34: 71–80. (In Swedish.)

Malmström, C. 1939. Hallands skogar under de senaste 300 åren. (Halland's forests during the past 300 years.) Meddelande från Statens Skogsförsöksanstalt 31: 171–300. Stockholm. (In Swedish.)

Miettinen, M. 1991. Skogens kulturhistorie i Finland. (The cultural history of forests in Finland.) Lommen – Kontaktblad för Nordiska Förbundet för Kulturlandskap 4: 24–25. (In Swedish.)

Nielsen, P.C. 1986. Mindesmærker og nyere litteratur om Johann Georg von Langen. (Memorials and newer literature about Johann Georg von Langen.) Årbok for Norsk Skogbruksmuseum, Nr. 11. Elverum. p. 37–62. (In Danish.)

Nielsen, P.C. 1990. Skovhistorie 1. (Forest history 1.) Kgl. Veterinær– og Landbohøjskole, København. (New edition by P. Søndergaard). 378 pp. (In Danish.)

Nordström, O. 1991. Resursen historikerna glömt. (The resource forgotten by historians.) Skog & Forskning 1: 30–36. (In Swedish.)

Odgaard, B. 1988. Heathland history in western Jutland, Denmark. In: Birks, H.J.B. (ed.). The cultural landscape – past – present – future. Cambridge. p. 311–319.

Ödum, S. 1991. Choice of species and origins for arboriculture in Greenland and the Faeroe Islands. Dansk Dendrologisk Årsskrift bd. IX: 1–78.

Oksbjerg, E. 1989. Alminding, fællesskov og overdrev. (Common agricultural land, common forest and common pasture.) Museerne i Viborg Amt 15: 73–80. (In Danish.)

Rackham, O. 1980. Ancient woodland: its history, vegetation and uses in England. Edward Arnold, London. 402 pp.

Reunala, A. 1984. Some aspects of social and psychological importance of forests in Finland. In: Kaiser, F., Schweitzer, D. & Brown, P. (eds.). Proceedings for economic value analysis of multiple–use forestry. Oregon State University, Corvallis. p. 55–64.

Romell, L.G. 1966. Röjningsbruket och dess hemlighet. (Clearing forestry and its secret.) Ymer Årsbok. p. 183–195. (In Swedish.)

Rubner, K. 1960. Grundlagen des Waldbaues. Neumann Verlag, Berlin. 620 pp.

Rydbo, F. & von Hofsten, E. 1984 (1979). Skoven og mennesket på kollisionskurs? (Forest and people coming into collision?) København. 189 pp. (In Danish.)

Selander, S. 1955. Det levande landskapet i Sverige. (The living landscape in Sweden.) Stockholm. (In Swedish.)

Sjöbeck, M. 1927. Bondskogar, deres vård och utnyttjande. (The management and utilization of farmforests.) Skånska Folkminnen Årsbok. p. 36–62. (In Swedish.)

Sjöbeck, M. 1933. Lövängskulturen i Sydsverige: dess uppkomst, utveckling och tilbagegång. (Broadleaf meadow culture in southern Sweden: its development and decline.) Ymer 53: 33–66. (In Swedish.)

Smith, T. 1960. Ivy, Mistletoe and Elm–Climate Indicators Fodder–Plants. D.G.U.. Ser. II. 66: 1–65.

Stridsberg, E. 1988. Den skogshistoriska forskningens objekt och problemställningar. (Definitions of objectives and problems in forest history research.) Kungl. Skogs– och Lantbruksakademiens Tidskrift 127(3): 183–193. (In Swedish.)

Stridsberg, E. & Mattsson, L. 1980. Skogen genom tiderna: dess roll för lantbruket från forntid till nutid. (Forestry through the ages: its importance to agriculture in the past and today.) Stockholm. (In Swedish.)

Sveli, A. 1987. Skogbruk i Nord–Norge: strejftog gjennom historien. (Forestry in northern Norway: a journey through the history.) Nord–Norges Skogmanns-forbund, Mosjøen. 504 pp. (In Norwegian.)

Tenow, O. 1974. Det nordiska skoglandskapets och skogbrukets utveckling fram till 1900–talet: en kort översikt. (The development of the Nordic forest landscape and forestry up till 19th century: a short overview.) The Swedish Coniferous Forest Project "Barrskogslandskapets ekologi", Internal Report 2. Uppsala. 34 pp. (In Swedish.)

Thorarinsson, Th. 1974. Thjodin lifdi in skogurin dó. (The people lived but the forests died.) Arsrit Skogræktar félags Islands. p. 16–29. (In Icelandic.)

Tuxen, O. & Hellesen, J.K. (eds.). 1988. Historisk Atlas Danmark. (Historical atlas of Denmark.) Copenhagen. (In Danish.)

Tveite, S. 1960. Engelsk–Norsk Trelasthandel 1640–1710. (English–Norwegian timber trade 1640–1710.) Bergen–Oslo. (In Norwegian.)

Tveite, S. 1964. Skogbrukshistorie. (History of forestry.) In: Seip, H.K. (ed.). Skogbruksboka bd. 3. Skogekonomi. Oslo. p. 16–76. (In Norwegian).

Tveite, S. 1964. Skogbrukshistorie. Kompendium. (History of forestry. A handbook.) Norges Landbrukshøgskole, Ås. 125 pp. (In Norwegian.)

Tvengsberg, P.M. 1988. Finnskogen brukes. Bönder, finner og godseieres utnyttelse av granskogsområdene på Östlandet i Norge 1600–1900. (Forestry in Finnskogen. Peasants, Finns and farmowners' utilization of spruce forest areas in Östlandet in Norway 1600–1900.) Nord Nytt 33/34: 59–70. (In Norwegian.)

Vestergaard, P. (ed.). 1987. Kulsvierlandet. København. 191 pp. (In Danish.)

Vevstad, A. 1992. Norsk Skogpolitikk: streiftog i det 20. århundre. (Norwegian forest policy: developments in the 20th century.) 296 pp. (In Norwegian.)

Weismann, C. 1900. Skove og skovbrug på Fyn i det nittende århundrede. (Forests and forestry in Fyn in the 19th century.) Odense. 172 pp. (In Danish.)

Worsøe, E. 1979. Stævningsskovene. København. (In Danish.)

Zackrisson, O. 1977. Influence of forest fires on the North Swedish boreal forest. Oikos 29: 22–32.

Zackrisson, O. 1979. Dendroekologiska metoder att spåra tidigare kulturinflytande i den norrländska barrskogen. (Dendroecological methods for defining former impacts of culture on the coniferous forests of Norrland.) Fornvännen 74: 260–268. (In Swedish.)

3 History, evolution and significance of the multiple–use concept

Marjatta Hytönen[1]

Abstract

Origins of the multiple–use approach can be found in Germany and in the United States of America. The concept became frequent in Scandinavian forestry vocabularies after the Fifth World Forestry Congress held in Seattle in 1960. The main part of the article describes interpretations of the concept in the Nordic countries from the 1960s. A general trend in the definitions has been a move away from the listing of products and functions to defining problems and setting goals. Also the components of multiple use have changed. The new objectives of forest management include carbon sequestration and preservation of biological diversity. Nowadays, multiple use is regularly mentioned in policy programs and action plans. Recently, international agreements on sustainable forestry have given a new demanding context to the experience obtained in the field of multiple use. Realization of the ideals of ecologically, economically and socially sustainable multiple–use forestry requires adjustments in practical forestry as well as development of new forest policy practices.

Keywords: multiple–use definitions, history, research, forest policy, debate, sustainable forestry.

---

[1] Finnish Forest Research Institute
Unioninkatu 40 A, FIN–00170 Helsinki, Finland

3.1 Too abstract to be useful?

The concept "multiple–use forestry" has been frequently used in the Nordic countries for about 30 years. However, there still exists confusion as to its meaning. Hultman (1984) argues that in Sweden the concept has not been defined and that is why it is not applied either. According to Koch and Kristiansen (1991), foresters have been using the concept as a kind of magic excuse ("trylleformular") to justify almost any decision concerning forests in Denmark. Hoen (1991b) claims that the concept is still hanging in the air ("henger i lufta") in Norway. A Finn, Lindqvist (1979), warns that the multiple–use concept, if used carelessly, may lead to quite unexpected results which run contrary to the intended purposes.

Despite problems in defining the concept, multiple use has established itself in forestry and environmental vocabularies. Nowadays it is regularly mentioned in policy programs, action plans and even in law texts. The reasons behind the need for such a multi–dimensional approach include the diverse demands for material and non–material forest products and the change in the use of forest resources from uncoordinated utilization to modern planning and management oriented forestry.

This chapter aims to clarify the contents and use of the "multiple–use forestry" –concept. First there is a short description of the development of multiple–use principles in Germany and in the United States of America. The main part of the text deals with interpretations of the concept in the Nordic countries from the 1960s. The article is completed with a discussion on the relationship of multiple use and the new concept of sustainable forestry.

3.2 Origins of the concept

Germany

The scientific basis for multiple use was clarified significantly in Germany in the 1950s. Viktor Dieterich, professor of forestry in the University of Munich, elaborated a theory about forest functions which he published in 1953 in a book called "Forstwirtschaftspolitik" (Forestry policy). According to him, a prerequisite of succesful forestry is knowledge of the complex interrelationships between forests and people as well as an understanding of the importance of forest to economy and welfare (Dieterich 1953). Consequently, forest policy should be based on the functions of forests. These functions have, after Dieterich's time been arranged in many connections into three main categories: utilization (Nutz), recreation (Erholung) and protection (Schutz) (e.g. Hasel 1971). The German forestry law of 1975 uses the same categories (Welp 1993).

Hasel (1971) states that Dieterich replaced the previous static way of practising forestry characterized by unintegrated measures by a new dynamic approach. Niesslein (1985) claims that Dieterich gave German forestry a new social dimension. Niesslein himself uses the multiple–use concept ("Mehrzweckforstwirtschaft"). According to him it refers to the relationship between the multiple functions of forest and the expectations of the various interest groups in society. These two interactive factors often lead to conflict situations.

The German forestry ideas were well known in the Nordic countries before the Second World War. After that, contacts of Nordic foresters with German forestry decreased and North American forestry ideas started to become more influential (e.g. Saastamoinen 1988).

United States of America

The philosophy of multiple use started to evolve in the USA in the beginning of the 19th century. In 1905, the Forest Service was established by transferring the administration of national forests from the Department of the Interior to the Department of Agriculture. Gifford Pinchot was the first chief of the Forest Service. His ideas accelerated the evolution of the present multiple–use approach. He considered wood, water and forage to be the main products of forests and wrote that "where conflicting interests must be reconciled the question will always be decided from the standpoint of the greatest good of the greatest number of people in the long run" (Gregory 1987).

After the establishment of the Forest Service, the multiple–use concept was repeatedly debated in forestry periodicals (e.g. Dana 1943, Pearson 1944, McArdle 1953, Gregory 1955). Leary (1987) calls the concept "perennially troublesome" in his short review dealing with the establishment of the term in the North American vocabulary.

Despite the controversies, multiple use became an official concept in 1960, when the Multiple Use Sustained Yield Act (MUSY–Act) was passed. The initiator of the law was the Forest Service. It had been under increasing pressure from conflicting timber production and conservation interests. Because of this pressure, working principles had to be clarified. The law requires the Forest Service to afford "due consideration" to non–wood forest–based resources such as recreation, water, wildlife, range, and fisheries in its management efforts. According to Wilkinson and Anderson (1987), the law has been implemented mainly through increasing the share of recreation, wildlife and watershed management in planning.

The MUSY–Act defines multiple use in the following way: "Multiple use means the management of all the various renewable surface resources of the national forests so that they are utilized in the combination that will best meet the needs of the American people, making the most judicious use of the land for some or all of these resources or related services over areas large enough

"Multiple Use of Forest Lands" is the theme of the Fifth World Forestry Congress. As symbolized by the official emblem, this concept has worldwide application. Forest and related areas should be managed in a manner that will conserve the basic land resource while at the same time producing high-level sustained yields of water, timber, recreation, forage, and wildlife harmoniously blended for the use and benefit of the greatest number of people.

RICHARD E. MCARDLE
President

Figure 3.1 The theme of the Fifth World Forestry Congress in Seattle, 1960 (Proceedings... 1962).

to provide sufficient latitude for periodic adjustments in the use to conform to changing needs and conditions; that some land will be used for less than all of the resources; and harmonious and coordinated management of the various resources, each with the other, without impairment of the productivity of the land, with consideration being given to the relative values of the various resources, and not necessarily the combination of uses that will give the greatest dollar return or the greatest unit output" (Gregory 1987).

Two months after the passing of the MUSY–Act, the Fifth World Congress of Forestry was held in Seattle under the title "Multiple use of forest lands" (*Figure 3.1*). Through the congress, the principles of the MUSY–Act reached a wider public. The meeting increased the interest towards multiple use considerably also in the Nordic countries (Saari 1962, Helles 1983a, Hultman 1984, Saastamoinen 1988, Koch & Kennedy 1991).

3.3 Nordic countries

Denmark

Multiple–use theories were introduced to Denmark by forest economist Finn Helles. In 1977, he published an article in which he discusses the theory of determining the optimal combination of forest products to maximize net revenue (Helles 1977). In 1978, he wrote a paper about the external effects of forestry, which he considers to be essential components of multiple use (Helles 1978). Five years later he tells about the Seattle Congress and discusses the applicability of the US concept to Denmark in a periodical "Skoven". He concludes that the most prominent forest use in Danish conditions, besides timber production, is outdoor recreation. Protective functions of forests and forest grazing are less important. He also emphasizes and analyzes interest conflicts in multiple use (Helles 1983b).

In another article from 1983, Helles states that the main idea in multiple use is to coordinate different interests which in principle are equal from society's viewpoint. In older times, many functions of forests (e.g. recreation environments and protection provided by forests) were abundant and taken for granted. Recently they have become deliberately produced services. As a consequence of this change, it is necessary to include new branches in the decision making of forestry and resource allocation in society. In this socially and politically oriented article, Helles divides the multiple–use functions which are important in Denmark into six main categories. They are 1) impact of forests on climate, 2) protection of water supply, 3) protection against erosion, 4) protection against pollution, 5) forest recreation, and 6) the positive and negative external effects of forestry practices (Helles 1983a).

The contents of the present multiple–use forestry definition started to take shape in 1987, when the Danish Forest Service merged with the Agency for Protection of Nature, Monuments and Sites to form the National Forest and Nature Agency. The new organization published a report dealing with the relationship between forestry and recreation. The report defines multiple use as the conscious and deliberate use of every area of the forest in order to produce at the same time several goods and services, the values of which are not necessarily measured economically (Koch & Canger 1987).

The same definition is repeated in a seminar report which was published by the National Forestry and Nature Agency to create debate on conservation of plants and animals in modern forestry. In their article, Billeschou and Koch, both employed by the National Forestry and Nature Agency at that time, elaborate the concept further. According to their interpretation "multiple–use forest management implies, from an idealistic viewpoint, that the decision–maker at each decision – for each piece of forest land – takes into account all the present and future goods and services which forest lands could provide, aiming at the optimal combination – the optimal compromise" (Billeschou & Koch 1989).

The present Danish Forest Act was passed in 1989. Compared to the previous laws, it puts more emphasis on non–market forest values. One of the four main purposes of the Act is to strengthen advisory and information activities concerning "good and multiple–use forest management" which is defined in the following way: "In good and multiple–use forest management it shall be endeavored to manage forests in order to increase and improve wood production and to protect landscape amenity, nature conservation, cultural heritage and environmental protection interests, as well as recreational activity interests" (Act of Forests... 1989).

In September 1991, the periodical Skoven published a special issue on multiple use. The publication was produced in joint cooperation by the National Forest and Nature Agency, the Ministry of Agriculture, the Danish Forestry Society, the Federation of Danish Forest Owners' Associations and the Danish Land Development Service (Hedeselskabet). The handbook repeats the definition introduced in 1987 (Koch & Canger 1987). In addition, the writers, Koch and Kristiansen, state that more important than the law text is how foresters respond to the demands concerning multiple use which society imposes on forestry through parliament (Koch & Kristiansen 1991).

Koch and Kristiansen (1991) divide forest functions in the following way. Forests produce wood and wood products, Christmas trees and decorative greenery. They serve as recreation sites, for example, for hiking, hunting and orienteering. They contain considerable landscape aesthetic, natural history and cultural history values. They stabilize climate, regulate water, bind soil and sand, and reduce soil–, water–, air– and noise pollution. Furthermore, people can be employed in forestry.

Multiple–use principles in Danish forestry have recently been adapted and supported by various policy statements of the government. The forest policy of 1987 states that forest production should be diversified and should not include only wood products but also a wide range of nature, environment and leisure products. It also stresses that diversified production of forests should be actively promoted. According to the 1988 action plan for environment and development of the government, multiple–use forestry should be practised so that the integration of timber production and different environmental measures is strengthened. The integration should be supported by education of forest owners. The plan also emphasizes the need for economic incentives and the strengthening of research and extension (Billeschou & Koch 1989).

In the beginning of the 1990s, the concept was geared towards a more socially oriented approach. In the article "Multiple use for social values", Koch and Kennedy define social values as "those goods, services, or ideals that large groups of people will make sacrifices to achieve (e.g. recreation, wildlife, wood products, scenery, etc.)". Consequently, they propose a new definition for forestry as "the management of forest resources to provide a satisfactory amount and mix of multiple–use social values for living clients in the same time protecting forest values and use options for future generations" (Koch & Kennedy 1991). Recently, the concept has been incorporated in the sustainable forestry discussion (e.g. Koch 1992, Strategy for sustainable... 1994).

Multiple–use rhetoric has also raised anger in Denmark. Madsen warns in his spicy article from 1991 about the dangers connected to the application of the concept. He maintains that in the long run private forest management and market forces are able to produce a balanced mixture of multiple forest benefits in society. Multiple–use ideology instead is paving the way for a bureaucratic system where private forest owners lose their freedom and the official sector regulates land use through restrictions and sanctions (Madsen 1991). This danger has also been admitted by other foresters and studies have been started to create methods to integrate private forest owners' and other interest groups' interests in a harmonius way in areas consisting of or including private properties (e.g. Clausen 1991, Kristiansen 1991) (*Figure 3.2*).

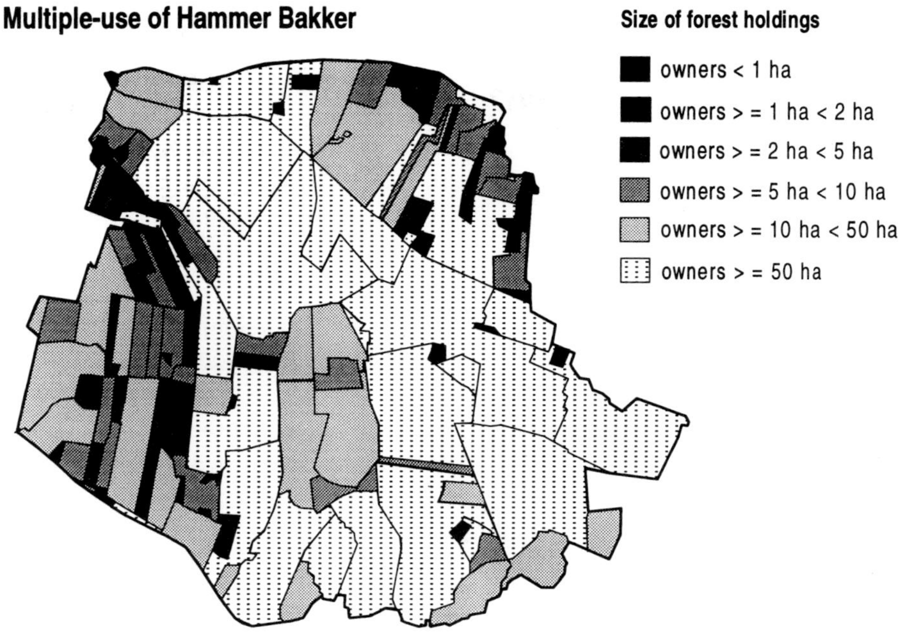

Figure 3.2 A map describing the ownership structure of the case study area (2,000 ha, 140 owners) of the project "Flersidig udnyttelse af Hammer Bakker" (Clausen 1991).

Finland

The concept of multiple–use forestry came to Finland from the USA. The first one to discuss the concept and its US interpretation in Finland was professor of forest economics Eino Saari in a forestry periodical "Metsätaloudellinen aikakauslehti" in 1962 (Saastamoinen 1988). He compares the contents of multiple use in the two countries. He considers the most important forest products in Finland, in addition to timber, to be water, outdoor recreation and wildlife (Saari 1962).

In the 1960s and 1970s, professor of forest biology Peitsa Mikola continued the discussion. He regarded timber production and outdoor recreation as the most important forest uses in Finland. He elaborated multiple–use principles for forestry planning, for example, concerning land–use classification and silvicultural criteria for improving the amenity of forest. Mikola also emphasized the societal and normative content of the concept. He introduced to Finland the idea of integrating forest uses in order to serve all people the best possible way and the idea presented in the Seattle Congress that multiple use requires conscious and coordinated management of the various renewable resources (McArdle 1962, Mikola 1969).

The contents of the concept were disputed in forestry periodicals. On some occasions, multiple use and timber production were considered to exclude each other. Professors Saari and Mikola often corrected the presented misconceptions. However, neither could they agree on a definition. Saari considered Mikola's interpretations to be idealistic. He claimed that the ideas presented by Mikola, such as the best possible serving of all people, the largest possible benefit and the coordination of uses so that the combination is the best possible, were unrealistic and utopian. Saari argued that such statements can describe the goals of multiple use but they can not be included in a definition of the concept (Saari 1970).

Since then, Finnish forest literature has contained both neutral and normative definitions of multiple use. In the 1970s, technical and value–free definitions were common. In the 1980s, definitions presenting societal goals became clearly more frequent.

In 1976, Jaatinen and Saastamoinen (1976) state that the concept includes the principle of producing many products and services instead of one. They divide forest uses into the following subgroups: 1) timber production, 2) game management and hunting, 3) picking of berries, mushrooms and other minor products, 4) reindeer husbandry and grazing, 5) recreation, 6) landscape aesthetics and management, 7) nature conservation, 8) environmental impacts (i.e. protective functions of forests), and 9) utilization of peat and gravel.

Saastamoinen uses the above division in his handbook of multiple–use forestry, which has been widely used as a textbook in vocational training in forestry schools. He also presents the definition used by the National Board of Forestry which divides the forest products into two main groups. According to this, multiple–use forestry means the use of the material and non–material forest products for more than one purpose simultaneously (Saastamoinen 1980).

In 1984, Seppo Kellomäki (1984a) elaborates further Jaatinen's and Saastamoinen's definition. He ranks the production potentials of forests according to their importance from the point of view of the needs of people: 1) material goods provided by forests (e.g. wood, game, reindeer, lichen, berries, mushrooms), 2) the indirect impacts of forests (e.g. protective functions of forests on air, soil and water), 3) recreation, 4) educational and scientific services of forests, 5) landscape and cultural functions of forests, and 6) nature conservation because of intrinsic value. He states that the ranking of the uses according to hierarchical needs helps one to understand the dependence of people on the ecological system (*Figure 3.3*).

The same year, Kellomäki presents a clearly socially oriented definition of the concept in his handbook on environmental management of forestry (1984b), which is based on his lectures held in the forestry faculty of the University of Joensuu. According to him, multiple–use forestry means the conscious use of forests so that the objectives and means are well known and the practices are based on the sustained production capacity of the ecosystem. Multiple use refers to such utilization and management of forest resources which, taking into account the ecological and economic limitations, aims "holistically, impartially and consciously" to satisfy the changing needs of society.

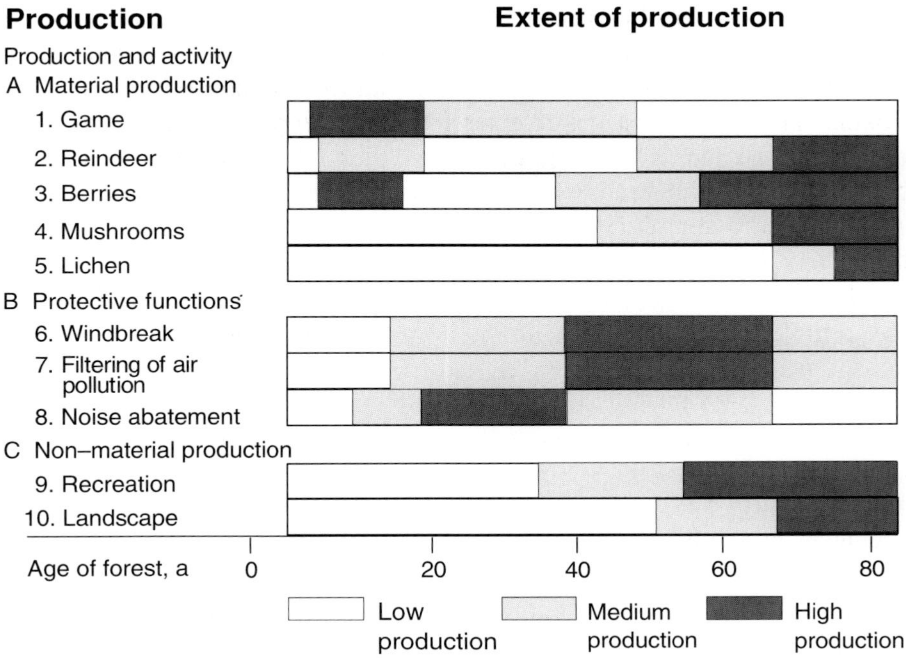

Figure 3.3 Potential production of a forest ecosystem in different stages of succession (Kellomäki 1984a).

The multiple–use forestry working group of the Forest 2000 Programme, which is a long–term plan for developing forestry and forest industry in Finland, presents the same type of ideas as Kellomäki. According to it, "multiple–use forestry means the planned utilization of forest resources in such a way that the various needs of people are satisfied sustainably and that the total of material and non–material benefits provided by the forests for society is as large as possible". The working group arranges the benefits provided by forests to society and individuals in four categories: material benefits, protective functions, recreation and other benefits (e.g. forest as landscape elements, as nature conservation areas and as a study and educational object). The group also mentions that reindeer husbandry and tourism are livelihoods which utilize forests (Metsien moninaiskäytön... 1985).

Environmental policy action plans contain objectives related to the above presented definitions of multiple use. The Finnish Commission of the Environment and Development proposes, among other things, that forestry should aim in the long term to direct development towards natural silviculture and regeneration, towards rehabilitating forests as healthy ecosystems including all the indigenous species and also towards the re–establishment of the cultural, educational and social uses of forests (Ympäristön... 1989).

The report "Sustainable Development and Finland" prepared by the Council of State for Parliament emphasizes the social and recreational functions of forests. Furthermore, the report states, that when planning the utilization of forests, attention should also be paid to non–timber production (Kestävä kehitys... 1990).

Recently, the multiple–use concept has been frequently included in new environmental and silvicultural studies and guidelines. Attempts to define it have become rare. Instead, it is problematized through practical contexts and accompanied by new viewpoints like biodiversity, sustainability and social values (e.g. Metsätalouden ympäristöopas 1993, Kellomäki 1994).

Iceland

The objectives for planting forests in Iceland, in order of priority, are soil protection, recreation and wood production (Loftsson 1993). The concept of multiple use does not occur in the English language papers dealing with Icelandic forestry. This may be due to the relative simplicity of forestry problems on the island. However, the functions of forests have been specified in ways which are comparable to the multiple–use thinking in the other Nordic countries.

Blöndal et al. (1986) divide the most important protective functions of forests in Iceland into four subgroups: 1) erosion control, 2) water management, 3) climate regulation (shelterbelts) and 4) nature conservation. In another context, the Icelandic forests have been grouped in four main categories: 1) protection forests for soil stabilization and land reclamation, 2) production forests for timber, firewood and Christmas trees, 3) recreation forests for short term visits close to urban areas and for longer visits for camping, and 4) shelterbelts and woodland for crop cultivation, domestic animals, farmsteads and urban areas (Blöndal 1993).

At the moment Iceland is almost treeless but afforestation is being carried out in many parts of the country (*Figure 3.4*). Blöndal (1993) writes about the future: "Intangible benefits worth mentioning are new landscapes, diversified environments, vegetation (including herbs, mosses, lichens, and fungi), and especially the new bird– and wildlife that accompany forests. The pride of living in a forested country which was previously barren and the opportunity to spend more leisure time outdoors when an amenity forest is nearby are important intangible benefits of afforestation".

Norway

After the Seattle Congress, the English language concept was used in Norway without translation. Børset (1970) writes in an article "If we want to participate in the multiple–use dance, we have to listen to the people" (Skal vi være med i "multiple–use" –dansen, må vi lytte til folket). Frederiksen (1970) refers with the "multiple–use –oppgaver" to nature conservation and outdoor recreation in forests. The Norwegian translation of the concept, "flerbruk",

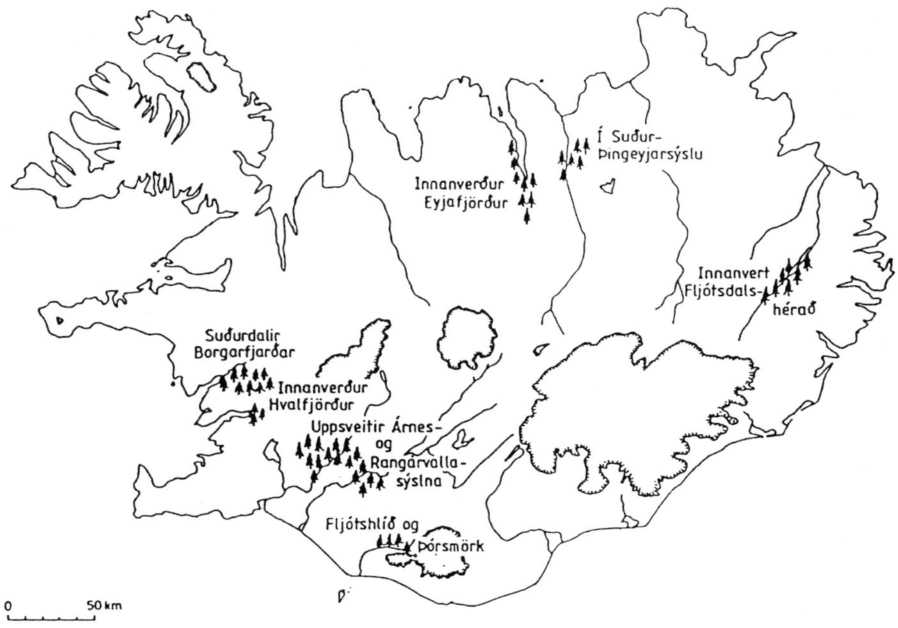

Figure 3.4 Areas potential for wood production in Iceland (Blöndal et al. 1986).

was registered for the first time by the Norwegian Lexicographic Institute in 1973 when the newspaper Dagbladet (12.5.) wrote about the multiple–use plan for Oslomarka, the forest north of Oslo (Fernand 1989).

In 1986, the journal Norsk Skogbruk published a special issue on multiple use. The Ministry of Agriculture and the Norwegian Forestry Society supported the publication. In the introduction of the issue, it is said that the booklet can be regarded as an expression of the intentions of the Ministry of Agriculture and Parliament as regards the practical recommendations concerning recreation, nature conservation and fauna management in forestry. When defining the concept, the editors refer to the Forestry Act, which was revised in 1976 to include the following sentences: "The objective of this act is to encourage forest production, afforestation and forest protection. The aim will be that forestry through rational management can give a satisfactory result for the people connected to the forest enterprise and secure an effective and regular supply of raw material for the industry. Further, there should be emphasized the importance of the forest as a source for recreation for the population, as an important part of the landscape, as environment for plants and animals, and as areas for hunting and fishing" (Flersidig skogbruk... 1986, Haveraaen 1991).

In 1987, the Ministry of Agriculture set up a working group to clarify the relationship of forestry to natural environment and recreation and to promote new measures to carry out the objectives presented in the Forestry Act. The working group states that multiple–use forestry is achieved by adjusting forestry practices so that the objectives mentioned in the Act are reached (Flersidig skogbruk... 1989).

The working group makes a clear distinction between multiple use and nature protection, because they have different objectives. Nature protection aims at preserving natural conditions of special and rare sites through establishing national parks, nature reserves, natural heritage sites and landscape–protection areas. Multiple–use forestry means that attention is paid to natural environment and recreation when practising economically oriented forestry. In other words, protection has more ambitious objectives regarding the preservation of nature than forestry, which aims at ensuring the long–term utilization possibilities of forest resources (Flersidig skogbruk... 1989).

The Norwegian Forest Research Institute has published its own definition. According to it, multiple–use forestry means the diversified use of forest resources, originating both from the various interests of society and from the economic objectives connected to forests. Important interests of society are: hunting, fishing, outdoor recreation, nature experiences, employment, economic profit and the protection of species, biotops, landscape and drinking water. Important economic objectives of forest utilization are: production of timber, Christmas trees, decorative greenery, game, fish, forage for wild and domesticated animals and investment in income generating activities such as tourism and outdoor recreation (Petersen 1989).

Researcher Knut Solbraa from the Norwegian Forest Research Institute emphasizes the social and political aspects of multiple–use forestry. He also brings forest managers and their responsibility into the definition. According to him, the aim of multiple–use forestry is to solve or reduce conflicts between different users and to result in maximum public utility of forests. Differing opinions about the importance of each group's needs, and thereby the content and extent of multiple use, are likely to show up. Consequently, managers who are evaluating the needs of two or more groups of forest users or inhabitants and who pay attention to these needs are practising multiple use. Furthermore, the framework set by laws, rules, and state investment contributions, as well as the influence exercised by state and private service organizations and voluntary organizations, considerably influence the choice of alternatives (Solbraa 1989).

Solbraa divides the multiple–use approach into four main subfields according to whom or to what attention is paid: 1) attention to species, biodiversity, biotopes, vegetation societies, research and education, 2) attention to climate, water regulation/quality, air cleaning, aesthetic effects, other environmental effects and cultural heritage, 3) attention to recreation (other than in points 1 and 2, and 4) attention to employment, settlements, private and public economic relations and raw material supply. Solbraa's opinion concerning the relationship between forestry and nature protection is that when protection serves other purposes, for example tourism, it can be regarded as part of multiple use, otherwise not (Solbraa 1989).

Hans Fredrik Hoen criticizes the Forestry Act of 1976, because it creates many questions for practical forestry and leaves them unanswered (Hoen 1991b). He defines multiple use the following way: multiple–use forestry practice, i.e. harvesting and investments, is formed on the basis of a deliberate choice from a set of possible handling alternatives. In other words, the chosen treatment defines whether the activity is multiple use or not (Hoen 1991a). He emphasizes that the choice between different alternatives is always value dependent. According to him, the overall objective of multiple use is to contribute to the maximization of the welfare of society (Hoen 1991b).

Hoen divides the forest products into goods and services. He gives some examples (1991a): wood and other flora, fauna, genetic material, pasture, berries, mushrooms, moss, hunting, drinking water, climate regulation, absorption of pollution from air and water, CO_2 –assimilation, landscape aesthetics and recreation. These goods and services can be produced in various combinations.

In 1991, there was lively debate in the periodical "Norsk Skogbruk" concerning the significance of the concept. Knut Solbraa states (1991b) that the concept should be abandoned, because when forestry is practised in accordance with the Forestry Act, forestry can be said to cover the whole field of multiple use. Petter Nielsen (1991) regards Solbraa's ideas as too simplistic. The objectives of the Act are on too general a level to ensure that multiple use is practised. He argues that the concept is needed to defend the non–timber production interests. In his response, Solbraa (1991a) claims that forestry is already on its way to being multifunctional. That is why multiple use is only a "smør–på–flesk" (butter–on–fat) –concept, which can soon be forgotten. Liv Sølland (1991) agrees with Nilsen that the concept is useful but criticizes the quality of the popularization of research results. According to her, the main problems can be found in the inadequacy of the dissemination of existing knowledge of multiple–use forestry.

Jann Fernand is elaborating a new definition for the concept on the basis of the old German doctrine of functions. According to him, multiple use includes and covers four main functions of forests: utility, protection, perception and human self–realization. Multiple use is thus given a prime position in relation to forestry and nature protection (Fernand 1989, 1992). Fernand is not the first one to develop function thinking in Norway. For example, Sigmund Huse wrote about the various functions of forests and their interrelationships in 1973 (*Figure 3.5*).

Fernand also participates in the debate on the significance of the concept. He criticizes Knut Solbraa's willingness to omit the concept, because he regards it as a valuable means for widening the forestry discussion in Norway by giving space to the arguments promoting various, often conflicting, forest uses. He would like the next phase in the debate to be a discussion of the position of the concept in a wider societal context and the objectives of the use of the concept (Fernand 1991).

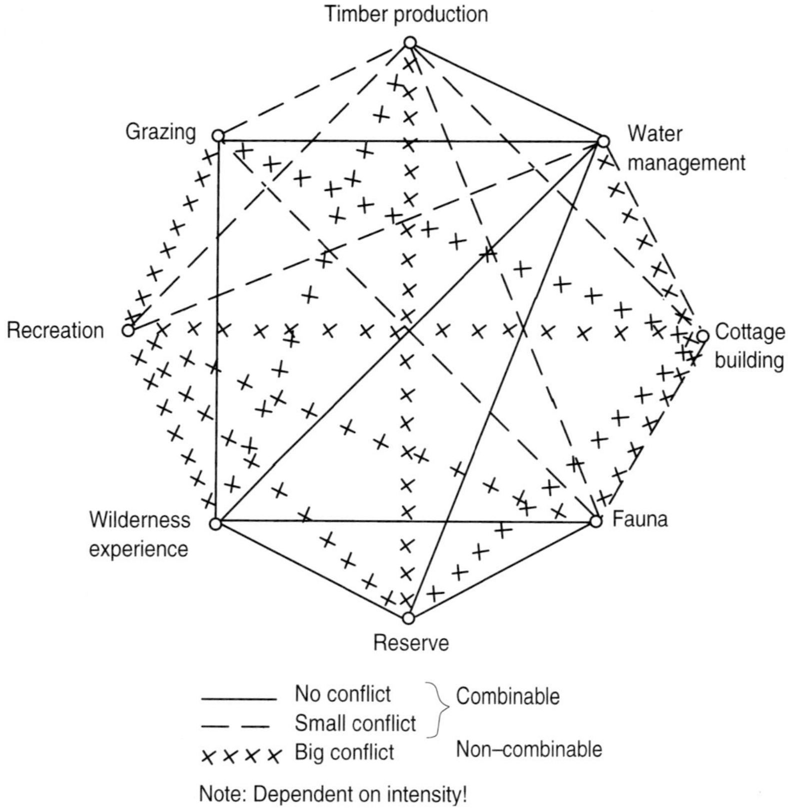

Figure 3.5 Forest functions and their interrelationships (Huse 1973).

Since the heated debate, multiple–use has become a common term in Norway. Outside scientific circles, it is used in a simple and straightforward manner. A brochure published by the Ministry of Agriculture puts it in the following way: "Forestry aims to combine economic, ecological and outdoor life considerations. The management of the forests is based on this multiple–use concept" (Agriculture in Norway 1993).

Sweden

The first person to mention multiple use frequently in Sweden was Erik Höjer, former chief of the Swedish Forest Service. He started to use the concept in his presentations after returning from the 1960 World Forestry Congress (Heden 1984, Hultman 1984).

In the 1970s, multiple–use forestry was mentioned occassionally by foresters. According to Hultman (1984), timber production was clearly considered to be the main purpose of forestry and it was claimed that there are no conflicts between wood production and other forest uses. It was argued that multiple use is already practised in Sweden, because there exists an extensive network of laws, "which take care of all important conflicts that could possibly arise between the use of forests for wood production and other purposes" (Hultman 1984).

However, at the same time there was lively debate concerning forestry activities in Sweden. In 1978, The Swedish University of Agricultural Sciences decided to make a study of the scientific basis of forestry practices together with forestry organizations and other related interest groups. The aim of the study was to improve the preconditions of making land–use decisions concerning timber production, nature conservation and recreation. The multiple–use concept is not mentioned in the book but its outline is based on the idea that forest is a producer of many material and non–material benefits and values. The book classifies the functions of forests into eight categories: 1) existence value, 2) ecological stabilization, 3) timber production, 4) forage production, 5) source of knowledge, 6) culture, 7) environment for recreation and (8) potential for the future (Andersson & Hultman 1980).

In 1979, Birger Andersson wrote an article about the debate dealing with forest utilization in Sweden. He states that there are no self–evident goals for utilizing forest resources. According to him, forests can be used to produce both wood and aesthetic experiences. They can serve as an environment for many activities which are important to people's feeling of security and self–confidence. Forest organisms also have an intrinsic value. Because of the many goods and services produced by forests, various kinds of demands and combinations of demands are possible. The choosing of utilization alternatives is value dependent and often leads to conflicts. Power relationships are central factors in the analysis of interest conflicts. Finally, Andersson suggests that the most essential issues in the whole debate may be the questions dealing with cultural identity and ethics (Andersson 1979).

In 1991, Mattsson and Li wrote an article about the multiple benefits of forests (skogens mångnytta). They state that forests produce, in addition to timber, an environment for animals and plants, recreation, hunting and picking of berries and mushrooms. The various interests concerning forest utilization create different kinds of demands on the age structure, tree species combination and other forest qualities. The writers emphasize that when defining the combination of benefits that will be produced or when prioritizing different interests one should attempt to use forest resources so that the total benefit for society is as large as possible. When doing this, it is of fundamental importance to take into account the external impacts which follow the production of certain benefits, affecting negatively or positively the production of other benefits. Finally, a definition of the ideal utilization of forested areas requires an analysis of the means and regulations which enable a balancing of different alternatives and interests so that good results are reached in practice, out in the forests. Such means and regulations include, for example, legislation, subsidies and provision of information (Mattsson & Li 1991).

In 1991, acting professor of silviculture Jan Falck refers with multiple use to walking, jogging, picking berries and mushrooms, camping and hunting. These activities are common because all forest land is open to public use on the basis of the Right of Public Access. Multiple use is thus practised in Sweden although the main use is commercial wood production. According to Falck, the Swedish forestry policy is changing from strict governmental control, which aims at the highest possible production of timber, to a more flexible attitude to the use of forest land. Environmental aspects and recreation are coming into focus. This creates new requirements for foresters. A major problem for practical forestry is the lack of knowledge and experience of alternative silvicultural systems. New silvicultural methods are needed first of all in forests with special uses. Those areas can be classified the following way: 1) forests close to cities and villages, 2) forests of high altitudes in the mountains, 3) forests on islands, by lakes and in sea archipelagoes, 4) forests close to nature reserves, 5) forests on water catchments, 6) forests, fauna and flora dependent on a particular density or species composition, 7) grazing areas for reindeer, and 8) small forest estates (Falck 1992).

In an outline of the extension of the national account of income from forest resources in Sweden, Lars Hultkrantz divides the yields of forests into the following eleven categories: 1) timber, 2) berries, 3) mushrooms, 4) meat, 5) recreation, 6) biodiversity, 7) effects on hydrological flows, 8) fixing of carbon 9) buffering of acid rain 10) nitrogen leaking and 11) reindeer forage. He also defines the respective stocks and maintenance activities (*Table 3.1*). The reasons for revising the accounting of forest products are related to the inadequacy of national income as a measure of welfare and to the need of taking into account environmental degradation when pursuing sustainability of income (Hultkrantz 1992).

A typical feature in the Swedish discussion is that the concept "multiple use" has been used very seldom. Still, there has been lively debate about the problems and principles of forestry and many remarkable books which promote multiple–use forestry have been published. They include the publications "Handbook of ideas for a private forest owner" (Idékatalog... 1988), "Richer forest: knowledge of nature management and ecology of the 1990s" (Rikare skog... 1990), and "Management of cultural heritage in forests" (Kulturmiljövård... 1992).

In 1990, the Department of Agriculture set up a committee to revise Swedish forest policy and legislation. Its report contains a definition of multiple use (mångbruk). According to it, multiple–use forestry includes all the functions of forests which provide renewable raw material, habitat for plants and animals, environment for recreation, cultural and aesthetic values, pasture for reindeer husbandry and also such products as berries, mushrooms and game animals. Attention should be paid to all these components on all forested lands. The balance between the different functions is dependent on the natural conditions and the needs of the population and industries. The report empha-

Table 3.1 Natural stocks of forest resources, the benefits and disutilities that they yield and the associated maintenance activities (Hultkrantz 1992).

Yield	Maintenance	Stock
Timber	Silviculture	Forest inventory
Harvest of berries		Berry–yielding herbs
Mushroom		Mycelium
Meat (bag)	Game protection	Game populations
Recreation	Various activities	Various features
Biodiversity	Fauna and flora protection	Conditions for the survival of species
Effects on hydrological flows	Measures that affect runoff	Forest inventory, bare lands, ditches, etc.
Fixing of carbon	Silviculture and hoarding (non–harvesting)	Carbon pools
Buffering of acid rain, tree nutrition	Liming, fertilization	Content of exchangeable base cations in soil and vegetation
Nitrogen leaking	Construction of nitrogen sinks	Nitrogen–fixing capacity
Reindeer forage		Lichen stocks

sizes that all utilization should be multiple–use oriented ("Nyttjandet skall präglas av mångbruk") (Skogspolitiken inför... 1992).

The Swedish parliament approved the new forest policy in May 1993. In a brochure describing the main principles of the new strategy, the National Board of Forestry declares: "The forests shall be managed so, that the needs for both high timber production and other functions of the forests are satisfied, in principle, in every hectare of forest land. This is a multiple use approach to forest management" (Sweden's new... 1994).

3.4 Sustainable multiple–use forestry

The multiple–use definitions presented by Nordic researchers have emphasized different combinations of forest benefits. A general trend has been a move away from the listing of products and functions to defining problems and setting goals. However, the lists of components have also changed. Old functions such as the production of forage for domestic animals have been left out and new ones such as carbon sequestration and preservation of biological diversity have become more frequently mentioned. Moreover, the interrelationships between uses and functions are increasingly dealt with. The need to cope with conflicting interests has often been mentioned recently in the multiple–use context. Conflict situations have arisen, for example, when the Sámi people herd their reindeer on state forest land affected by logging operations, when the aims of private forest owners go against the desires of recreationists, and when local and national needs do not meet.

Today, the ideas of multiple use are merging with the new concept of sustainable forestry. The following widely quoted definitions of sustainable forestry are clearly normative in their contents and also because they are presented in international agreements. According to the Forestry Principles approved in the UNCED (United Nations Conference on Environment and Development) Conference in 1992, "Forest resources and forest lands should be sustainably managed to meet the social, economic, ecological, cultural and spiritual human needs of present and future generations. These needs are for forest products and services, such as wood and wood products, water, food, fodder, medicine, fuel, shelter, employment, recreation, habitats for wildlife, landscape diversity, carbon sinks and reservoirs, and for other forest products" (Non–legally binding... 1992). The General Guidelines for the Sustainable Management of Forests in Europe which were agreed upon in the Ministerial Conference on the Protection of Forests in Europe in 1993 states: "sustainable management means the stewardship and use of forests and forest lands in a way, and at a rate, that maintains their biodiversity, productivity, regeneration capacity, vitality and their potential to fulfil, now and in the future, relevant ecological, economic and social functions, at local, national, and global levels, and that does not cause damage to other ecosystems" (Ministerial conference... 1993).

The abovementioned agreements give a new and demanding context to the experience obtained in multiple–use forestry. An essential component in the follow–up to these agreements is the defining of criteria and indicators for sustainable forestry. The creation of a comprehensive system of indicators reflecting the level of sustainability in forestry requires widening of the information basis and improvements in the statistical and other documentation concerning forest resources and their utilization. The research carried out within the scope of multiple use, for example, on the production and valuation of non–wood forest products and benefits provides valuable information for the elaboration of the criteria and indicators (see e.g. Hytönen 1992).

The real value of multiple–use rhetoric and theory will ultimately be tested in forests and by people using and enjoying them. Many examples of well functioning multiple use can be found in the Nordic countries, but there are still a lot of problems to be solved. Ecologically, economically and socially sustainable multiple–use forestry requires adjustments in assessments of forest resources and their societal connections, in the choice of silvicultural methods, and in the structure of decision making. The fair and equitable satisfaction of forest–related needs of the various organized and unorganized groups of people is one of the biggest challenges emphasized in recent sustainability definitions and international agreements. Furthermore, planning and realization of forest policy measures according to the ideals of multiple use and sustainability require new types of systematic and integrating methods of analyzing the interdependencies between the various uses and functions of forests. Such methods could also help to clarify the relationship of forestry with other forms of land and resource use.

References

Agriculture in Norway. 1993. Royal Ministry of Agriculture, Oslo. 14 pp.

Andersson, B. 1979. Vad handlar debatten egentligen om? (What is the debate really dealing with?) Sveriges Skogsvårdsförbunds Tidskrift 77(1): 75–82. (In Swedish.)

Andersson, B. & Hultman, S.–G. 1980. Skogens värden: skogsbrukets roll. (The values of forests: the role of forestry.) Sveriges lantbruksuniversitet och LTs förlag, Kristianstad. 233 pp. (In Swedish.)

Billeschou, A. & Koch, N.E. 1989. Flersidigt skovbrug. (Multiple–use forestry.) In: Naturen i skoven. (Nature in Forest.) Miljøministeriet/Skov– og Naturstyrelsen, Naturovervågning/89. p. 69–76. (In Danish.)

Blöndal, S. 1991. Socioeconomic importance of forests in Iceland. In: Alden, J., Mastrantonio, J.L. & Odum, S. (eds.). Forest development in cold climates. Plenum Press, New York. 13 pp.

Blöndal, S., Benedikz, P. & Ottósson, J.G. 1986. Forestry in Iceland. Iceland Forest Service, Reykjavik. 32 pp.

Børset, O. 1970. Norsk skogbruk: Oslomarka. (Norwegian forestry: Oslomarka.) Norsk Skogbruk 16(9): 230–231. (In Norwegian.)

Clausen, J.T. 1991. Case: Hammer Bakker. Ugeskrift for Jordbrug 51/52: 830–831. (In Danish.)

Dana, S.T. 1943. Multiple use, biology, and economics. Journal of Forestry 41(9): 625–626.

Dieterich, V. 1953. Forstwirtschaftspolitik: eine Einführung. Verlag Paul Parey, Hamburg und Berlin. 398 pp.

Driver, B.L. & Peterson, G.L. 1992. Evaluation of the multiple–use research program of the Finnish Forest Research Institute. The Finnish Forest Research Institute, Research Papers 438. 71 pp.

Eckerberg, K. 1987. Environmental protection in Swedish forestry. Avebury Studies in Green Research, Aldershot. 179 pp.

Falck, J. 1992. New aspects of multiple–use in education and research at the Department of Silviculture in Sweden. Norwegian Journal of Agricultural Sciences, Supplement 8: 25–30.

Fernand, J. 1989. Refleksjoner omkring flerbruk og flerbruksbegrepet: fusjon og konfusjon ? (Reflections around multiple use and multiple–use concept: fusion and confusion?) M.Sc.thesis. Norges landbrukshøgskole, Institutt for skogskjøtsel, Ås. 110 pp. (In Norwegian.)

Fernand, J. 1991. Flersidig kaos. (Multiple chaos.) Norsk Skogbruk 10: 30–31. (In Norwegian.)

Fernand, J. 1995. Multiple–use forestry: a concept of communication. In: Hytönen, M. (ed.). Multiple–use forestry in the Nordic countries. The Finnish Forest Research Institute, Helsinki. p. 67–80.

Flersidig skogbruk: skogbrukets forhold til naturmiljø og friluftsliv. (Multiple–use forestry: the relationship of forestry with natural environment and outdoor recreation.) 1989. Norges offentlige utredninger, NOU 1989:10. 139 pp. (In Norwegian.)

Flersidig skogbruk: veiledende retningslinjer for det praktiske skogbruk. (Multiple–use forestry: guidelines for practical forestry.) 1986. Norsk Skogbruk 5: 1–40. (In Norwegian.)

Frederiksen, J.M. 1970. Ragnvald Aaheim og Oslos turgåere. (Ragnvald Aaheim and the hikers of Oslo.) Norsk Skogbruk 16(10): 255–256. (In Norwegian.)

Gregory, G.R. 1955. An economic approach to multiple use. Forest Science 1(1): 6–13.

Gregory, G.R. 1987. Resource economics for foresters. John Wiley & Sons, New York. 477 pp.

Hasel, K. 1971. Waldwirtschaft und Umwelt: eine Einführung in die Forstwirtschaftspolitischen Probleme der Industriegesellschaft. Verlag Paul Parey, Hamburg und Berlin. 322 pp.

Haveraaen, O. 1992. Multiple–use forestry in Norway. Norwegian Journal of Agricultural Sciences, Supplement 8: 7–9.

Heden, E. 1984. Multiple–use forestry practiced by the Swedish Forest Service. In: Saastamoinen, O., Hultman, S.–G., Koch, N.E. & Mattsson, L. 1984. Multiple–use forestry in the Scandinavian countries. Communicationes Instituti Forestalis Fenniae 120: 23–26.

Helles, F. 1977. Om teorien bag flersidig produktion i skovbruget. (The theory behind multiple production in forestry.) Dansk Skovforenings Tidsskrift 62(3): 179–198. (In Danish.)

Helles, F. 1978. On the theory of external effects: as applied to forestry. Royal Veterinary and Agricultural University, Reports from Department of Forestry 6. 12 pp.

Helles, F. 1983a. Flersidig brug af skovene. (Multiple–use forestry.) Den Kgl. Veterinær– og Landbohøjskole, Skovbrugsinstituttet, Arbejdsnotat 16. 47 pp. (In Danish.)

Helles, F. 1983b. Flersidig udnyttelse af skovene. (Multiple utilization of forests.) Skoven 12: 350–353. (In Danish.)

Hoen, H.F. 1991a. Planlegging for flersidig skogbruk: om bruk av kvantitative modeller. (Multiple–use forestry planning: the use of quantitative models.) Landbruksøkonomisk Forum 4: 59–76. (In Norwegian.)

Hoen, H.F. 1991b. Hvordan drive (flersidig) skogbruk? (How to manage (multiple–use) forestry?) Norsk Skogbruk 7–8: 26–27. (In Norwegian.)

Hultkrantz, L. 1991. National account of timber and forest environmental resources in Sweden. Environmental and resource economics 2: 283–305.

Hultman, S.–G. 1984. Multiple–use forestry in Sweden: undefined and non–existent? In: Saastamoinen, O., Hultman, S.–G., Koch, N.E. & Mattsson, L. 1984. Multiple–use forestry in the Scandinavian countries. Communicationes Instituti Forestalis Fenniae 120: 27–32.

Huse, S. 1973. Flerbruksbegrepet og skogens funksjoner. (Multiple–use concept and the functions of forests.) Norsk Skogbruk 22: 395–397. (In Norwegian.)

Hytönen, M. 1992. Metsien monikäytön tutkimus Suomessa 1970–1990: tiivistelmäbibliografia. (Multiple–use forestry research in Finland 1970–1990: an annotated bibliography). Metsäntutkimuslaitoksen tiedonantoja 430. 395 pp. (Partly in English.)

Idékatalog för skogsbrukare. (Handbook of ideas for forest managers.) 1988. LTs förlag, Stockholm. 180 pp. (In Swedish.)

Jaatinen, E. & Saastamoinen, O. 1976. Metsien moninaiskäyttötutkimuksen perusongelmat. Summary: Multiple use of forests: basic research tasks. Silva Fennica 10(2): 141–147.

Jespersen, C. 1991. Hvordan kommer vi videre. (How can we go further.) Ugeskrift for Jordbrug 51/52: 827–829. (In Danish.)

Kellomäki, S. 1984a. Metsien sivutuotteet. Summary: By–products of the forests. Silva Fennica 18(4): 382–387.

Kellomäki, S. 1984b. Metsätaloudellinen ympäristönhoito. (Environmental management in forestry.) Silva Carelica 1. 200 pp. (In Finnish.)

Kellomäki, S. 1994. Metsätalous. (Forestry.) In: Kurki–Suonio, I. & Heikkilä, M. (eds.). Kestävän kehityksen edellytykset Suomessa. (The prerequisites for sustainable development in Finland.) Tammi, Helsinki. p. 417–482. (In Finnish.)

Kestävä kehitys ja Suomi. (Sustainable development and Finland.) 1990. Valtion painatuskeskus, Helsinki. 97 p. (In Finnish.)

Koch, N.E. 1990. Flersidigt skovbruk i går, i dag og i morgen. (Multiple–use forestry yesterday, today and tomorrow.) In: Landet og loven. Miljøministeriet/Skov– og Naturstyrelsen, Hørsholm. p. 109–113. (In Danish.)

Koch, N.E. 1992. Integrated multiple–use forest planning and management. In: Proceedings from IUFRO international conference "Integrated sustainable multiple–use forest management under the market system", September 6–12, 1992, Pushkino, Moscow Region, Russia. p. 305–314.

Koch, N.E. & Canger, S. 1987. Skovopbygning til glæde for friluftslivet. (Forestry for joyful outdoor recreation.) Skov– og Naturstyrelsen, Hørsholm. 239 pp. (In Danish.)

Koch, N.E. & Kennedy, J.J. 1991. Multiple–use forestry for social values. Ambio 20(7): 330–333.

Koch, N.E. & Kristiansen, L. 1991. Flersidigt skovbrug: et idékatalog. (Multiple–use forestry: a handbook of ideas.) Skov– og Naturstyrelsen, Hørsholm. 39 pp. (In Danish.)

Kristiansen, L. 1991. Case: Hov Skov. Ugeskrift for Jordbrug 51/52: 832–833. (In Danish.)

Kulturmiljövård i skogen. (Management of cultural heritage in forests.) 1992. Skogsstyrelsen, Jönköping. 259 pp. (In Swedish.)

Leary, R.A. 1985. Interaction theory in forest ecology and management. Martinus Nijhoff/Dr. W. Junk Publishers, Dordrecht. 219 pp.

Li, C.–Z., Mattsson, L. & Söderberg, U. 1990. Forests for timber production and environmental services: a conceptual economic analysis. Sveriges lantbruksuniversitet, Institutionen för skogsekonomi, Arbetsrapport 128. 36 pp.

Lindqvist, O. 1979. Pelastaako luonnon moninaiskäyttö luonnon? Summary: Will the multiple use of nature save nature? Silva Fennica 13(2): 132–135.

Loftsson, J. 1993. Forest development in Iceland. In: Alden, J., Mastrantonio, J.L. & Odum, S. (eds.). Forest development in cold climates. Plenum Press, New York. p. 453–461.

McArdle, R.E. 1953. Multiple use: multiple benefits. Journal of Forestry 51(5): 323–325.

McArdle, R.E. 1962. The concept of multiple use of forest and associated lands: its values and limitations. In: Proceedings of the Fifth World Forestry Congress "Multiple Use of Forest Lands", August 29–September 10, 1960. Vol. I. p. 143–145.

Madsen, E.M. 1991. Et moderne misfoster. (A modern mongrel.) Ugeskrift for Jordbrug 51/52: 836–837. (In Danish.)

Mattsson, L. & Li, C.–Z. 1991. Hur bör skogen användas? Ett samhällsekonomiskt problem. (How should forests be utilized? A socioeconomic problem.) Skog & Forskning 1: 43–53. (In Swedish.)

Metsätalouden ympäristöopas. (Environmental guidelines for forestry.) 1993. Metsähallitus, Vantaa. 112 pp. (In Finnish.)

Metsien moninaiskäytön työryhmän raportti. (Report of the multiple–use forestry working group.) 1985. Talousneuvosto, Metsä 2000–ohjelmajaosto, Helsinki. 59 pp. (In Finnish.)

Mikola, P. 1969. Monikäyttöinen metsä. (Multiple–use forest.) Metsä ja puu (7–8): 4–6.

Ministerial conference on the protection of forests in Europe, 16–17 June 1993 in Helsinki: documents. 1993. Ministry of Agriculture and Forestry, Helsinki. 56 pp.

Niesslein, E. 1985. Forstpolitik: ein Grundriss sektoraler politik. Pareys Studietexte 47. Verlag Paul Parey, Hamburg und Berlin. 150 pp.

Nilsen, P. 1991. Flersidig skogbruk ≠ skogbruk. (Multiple–use forestry ≠ forestry.) Norsk Skogbruk 4: 39. (In Norwegian.)

Non–legally binding authoritative statement of principles for a global consensus on the management, conservation and sustainable development of all types of forests. 1992. United Nations Conference on Environment and Development, Rio de Janeiro, 3–14 June 1992. 7 pp.

Pearson, C.A. 1944. Multiple use in forestry. Journal of Forestry 42(4): 243–249.

Petersen, K. 1989. Skoglig flerbruk: en litteraturstudie. (Multiple–use forestry: literature survey.) Norsk institutt for skogforskning, Ås. 80 pp. (In Norwegian.)

Rikare skog: 90 –talets kunskaper om naturvård och ekologi. (Richer forest: knowledge of nature protection and ecology in the 90s.) 1990. Skogsstyrelsen, Jönköping. 133 pp. (In Swedish.)

Saari, E. 1962. Metsän monikäyttö. (Multiple–use forestry.) Metsätaloudellinen aikakauslehti 79(7–8): 255–256. (In Finnish.)

Saari, E. 1970. Monikäyttö–sanan merkitys: lisäselvennystä. (The meaning of the multiple–use concept: additional clarification.) Metsänhoitaja 10: 336. (In Finnish.)

Saastamoinen, O. 1974. Metsien moninaiskäytön käsite ja perusteet. (Multiple–use forestry concept and principles.) Metsäntutkimuslaitos, Rovaniemen tutkimusaseman tiedonantoja 6: 42–50. (In Finnish.)

Saastamoinen, O. 1980. Metsien moninaiskäyttö. (Multiple–use forestry.) Metsäalan perusoppi 8. Otava, Keuruu. 108 pp. (In Finnish.)

Saastamoinen, O. 1988. Metsien monikäytön kehitys: arvioita menneestä ja nykytilasta. (The development of multiple–use forestry: evaluation of the past and present situation.) Metsäntutkimuslaitoksen tiedonantoja 288: 145–157. (In Finnish.)

Skogspolitiken inför 2000–talet: huvudbetänkande av 1990 års skogspolitiska kommitté. (Forest policy for the 21st century: main report of the 1990 forest policy committee.) Jordbruksdepartementet, Statens offentliga utredningar, SOU 1992:76. 343 pp. (In Swedish.)

Solbraa, K. 1989. Flersidig skogbruk. Summary: Multiple–use forestry. Norsk institutt for skogforskning, rapport 7. 35 pp.

Solbraa, K. 1991a. Skogbruk skal være flersidig. (Forestry will be multifunctional.) Norsk Skogbruk 6: 30. (In Norwegian.)

Solbraa, K. 1991b. Vekk med flersidig skogbruk. (Getting rid of multiple–use forestry.) Norsk Skogbruk 2: 25–26. (In Norwegian.)

Sølland, L. 1991. Flersidig skogbruk er forskjellig fra skogbruk? (Multiple–use forestry is different from forestry?) Norsk Skogbruk 6: 31. (In Norwegian.)

Strategy for sustainable forest management. 1994. Ministry of the Environment, Copenhagen. 64 pp.

Sweden's new forest policy. 1994. The National Board of Forestry, Jönköping. 13 pp.

UNCED: YK:n ympäristö– ja kehityskonferenssi, Rio de Janeiro 3.–14.6.1992. (UNCED: United Nations Conference on Environment and Development, Rio de Janeiro 3.–14.6.1992.) Ympäristöministeriö/Ulkoasiainministeriö, Helsinki. 239 pp. (In Finnish.)

Welp, M. 1993. Metsälait ja metsäpolitiikka Saksassa. (Forest laws and forest policy in Germany.) In: Palo, M. & Hellström, E. (eds.). Metsäpolitiikka valinkauhassa. (Forest policy in a melting pot.) Metsäntutkimuslaitoksen tiedonantoja 471: 23–73.

Wilkinson, C.E. & Anderson, H.M. 1987. Land and resource planning in the National Forest. Island Press, Covelo, California. 396 pp.

Ympäristön ja kehityksen Suomen toimikunnan mietintö. Summary: Report of the Finnish Commission on Environment and Development. Ympäristöministeriö, Komiteamietintö 1989:9. 264 pp.

4 Multiple–use forestry – a concept of communication

Jann Fernand[1]

Abstract

The multiple–use forestry concept has been interpreted as a new sector within forestry, as a synonym to forestry and as a notion superior to the interests related to forests. Different professional and other interest groups have a tendency to define the concept to suit their own purposes. There has been debate, for example, on the relationship between nature protection and multiple use. The German theory of forest functions is elaborated further by specifying the social dimension more distinctly. This is done by widening the scope of the recreation function by emphasizing the self–realization aspect of recreation. Also a new function of perception is discussed in addition to the traditional utility and protection functions. Multiple use arises from a communicative process between the interests related to these four functions. Tying multiple use to the concept of communication results in a process of creativity which may lead to better economy, protection and recreation.

Keywords: forest functions, concepts, foresters, environmentalists, communication, philosophy.

[1] Agricultural University of Norway
P.O.Box 5044, N–1432 Ås, Norway

4.1 Introduction

Although foresters in Germany were using terms related to multiple use as early as the 19th century ("Waldnebennützungen" (Cotta 1860) and "Zweck des Waldes" (Gayer 1882)), the prevailing discussion concerning multiple use in the Nordic countries seems to have originated with "The Multiple Use Sustained Yield Act", approved in 1960 in USA. This act was presented at the Fifth World Forestry Congress in Seattle the same year. In Denmark and Norway this has been said to be the starting point of the current debate (Koch & Kristiansen 1991, Samset 1988). The term multiple use was first used in Scandinavia at the end of the 1960s and had become well known by the first half of the 1970s. This did not mean that the notion multiple use – or its various translations – was unambiguous. On the contrary, many used it to legitimize their own views or understandings because of its relative blankness in relation to other notions. That this lack of agreement was more than purely semantic became obvious during the power struggle over, for example, budget dispositions. The need for an unambiguous definition became clear in all the Nordic countries, and almost all organizations involved sought to create such a definition. The formulations used were understood to be descriptive by those who looked at multiple use as an interaction between different goals, in which forestry always had been involved. The normative interpretation was imposed by those who saw the possibility of gaining acceptance of their own ideas on what forestry ought to be like.

The antagonism which arose, connected to how foresters interpret multiple use, is typical for people with a natural scientific approach to problem solving. When we start with a definition, we can soon move into a natural scientific world of registration and techniques. There is no scientific method for formulating questions, because there can never be a complete description of the reality. Our interpretation of the reality will still be our point of departure when we pose questions, but our interpretation has to be developed continuously. Problems connected with formulating questions are therefore not something we as foresters are socialized to be aware of. If we had understood why the new notion was necessary, our entire experience of the reality would have been different. Our framework has been to take reality for granted, in order to get to work. That other questions may be better than our own has not been seen as a problem, as long as our interpretation of reality satisfies a certain definition.

> "When we talk about "multiple–use forestry", it is a common experience that the notion is very well received by all of the interest groups. Maybe this is because the different groups each imagine that "multiple use" signifies that exactly their interests will be promoted at the cost of other considerations which must be included within multiple use." (Koch & Kristiansen 1991, translated by J. Fernand).

Hall (1963) has given a corresponding analysis of how the multiple–use act was perceived in the USA. Hall concludes that it is impossible to decide whether the concept was supposed to be apprehended as normative or descriptive.

The discussion of what multiple use is – what is the concept of multiple use – has continued in all the Nordic countries and has been revised several times during the years we have had this term in our vocabulary. The longest discussion has been on whether multiple use has to be comprehended as conflict or harmony, optimization or adaptation. Is man or nature the subject/object in our comprehension of multiple use and how is the concept of multiple use related to other areas of nature management? The latter will be the point of departure for this article.

4.2 Seeking a position for the notion multiple use

Multiple use as a sector within forestry

"A translation of the notion multiple use as "flerbruk" is possibly philologically acceptable. However, ...the general considerations we are trying to take into account have a long tradition in Norway and are completely covered by the roomy notion forestry." (Samset 1988, translated by J. Fernand).

Samset's statement can be interpreted in two ways. Firstly, multiple use can be a sector of forestry along with other forest activities such as economy, biology and technology (*Figure 4.1*). As multiple use, according to Samset, already has a long tradition in other forest activities, it will be a vacant box or needless sectorization.

Notions such as "new forestry", "urban forestry" and "landscape ecology" well satisfy the criteria for such a placement.

Multiple use as a synonym for forestry

The other interpretation equates the concept of multiple use with the concept of forestry. This was discussed before the 1976 amendments of the Norwegian Forest Act (Om lov om endringer... 1974/75). It was recently repeated by

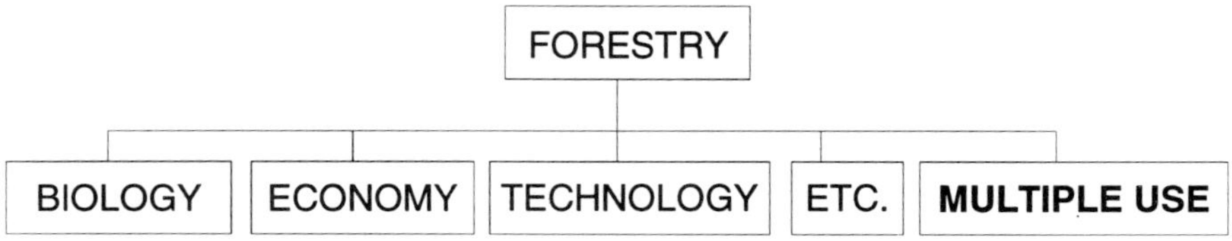

Figure 4.1 Multiple use as a new sector within forestry, equal to existing and future activities.

the coordinator of the Norwegian research programme "Forest Ecology and Multilateral Forestry", who equates multilateral forestry and forestry (Solbraa 1991a, 1991b).

"Multilateral forestry is used today in order to denote the intention to follow the Forest Act. Even if such a notion is useful during a transition stage, we do have to cross out multilateral as soon as possible." (Solbraa 1991a, translated by J. Fernand).

With no distinction between "multiple–use forestry" and "multilateral forestry" (forestry with multiple–use considerations), the passage above is a good example of those that equate forestry and multiple use (*Figure 4.2*) (Fernand 1991).

The models shown in *Figures 4.1* and *4.2* legitimize, but do not stipulate, that multiple use remains a problem to be solved internally among foresters. Multiple use is given the role of adapting forestry to take into consideration other interests.

Forest actions are not an isolated act among foresters. Modern forest activity can only be legitimized in a wider context, in an interaction with a larger society. Both of the models presented make a continued forest monopoly possible when it comes to posing the questions. Foresters can still seek internally among themselves what the reality is like. By this it becomes nearly impossible to evaluate the traditions on which forestry is based. In relation to what are we supposed to make evaluations?

It is not enough to register outside criticism and act in response to the critique. The criticism is most likely a reaction to something that is wrong. Those who criticize forestry probably do not know any more about the hidden values influencing their reactions, than foresters understand how they find themselves in their present tenuous positions.

If multiple use is put as a sector under forestry, people in charge of multiple–use management will hardly be able to question presuppositions in forestry policy. In a sectorized society, multiple use will need admission to groups with forest interests, including foresters. If multiple use is interpreted as a sector under forestry or as a synonym for forestry, the concept of multiple use will not make any contribution to the legitimization needed to be critical of presuppositions formed outside the sector of forestry. In order to be part of an interaction, multiple use needs admission to all those interested in forests.

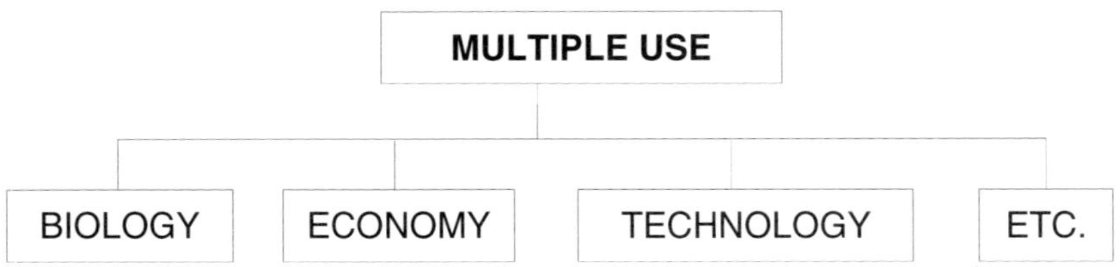

Figure 4.2 Multiple use as a synonym for forestry.

Multiple use as a superior notion

In an official report by the Norwegian Ministry of Agriculture, the term multiple use is elucidated as follows:

> "Multiple use of the forest is understood to mean several forms of use of forest resources on the same area or at the same time, as opposed to single use of the same resources. ...The concessions forestry has to make are basic presuppositions for the multiple use mentioned, but forestry alone does not direct multiple use. We therefore say that forestry takes multiple use into consideration or is multilateral. ...When forestry is supposed to adapt itself to other interests, it is considered multilateral...." (Flersidig skogbruk 1989, translated by J. Fernand).

Multiple use has thus, to official authorities in Norway, become a notion superior to forestry, while forestry is run multilaterally or with multiple–use considerations (*Figure 4.3*). Accordingly, it is possible to refer to protection with multiple–use considerations or adaptation for recreation with multiple–use considerations.

Some foresters want to abandon the notions "multilateral forestry" and "forestry with multiple use considerations" (e.g. Solbraa 1991a, b). These notions have, however, a major and positive mission. They are discussed in a lively way, and they are connected to a creative process within practical forestry. When a forester needs to point out the importance of new methods, it can be productive to have a notion showing that these methods take into consideration other forest–related interests.

If we see multiple use as a concept superior to forestry, foresters might feel their position and power threatened. Even nature conservationists might feel their power threatened, as multiple use would also be superior to protection.

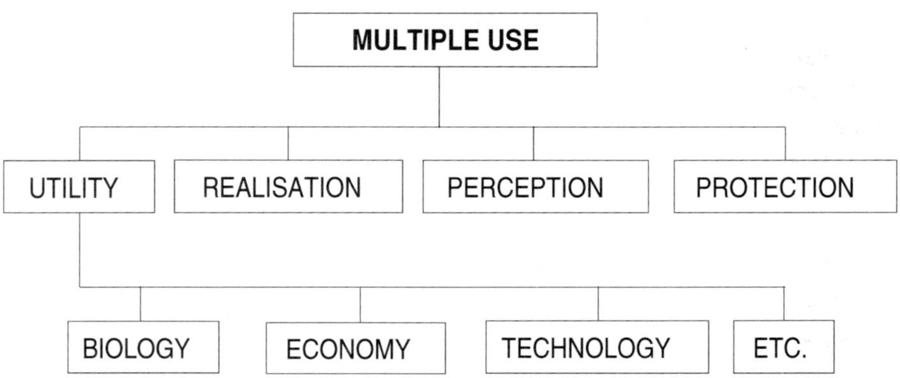

Figure 4.3 Multiple use as a notion superior to our interests in relation to forests.

One strategy used by conservationists, in order to increase the importance of protection, has been to separate the motives for multiple use and protection (e.g. Hågvar 1989). If, however, we break down multiple use according to motives, we very soon end in a situation where multiple use becomes equivalent to production of wood fibre and the concept of multiple use is unable to cover more than forestry. The reason for introducing the new notion then becomes purely strategic, in order to manipulate an opinion.

Multiple use as a notion superior to our interests in relation to forests can bring added complexity to decisions concerning foresters' and conservationists' activity. Their status can be reduced, and external financial aid for their system can be in danger. For individuals, a new career ladder can also be considered as threatening to one's own interests. Seeing multiple use in a superior position comes up against high barriers which the actors might not be conscious of.

4.3 Forest functions

Among concepts related to multiple use, forest functions play an important part. They are particularly used in Central Europe, but only sporadically in Scandinavia (e.g. Huse 1973, Andersson 1979, Koch 1990).

The theory of forest functions was introduced in Germany in the 1930s (Dieterich 1953), and discussion of this theme has been one of the main propellants of the central European approach to multiple–use forestry. A large number of forest functions have been described (e.g. Hasel 1971). Usually, the theory of forest functions is presented with three main functions: utility, protection and recreation (Nutz–, Schutz– und Erholungsfunktion). It is, however, desirable to specify the social dimension more distinctly (cf. Reunala 1987). I therefore propose four main functions or function groups: utility, realization, perception and protection. For readers familiar with the use of forest functions, it should be noted that the forest functions can be interpreted independently of a dialectic historical and conflict orientated frame.

Utility function

The forest is a source of raw materials and income. It represents a reserve of capital which can be realized by property transactions or by harvesting of resources. Forests also provide employment.

Traditional production and harvesting of timber are covered by the utility function. The forest will also have a function of utility for those who sell hunting or fishing rights, for those who work in the tourist industry or for others with an income connected to forest areas. The forest has a function of utility for those who harvest products for personal use (berries and mushrooms), provided that economy, not recreation, is the main motivation. The function of utility also covers nature as a resource for new medicines in the future.

Realization function

The forest has a function for those who want to spend their leisure time. How we realize our lives is dependent on norms, which differ from culture to culture and time to time, but still, human realization can be very conformist (i.e. it has a normative aspect). Sports for exercise and competition, hunting, fishing and berry picking are well covered by the function of realization, as long as leisure is the main motivation.

The word "realization" (or self–realization, which has to take place in the context of normative regulated interaction) has been chosen for several reasons. It allows the concept to be used in times or places where "leisure time" and "working time" overlap for forest users. Another reason is to make the theory of forest functions comparable to other sectors of society. For example, cinemas and theaters have a realization function, too. Thus, it is easier to define the importance of forest for recreation in a wider context.

Most Western societies are today in transition between a situation where realization and utility are woven together and a situation where "objective" work and "normative" self–realization are separated. In such a transitionary stage, it is not surprising that differences arise between the rural and urban population over interpretation of the forests.

Perceptive function

In contrast to the function of realization this function is oriented not toward the normative, but toward the subjective. This function covers the forest as a symbol for our emotions of height, time, light, spirit and wrath (see Bevan 1938), as an archetype (Reunala 1984), as well as how the forest binds us to history as a process and as a symbol for existing culture.

To facilitate a general survey, the function of perception can be divided in two. Firstly, as a medium for communication with what we can call our hidden values, or our intuitions of second order (Rawls 1971). The certainty of the forest's existence and continuity is significant for individual and social self–perception. Secondly, as experience in a forest stand. Different forms of the forest and its infrastructure give us associations, and thereby words, in our culture and emotional lives.

We all bob about in a sea of symbols, whether we are conscious of them or not. The function of perception has to do with how we interpret nature and human activity in relation to nature. Our creativity depends on our ability to sense, reconstruct and question our symbols. To describe our perception of a tree in terms of chemical formulas and cell structure is as genial as explaining a piano concert of Chopin by studying the tracks on a record.

I have intentionally not drawn aesthetics into this presentation. I find this difficult because most discussions of this concept end in a designation of good and bad. It is possible to give aesthetics a deeper content, but as long as aesthetics – by nature management – is known as pleasure of a different and superior quality, or that aesthetic experience is distinguished not by pleasure at all, but by a special aesthetic emotion, I can not see any reason for using aesthetics as a point of departure for our purpose. This does not indicate that I intend to forget about aesthetics, rather that I want to treat it more as action, creation and re–creation, and less as an attitude.

Protective function

There is no way in which we can suppose distinct notions. But even so, through all our descriptions we maintain the existence of a unity, a center ("das Sein" – Heidegger (1967)); as something which makes the reality static and stabilizes our description. In the theory of forest functions, nature plays the role of a center for understanding. The forest becomes a prerequisite for constructive communication about forests. In its most extreme consequence – if we lose the forests – it will soon be absurd referring to forests in communicative action (Habermas 1989, Fernand 1989).

The existence of forests provides the existence of humans and human relations to forests. The importance of forests follows from our descriptions. Forests have a protective function for species and ecosystems. Forests maintain water quality, prevent soil erosion, protect us from noise, improve air quality and are a filter for pollution. Forests can encourage humans in relating to nature, including relating to each other. The existence of forests insures the development of new ways for humans to relate to forests and to nature.

This interpretation of the protective function must however not be confused with an extensive use of strategic and instrumental actions referring to the functions of utility and realization, used by nature conservationists. The most dominating strategy nowadays is to legitimize protection by combining an interest for objects with an economic rational mentality. This strategy very soon brings nature conservation under the function of utility. A utilitarian strategy of conservation is not necessarily wrong, but it is incomplete, and has to be corrected through an interaction with other forest functions. One of the purposes of the introduction of a function theory is to establish a frame to encourage criticism, not only for foresters, but for everyone in society who is concerned about nature.

Negative functions

The existence of negative forest functions does not disturb the function theory. Examples from within the perceptive function group which might be perceived as negative are that forest darkens the landscape and impedes orientation. We shall take a closer look at another example, much discussed in some of the Scandinavian countries. One of the protective functions of forests is the function of conserving forest wildlife, which in certain parts of Scandinavia still includes large predators. If the problem predator/domestic animals is to be removed completely, this will mean extermination of our population of large predators. Such extermination will thereby be nothing but a denial of the validity of a value principle which no science can prove, but to which our actions will still be oriented (Weber 1971). This is, in respect of the creation of new perceptions and social norms, the opposite of what we are trying to establish through the theory of forest functions. To what extent we want to have a predator population is therefore a question of management, which is covered by multiple use. The theory of forest functions can help us here to create better analyses, and to question already established presuppositions. Because of this, I consider the phenomenon of negative functions – to the extent they exist – to strengthen the theory of forest functions.

4.4 An interpretation of the theory of forest functions

In describing forest functions, we have started to come to terms with the concept of multiple use. Multiple use covers the four functions, but also – and this is probably the most important – the relation between the functions. Multiple use is more than the sum of four functions, since the functions are also mutually related to each other. This mutual relation is very central in the theory of forest functions. How we affect one function will always be reflected and have an influence on the other functions. To formulate these connections is not just a good exercise for the brain, but also a good source of enlightenment concerning multiple use. It becomes easier for each of us to see our position and to orient ourselves in the surrounding world. We have now reached the following pictures of multiple use (*Figures 4.4* and *4.5*).

Looking closely at *Figure 4.4*, we see that when we refer to one or another of the functions, the forest will still be the subject, while the function becomes the object. The functions as objects always have the same subject, the forest. This has been interpreted by Glück and Pleschberger (1982) to mean that nature becomes the subject while humans remain as the objects. If Glück and Pleschberger's interpretation is right, the theory of forest functions would imply a revolutionary new view of the relation between man and nature. The opposite view was already established in antiquity by the transition to abstract notions (Hegge 1978). The same problem of subject/object arises if we try to understand the functions in a hierarchical structure, where protection is placed on the top. It has been argued by conservationists that their protective function is so basic that it has to be placed above multiple use in *Figure 4.5*. In order to attain the equivalent of what those representing protection are looking for, I say that the function of protection is communicative. Protection is a supposi-

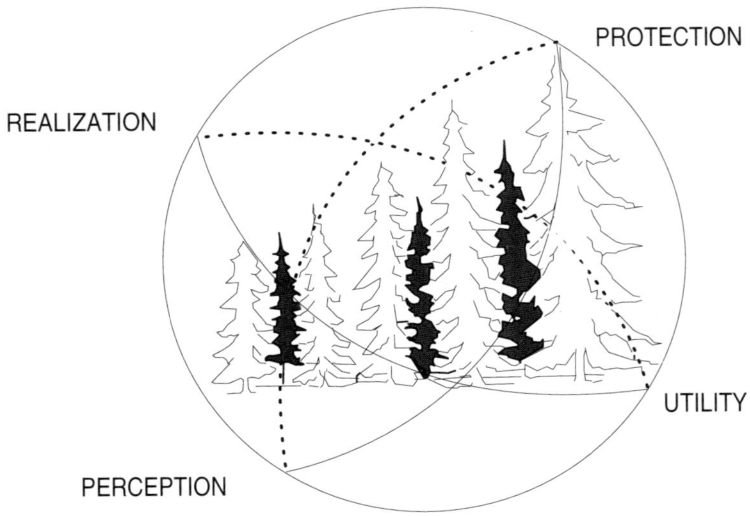

Figure 4.4 Forest functions.

tion which enables us to communicate intelligibly about forests. Without forests it makes no sense to talk about forests – and in its most extreme consequence – nobody to talk about them. Without language (the communicative), nobody is talking.

If we again take a look at *Figure 4.4*, it is possible to see that the theory of forest functions is not as revolutionary as is here suggested. You will see that the forest is the kernel, center or subject. What encloses the kernel is, however, not people, but the interests humans have in their relation to nature. As all the interests or functions are mutually related to each other, no humans will be able to neglect any of the functions. The gravity of human interests in forests will therefore always be within this sphere – in the forest. Whether the forest or the human is the subject is thereby a question which the theory of forest functions does not manage to answer. The relationship between individuality and sociality (for instance language, nature) within nature management is thereby a problem which the theory of forest functions is nowhere near solv-

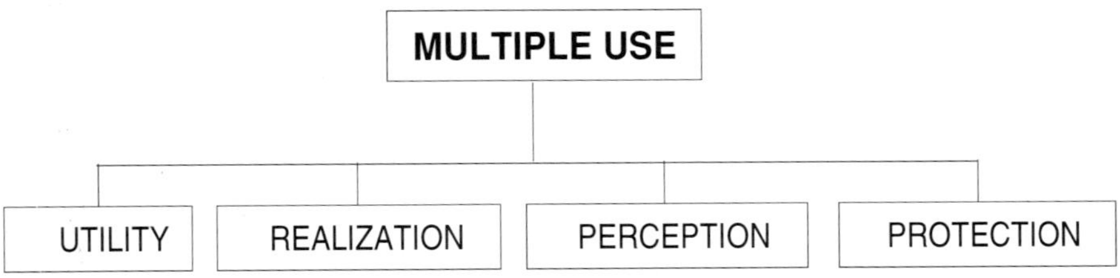

Figure 4.5 Forest functions.

ing. This does not prohibit the theory of functions from being a good foundation for further analyses.

It is necessary to point out that even if a representative invokes an interest, this interest does not represent human beings, but rather the interest the representative is "paid" to promote. Thus it is possible to see how the development of society toward a sectorized superstructure also has had its influence on nature management. On the other hand, it can be asserted that those who claim the interests of foresters claim the interests of the landowners, and they are humans. But does the forester protect the interests of landowners? According to official policy in Scandinavia today, the landowners are supplied with a vocabulary which expresses the forest's function of utility. The landowners are supposed to defend and legitimately interpret the forest economically. This is all very well, and many have done a major and important job in informing the landowners on this subject. But how does a landowner legitimize decisions which are correct for him as an individual – but not according to the textbook – when he lacks words and expressions for other interests in his forest? The forest has possibly the strongest function of perception for the landowner, at the same time as his management of the forest is of major importance for his socio–cultural position in the local community. Those who work with multiple use have a responsibility to promote words and argumentation which legitimize the whole spectrum of interests we have in relation to forests.

My theory of forest functions states that the forest has four main functions which are all mutually interrelated. One cannot, according to this theory, claim that one function is more important than another. There is no origin in perception, no priority given to vision, and no data of observation. Multiple use is not an account of rationalized movement from percept to concept. It begins and ends in concepts. The theory of forest functions gives us a basis for seeing ourselves and our actions in a new perspective. The theory of forest functions is a basis for reaching a deeper acknowledgement of the set of problems we are facing, but does not in itself make our actions legitimate.

4.5 What is multiple use?

Multiple use arises from a communicative process between the four functions or our interests in forests. This noticeable feature is valid whether the communication is between representatives for the different functions or between interests within each of us. Multiple use presupposes a substantial communicative rationality of value, which concentrates our attention on possibilities, possibilities of finding well–informed solutions to problems and values which arise and evolve in a communicative connection. This "production" of knowledge up through the system will later form the basis for new signals down through the system, signals which indicate where we lack knowledge and which strategy a representative ought to choose without disqualifying other representatives or interests. We can here study the relation between the four functions, interests and representatives. We can look at possibilities for coordination of activities which are separate today (for instance, registrations and infrastructure). We can come up with new concepts for education. We can start projects to analyze the individual sectors in fields where the sectors have not been willing or able to look at themselves in a new way. Multiple use can be an umbrella for case studies of the development of new procedures of settlements. In this way multiple use of forest becomes: philosophy, pedagogics, psychology, sociology, social anthropology and political science, which at the same time demand a basic knowledge of biology and economics (history) of the areas administered, also in order to communicate with those who are responsible for nature management today.

This can happen by analyzing and naming the context in which each sector is operating, or by replacing previously established understatements for one sector's activity. Such a change does not presuppose – as many seem to believe – that one sector will become less important. To the contrary, one sector can become increasingly important at the same time as other sectors become important. When we tie multiple use to a concept of communication we are getting into a process of creativity where we are also searching for a better economy, more protection and to do more in favour of recreation. This is in contrast to forestry with multiple–use contradictions, with its instrumental bureaucratic and strategic character, which creates more at the expense of another (strategic manipulation of a subject) or something else (instrumental manipulation of an object).

Interdisciplinary work, so much discussed these days, is not about confronting already constituted disciplines (none of which, in fact is willing to let itself go). To do something interdisciplinary it is not enough to choose a "subject" (a theme) and gather around it two or three sciences. Interdisciplinarity consists of creating a new object that belongs to no one (Barthes 1984).

References

Andersson, B. 1979. Vad handlar debatten egentligen om? (What is the debate really dealing with?) Sveriges Skogsvårdsförbunds Tidskrift 77(1): 75–82. (In Swedish.)

Barthes, R. 1984. Le bruissement de la langue. Le Seuil, Paris. p. 97–103.

Bevan, E. 1938. Symbolism and belief. George Allen & Unwin Ltd., London. 387 pp.

Cotta, H. 1860. Grundriss der Forstwissenschaft. 5. Auflage. Arnoldische Buchandlung, Leipzig. 348 pp.

Dieterich, V. 1953. Forstwissenschaftspolitik. Eine Einführung. Verlag Paul Parey, Hamburg und Berlin. 398 pp.

Fernand, J. 1989. Refleksjoner omkring flerbruk og flerbruksbegrepet: fusjon og konfusjon? (Reflections around multiple use and multiple–use concept: fusion and confusion?) M.Sc. thesis. Norges landbrukshøgskole, Institutt for skogskøtsel, Ås. 110 pp. (In Norwegian.)

Fernand, J. 1991. Flersidig kaos. (Multiple chaos.) Norsk Skogbruk 10: 30–31. (In Norwegian.)

Flersidig skogbruk: skogbrukets forhold til naturmiljø og friluftsliv. (Multiple–use forestry: the relationship of forestry with natural environment and outdoor recreation). 1989. Norges offentlige utredninger, NOU 1989:10. 139 pp. (In Norwegian.)

Gayer, K. 1882. Der Waldbau. Verlag Paul Parey, Berlin. 592 pp.

Glück, P. & Pleschberger, W. 1982. Das Harmoniedenken in der Forstpolitik. Allgemeine Forstzeitschrift 37: 650–655.

Habermas, J. 1989. The theory of communicative action: the critique of functionalist reason. Vol. 1–2. Policy Press, Cambridge.

Hågvar, S. 1989. Flerbruk kan ikke erstatte vern. (Multiple use can not replace protection.) Aktuelt fra Statens fagtjeneste for landbruket 4: 87–96. (In Norwegian.)

Hall, G.R. 1963. The myth and reality of multiple use forestry. Natural Resources Journal 3: 276–290.

Hasel, K. 1971. Waldwirtschaft und Umwelt. Eine Einführung in die forstwirtschaftspolitischen Probleme der Industriegesellschaft. Verlag Paul Parey, Hamburg und Berlin. 322 pp.

Hegge, H. 1978. Mennesket og naturen. Naturforståelsen gjennom tidene – med særlig henblikk på vår tids miljøkrise. (People and nature. Understanding of nature through the ages – with special emphasis on the present environmental crisis.) Universitetsforlaget, Oslo/Bergen/Tromsø. 162 pp. (In Norwegian.)

Heidegger, M. 1967. Sein und Zeit. Tübingen.

Huse, S. 1973. Flerbruksbegrepet og skogens funksjoner. (Multiple–use concept and the functions of forests.) Norsk Skogbruk 22: 395–397. (In Norwegian.)

Koch, N.E. 1990. Flersidigt skovbrug i går, i dag og i morgen. (Multiple–use forestry yesterday, today and tomorrow.) In: Landet og loven. Miljøministeriet/Skov– og Naturstyrelsen, Hørsholm. p. 109–113. (In Danish.)

Koch, N.E. & Kristiansen, L. 1991. Flersidigt skovbrug: et idékatalog. (Multiple–use forestry: a handbook of ideas.) Miljøministeriet/Skov– og Naturstyrelsen, Hørsholm. 39 pp. (In Danish.)

Om lov om endringer i lov av 21. mai 1965 om skogbruk og skogvern. Ot.Prop. Nr. 29 (1974–75). Forvaltningstjenestene, Statens trykningskontor, Oslo. 42 pp. (In Norwegian.)

Rawls, J. 1971. A theory of justice. The Belknap Press of Harvard University Press, Cambridge, Massachusetts. 607 pp.

Reunala, A. 1984. Forest as symbolic environment. Communicationes Instituti Forestalis Fenniae 120: 81–85.

Reunala, A. 1987. Forest benefits in Finland. In: Proceedings of the 18th IUFRO World Congress for Economic Value Analysis of Multiple–use Forestry. Oregon State University, Corvallis. p. 47–60.

Samset, I. 1988. Førti års samarbeid, tretti års naboskap. (Forty years of cooperation, thirty years of neighborhood.) Vi og Værket 33(2): 9–17. (In Norwegian.)

Solbraa, K. 1991a. Vekk med flersidig skogbruk. (Getting rid of multiple–use forestry.) Norsk Skogbruk 2: 25–26. (In Norwegian.)

Solbraa, K. 1991b. Skogbruk skal være flersidig (Forestry will have multiple objectives). Norsk Skogbruk 6: 30. (In Norwegian.)

Weber, M. 1971. Makt og byråkrati. (Power and bureaucracy.) Gyldendal Norsk Forlag, Oslo. 223 pp. (In Norwegian.)

5 Timber production and the forest industry

Marjatta Hytönen[1]
Sigurdur Blöndal[2]

Abstract

By the end of the 19th century, the forest resources of the Nordic countries were in a degraded state because of overutilization. The structure of forests was also shaped by fires, shifting cultivation and grazing. Silvicultural methods evolved fast in the first half of the 20th century and were intensified in the 1960s and 1970s, the main aim being timber production. Today, timber resources are growing faster than they are cut. In the 1990s, all the Nordic countries have started to look for new ways to manage forests more in harmony with nature and concern for interests other than timber production is increasing. The wood–processing forest industry is a prominent sector in the national economies of Finland, Norway and Sweden. In Denmark, and to some extent also in Iceland, wood processing is recognized as a significant local livelihood. Wood–based livelihoods will be important sources of income, employment and cultural identity also in the future in the Nordic countries.

Keywords: forest resources, timber, silviculture, forest industry, environmental awareness, trade, employment.

[1] Finnish Forest Research Institute
Unioninkatu 40 A, FIN–00170, Helsinki, Finland

[2] Kviaklettur, Hallormstadur, 701 Egilsstadir, Iceland

5.1 Introduction

Throughout history, wood has been one of the important natural commodities sought by man, and the Nordic countries have been no exception. Today, the forest and utilization of wood form one of the pillars on which the modern welfare society of at least three of the five Nordic countries relies. However, as the social structure gradually changes from rural to urban, the primary industries decrease in importance. The secondary manufacturing sector and services provide occupations for the majority of people. Consequently, forests are assigned a new role in which timber production plays a lesser part, and the importance of recreation and environmental protection increases. In despite of the changes, the wood–based livelihoods will be important sources of income, employment and cultural identity also in the future in the Nordic countries. Therefore, this chapter concentrates on describing the present trends and future challenges concerning Nordic timber resources and their management and utilization.

5.2 Timber resources

Most of the forests of Finland, Norway and Sweden belong to the boreal coniferous zone. The southern parts of these countries and the whole of Denmark belong to the temperate mixed forest zone (Kuusela 1990). The only forest–forming native tree species in Iceland is birch (Naturskogar i Norden 1994).

The structure of forests of the Nordic countries was heavily shaped by fire and overexploitation until the beginning of the 20th century. Since then forests have been affected by organized silviculture. Recently, various societal and forest management factors have brought about an expansion of timber resources.

Denmark

Forests cover 11 % (466,000 ha) of the total area of Denmark. About 70 % is arable land. The remaining areas include bogs, heaths, dunes, built–up areas and gardens (Brief glimpses... 1982, The forest resources... 1992).

Before man came on the scene, 90 % of Denmark was covered by trees. The majority of the original tree species were broadleaves. People started to utilize the forest around 2500 B.C. By the 18th century, tree resources had been used up and forest land cleared, leaving only 3 % forest coverage. In 1762, a German forester Johan Georg von Langen was invited to organize forest management in Denmark. Since then, Denmark's forests have been managed systematically and intensive afforestation efforts have been carried out, mainly using exotic conifers (*Table 5.1*) (Brief glimpses... 1982).

Table 5.1 Tree species distribution in Denmark according to the Nationwide Forest Inventory 1979–82 (The forest resources... 1992).

	1,000 ha	%
Conifers	316	68
Beech	87	19
Oak	33	7
Other broadleaved	30	6

Today, about two thirds of the forest area of Denmark is covered by conifers. The dominant species is Norway spruce (*Picea abies*). Sitka spruce (*Picea sitchensis*) has been planted in coastal areas because it tolerates salty air. Since 1864, large areas of heaths have been forested with mountain pine (*Pinus mugo*). Other exotic conifers used in afforestation include Douglas fir (*Pseudotsuga menziesii*), silver fir (*Abies alba*), grand fir (*Abies grandis*), Lodgepole pine (*Pinus contorta*), and larch (*Larix sp.*) (Brief glimpses... 1982, The forest resources... 1992, Scandinavia... 1993).

Beech (*Fagus sylvatica)* and oak (*Quercus robur*) are the most widespread broadleaved tree species in Denmark. Other common broadleaved species are ash (*Fraxinus excelsior*), sycamore (*Acer pseudoplatanus*), alder (*Alnus* sp.), elm (*Ulmus* sp.), aspen (*Populus* sp.), birch (*Betula* sp.) and lime (*Tilia* sp.). The broadleaved forest area is slowly increasing. The amount of oak has shown the largest increase. It has been used to make forests more resistant to winds. Elm, white alder, sycamore and common alder are increasingly planted in shelterbelts to protect the topsoil from drifting in sandy areas. They are replacing white spruce (*Picea glauca*), which was formerly preferred for this purpose (Brief glimpses... 1982, The forest resources... 1992, Scandinavia...1993).

In 1991, the Danish government decided to speed up reforestation of the country. It was decided to double the current forest area over the next 75 to 100 years. To achieve this goal the forest area must be increased by about 5,000 ha per year (Scandinavia... 1993).

Plum et al. (1990) list the following reasons for the decision to increase forest cover:

1) Farming has become less profitable. Because of the surplus production the European Union (EU) has started to regulate farming by quotas.
2) Growing concern about the environment has made it desirable to re–establish forests in open areas in order to absorb more carbon dioxide from the air, to prevent the movement of sand and to protect watercourses.
3) The need for increased tourism and for enhancing the amenity values of areas in general.

4) The benefit of planting woodland as a source of wood. Hardwoods are expected to be particularly profitable.

From 1976 to 1990, the annual planted acreage amounted to only 300–400 hectares. Several government and EU initiated programs encourage afforestation of agricultural land but the support has been insufficient to persuade forest owners to establish the desired amount of new forests (Scandinavia... 1993). To get government grants for reforestation, forest owners have to also agree to maintain the new forest areas in the future. This compulsory condition has created resistance (Grayson 1993).

However, Danish forest resources are growing (*Table 5.2*). Because of the extension of forest area and the uneven distribution of age classes, having a preponderance of young stands, harvesting potential will be doubled in the next thirty years (Strategy for sustainable forest... 1994). On the other hand, government policy will increase the area of forests that are not subject to commercial management. This policy will have a decreasing effect on the level of removals, especially from broadleaved forest (Outlook for the European... 1994).

Table 5.2 Growing stock, net annual increment and removals in exploitable forest in Denmark (figures over bark; the removals in 1990 have been converted to over bark figures by multiplying by 1.16).

	1950	1960	1970	1990
Growing stock, 1000 m^3	40 000	44 000	42 000	55 000
Growing stock, m^3/ha	91	118	113	132
Net annual increment, 1000 m^3	2 500	2 530	2 300	3 200
Total removals, 1000 m^3	2 180	1 970	2 470	2 100
Net annual increment, m^3/ha	5.7	6.8	6.2	7.7
Removals m^3/ha	5.0	5.3	6.6	5.0

Sources: years 1950, 1960, 1970 (European timber trends... 1986), year 1990 (Outlook for the European... 1994).

Finland

About 78 % (26 million ha) of the total area of Finland is forested. Lakes and other waterways cover 10 % (3 million ha). Exploitable forests grow on 57 % of the land area, while 13 % is arable land. One third of Finland is covered by peatlands. Of this, about 50 % (5 million ha) have been drained to increase timber production (Metsätilastollinen vuosikirja 1994).

The structure of Finnish forests has been affected by fire. It is estimated that, in the 18th and 19th centuries, dry forests burned at 50 year intervals and the more moist sites at 120 year intervals, on average. Fires diversified the landscape; old forests were able to develop in areas which the fire could not reach, but it also prevented natural succession and the formation of old forests in more accessible areas (Metsätalous ja ympäristö 1994).

Starting from the 16th century, the forest structure was also shaped by shifting cultivation, forest grazing and tar burning. Shifting cultivation increased the amount of broadleaved trees in the landscape. Forest pasture created open forest areas with retarded growth of trees. Tar burning consumed large amounts of pine and thus contributed to the spreading of spruce to areas where it would not grow otherwise (Metsätalous ja ympäristö 1994).

From the 1950s, forest structure has been modified by intensive silviculture. Widely used clearcutting and artificial regeneration methods have changed the forest landscape. They have resulted in even–aged stands, the decrease of mixed forests and stands dominated by broadleaved trees, and in excessive planting of pine at the cost of other tree species (Metsätalous ja ympäristö 1994). During the past ten years, silvicultural methods have been diversified significantly.

Today, Scots pine (*Pinus sylvestris*) is the main species in 65 % of the productive forest area. Norway spruce is the predominant species on 26 %, and birches (*Betula pendula* and *B. pubescens*) are the main species on 7 %. The rest is dominated by other broadleaves, mainly aspen and alder (Metsätilastollinen vuosikirja 1994). Exotic tree species have not been used in Finland on a large scale (*Table 5.3*).

Table 5.3 Total growing stock volume by tree species in Finland (Yearbook of Nordic statistics 1994).

	Million m^3	%
Scots pine	850	45
Norway spruce	690	37
Broadleaved	340	18

Finland's forest resources and their annual growth have increased from the beginning of the 1960s (*Table 5.4*). The annual growth is continuing to increase, primarily due to the relatively high percentage of young, fast–growing stands. At the beginning of the 1990s, only about 60 % of the total growth was harvested annually (Parviainen 1994). The situation is partly caused by intensive forest management and partly by the decreased dependence of forest owners on income from selling timber. In addition, attitudes opposed to large–scale cuttings have become common. These trends and the relatively low industrial demand for domestic roundwood have contributed to a situation where less timber is felled than grown (Seppälä 1994).

Table 5.4 Growing stock, net annual increment and removals in exploitable forest in Finland (figures over bark; the removals in 1990 have been converted to over bark figures by multiplying by 1.16).

	1950	1960	1970	1990
Growing stock, 1000 m^3	1 456 000	1 430 000	1 445 000	1 790 000
Growing stock, m^3/ha	80	79	77	92
Net annual increment, 1000 m^3	53 300	52 200	55 780	81 625
Total removals, 1000 m^3	44 890	50 290	50 270	51 040
Net annual increment, m^3/ha	2.9	2.9	3.0	4.2
Removals m^3/ha	2.5	2.8	2.7	2.6

Sources: years 1950, 1960, 1970 (European timber trends... 1986), year 1990 (Outlook for the European... 1994).

Iceland

Iceland is an island of 104,000 km^2. About 25 % of its surface is covered with vegetation. The center of the island is covered with glaciers, mountains and deserts. All productive farmland is found below 100 m above sea level (Lines 1990). Forests are situated in areas below 400 m above sea level (Naturskogar i Norden 1994).

At the time of the Viking settlement in the 9th century it is believed that vegetation covered about 75 % of the land area and it is estimated that birch occurred on 25 % of the total land area (Naturskogar i Norden 1994). The Vikings brought sheep with them which led to a big expansion in sheep farming in the 19th and 20th century. By the turn of the century, overgrazing and tree felling for firewood, charcoal and construction had destroyed almost all native birch woodlands and caused severe erosion problems (Lines 1990).

The native flora of Iceland consists of 440 species. Native trees are limited to *Populus tremula*, *Sorbus aucuparia*, *Betula pubescens*, *Betula nana*, *Juniperus communis* and 5 species of *Salix* (Lines 1990, Naturskogar i Norden 1994).

Today birch forests cover 125,000 ha, i.e. 1.2 % of the total land area (Naturskogar i Norden 1994). About 80 % of these forests are less than two meters high (Blöndal et al. 1986). Some 10–15 % of the natural birch forests are protected against sheep by fences. However, these fenced forests have often been thinned and exotic species have been planted among the domestic ones. Because of the decrease in the amount of sheep and grazing, birch forests are also recovering in non–fenced areas (Naturskogar i Norden 1994).

The first efforts to establish new forests in Iceland were made at the beginning of the 20th century. Exotic tree species including *Pinus mugo*, *Picea abies*, *Larix sibirica*, *Picea engelmannii* and *Pinus aristata* were planted in small groups in eastern and northern Iceland (Loftsson 1993). Trials with exotics were interrupted for the period 1908–1935 and efforts were concentrated on enclosing native birch woodlands. After the Second World War, the use of exotics started again. In 1950–1960, Scots pine was the main species. Sitka spruce was widely used between 1950 and 1963. Norway spruce was planted on a large scale between 1958 and 1972. Lodgepole pine (*Pinus contorta*) was first planted in 1940 and from 1960 it has been one of the most popular exotics. The first Siberian larch plantations were established in 1922 and it is now the most widely planted species (Lines 1990).

The main species planted for wood production are Siberian larch, Sitka spruce, Lodgepole pine and black cottonwood. Four species are planted for Christmas tree production: Engelman spruce, Norway spruce, subalpine fir (*Abies lasiocarpa*) and Lodgepole pine (Blöndal 1993).

A large afforestation project has been initiated in east Iceland, where, during the next four decades, 80 landowners aim to establish 15,000 ha of production forests. This forest development project is possible because the recent agricultural recession has released land for planting that otherwise would have been maintained for grazing (Loftsson 1993).

Norway

Productive forests cover 22 % (7 million ha) of the land area in Norway. Agriculture is practised on 3 %. The remaining 75 % consist of non–commercial forests, mountains, glaciers, lakes and built–up areas. About 70 % of the area is above the timberline and 25 % of the productive forest area is so steep that some form of cable transportation for timber extraction is required (Forestry in Norway 1991, Agriculture in Norway 1993, Landbruksdepartementet's... 1994).

The Norwegian forest landscape has been affected by grazing. On the west coast, large areas were burned 1,500–2,000 years ago to form pasture. Part of these areas has been planted with spruce while part is still treeless. The montane forests have been shaped by forest grazing, which was most widespread in the first half of the 19th century. In some places the pressure was so intense that the timberline moved downwards. Industries like mining, and salt and tar production have also affected the landscape through using large amounts of wood (Naturskogar i Norden 1994). By the end of the 18th century, the cutting of large diameter trees for sawmills and exports had been so intensive that such trees could not be found any more in southern Norway. Because of this, timber traders left for Sweden and Finland to look for untouched forests for cutting (e.g. Mattsson & Östlund 1992).

Intensive forest management (clearcutting, the following soil preparation, planting, cleaning and thinning) have affected the forests in Norway from the 1950s. Today, about half of the productive forests are younger than 50 years (except in the county of Finnmark). Forests older than 160 years comprise less than 0.6 % of the productive forest area but 36 % are older than 80 years. Plantations cover about 20 % of the forest area (Naturskogar i Norden 1994, Norwegian forest... 1994).

The predominant species in Norway are spruce and pine (*Table 5.5*). The main coniferous areas are situated in the southeast and central Norway. Elsewhere in the country the landscape is dominated by birch and scattered patches of conifers. Other important species are aspen and alder. The forests along the south coast contain a larger share of broadleaves (Forestry in Norway 1991, Agriculture in Norway 1993, Scandinavia 1993). About 8,000 ha are planted with 50 foreign species. Sitka spruce has been used to replace birch and pine on the west coast. Lodgepole pine has been planted in harsh areas instead of native pine, because it is more resistant to snow, cold and damage caused by fungi (Landbruksdepartementet's... 1994).

Table 5.5 Total growing stock volume by tree species in Norway (Yearbook of Nordic statistics 1994).

	Million m³	%
Scots pine	214	32
Norway spruce	314	47
Broadleaved	137	21

There is at present more standing timber than ever before in Norway. Forest resources have increased by 70 % from 1920. The biggest relative increase has been in broadleaves and pine. Still, spruce is the predominant species (Økt verdiskapning... 1994). The situation has changed remarkably compared to the turn of the century when cuttings were higher than annual growth (*Table 5.6*).

Table 5.6 Growing stock, net annual increment and removals in exploitable forest in Norway (figures over bark; the removals in 1990 have been converted to over bark figures by multiplying by 1.16).

	1950	1960	1970	1990
Growing stock, 1000 m³	436 000	465 000	516 000	630 000
Growing stock, m³/ha	68	72	79	95
Net annual increment, 1000 m³	14 370	14 720	15 770	20 721
Total removals, 1000 m³	11 317	10 065	9 660	13 524
Net annual increment, m³/ha	2.2	2.3	2.4	3.1
Removals m³/ha	1.8	1.6	1.5	2.0

Sources: years 1950, 1960, 1970 (European timber trends... 1986), year 1990 (Outlook for the European... 1994).

Sweden

Waterways cover 9 % of the total area of Sweden. About 56 % (23 million ha) of the land area is forested. Agricultural fields cover about 8 %, swamps 11 %, (4,6 million ha) and mountains 8 %. The rest includes protected and built–up areas (Skogsstatistisk årsbok 1994).

Many factors which have affected the structure of Swedish forests are very similar to those that have shaped the forests in Finland. They include forest fires, shifting cultivation, forest grazing and tar burning. However, in the 17th century ore mining was much more widespread than in Finland. This industry used large amounts of timber, mainly in the form of charcoal. Also, production of potash from broadleaved trees was more common in Sweden (Mattsson & Östlund 1992, Kulturmiljövård... 1992).

Because of intensive use by the rural population, the forest resources of southern and central Sweden by the middle of the 19th century are estimated to have been about 50 % smaller than today (Naturskogar i Norden 1994). In the latter half of the 19th century timber exports from Sweden led to overexploitation and impoverishment of forests, especially in the northern parts of the country. According to Remröd (1991), "Old and thick–trunked trees were chopped down and floated to the coast. Only the timber that was thin, rotten and valueless was left behind. Much of this devastation was financed with foreign capital". When the big trees in accessible areas had been cut, the sawmills started to utilize smaller timber. The already high–graded forests were cut selectively again. As a result of the intensive and unorganized utilization, the volume of timber resources in Swedish forests by the turn of the 20th century was smaller than ever before (Mattsson & Östlund 1992).

Since the 1950s, clearcutting has dominated Swedish forestry. Natural forests have often been replaced with cultivated ones. Many species of plants and animals that require long–term continuity have become threatened by this development, and a number of species are already extinct. The biological production capacity of forest soil has been affected by drainage, clearcutting and soil preparation with the consequent loss of nutrients. These problems have also been admitted by the government and the forest industry. During the past ten years, a number of measures have been implemented to increase the extent to which the environment is taken into consideration in forestry, both by voluntary means and by legislation (A search... 1994, Strategy for sustainable development... 1994).

Two thirds of Sweden belong to the boreal coniferous zone, the southern part belongs to the temperate mixed forest zone. Pine is the predominant species in the north, spruce in the south. Birch occurs as a natural interspersion in coniferous forests (*Table 5.7*). Over the past few years there has been an increase in the production of broadleaved trees to meet the rising demand for hardwood from pulp and paper mills (Scandinavia 1993).

Table 5.7 Total growing stock volume by tree species in Sweden (Yearbook of Nordic statistics 1994).

	Million m^3	%
Scots pine	1 078	39
Norway spruce	1 244	44
Broadleaved	426	15

The Lodgepole pine plantations in the northern part of Sweden form the largest use of exotics in Scandinavia. At the end of the 1960s, the large Swedish forest companies announced their great expectations for Lodgepole pine, based on experimental plantations. Planting on a commercial scale started with full impetus during the 1970s. By 1991, the plantations exceeded 500,000 hectares. A fungal attack in the late 1980s contributed to a reduction in planting. Since 1985, the annual planting of Lodgepole pine has been decreasing. In 1992, the total planted area was 13,700 ha (Contortatallen i Sverige... 1992, Skogsstatistisk årsbok 1994).

The timber resources of Sweden are now 60 % larger than they were at the beginning of the 1920s (Wibe 1992). Annual removals reached a peak in the 1970s. Since then they have been decreasing (*Table 5.8*). The increase has been particularly high in the southern parts of the country and for trees with a diameter above 25 cm. Since the period 1968–1972, growth has risen by 29 % (25 % for conifers and 49 % for broadleaved trees). Because of insufficient regeneration in the past, Sweden now has a shortage of middle–aged stands, but the share of forests mature for final felling is remarkably large (Skogsstatistisk årsbok 1994).

Table 5.8 Growing stock, net annual increment and removals in exploitable forest in Sweden (figures over bark; the removals in 1990 have been converted to over bark figures by multiplying by 1.16).

	1950	1960	1970	1990
Growing stock, 1000 m^3	1 820 000	2 089 000	2 288 000	2 557 000
Growing stock, m^3/ha	79	104	98	116
Net annual increment, 1000 m^3	57 000	65 100	63 300	91 300
Total removals, 1000 m^3	43 500	48 910	66 620	62 870
Net annual increment, m^3/ha	2.5	3.3	2.7	4.1
Removals m^3/ha	1.9	2.4	2.8	2.9

Sources: years 1950, 1960, 1970 (European timber trends... 1986), year 1990 (Outlook for the European... 1994).

5.3 Silviculture

By the end of the 19th century, the forest resources of the Nordic countries were in a degraded state because of overutilization. Silvicultural methods evolved fast in the first half of the 20th century and were intensified in the 1960s and 1970s, the main aim being timber production. This trend continued until environmental movements and the public reacted and started to demand more diverse forest environments. Today, all Nordic countries are looking for new ways to manage forests more in harmony with nature and concern for other interests than timber production is increasing.

Denmark

The forests of Denmark are artificial and have been managed very intensively. Denmark has a long tradition of heavy thinnings which have been carried out at short intervals (Brief glimpses... 1982). A typical Danish forest is divided into small parts, in each of which there grows only one tree species. Conifers are planted manually. Broadleaves are regenerated by self–seeding or by artificial sowing. Soil preparation consists of light surface treatment (Jespersen 1989).

Forest management in Denmark is nowadays clearly less intensive than 50 years ago partly due to the higher salaries which make silvicultural work less profitable. In addition, people's attitudes have turned against such intensive methods as fertilization and the use of pesticides (Honoré & Jespersen 1989, Jespersen 1989).

The Forest Act of 1989 and the Nature Management Act of the same year define Denmark's forestry policy and silvicultural objectives. The Forest Act describes "good and multiple–use forest management" in detail. The Act also contains prescriptions for the enhancement of biodiversity, for example concerning the creation and management of broadleaved forests. The Nature Management Act emphasizes the need to increase the amount of forest (Grayson 1993). Both these and a few other Acts contain prescriptions for subsidies for various forest operations from nature protection to intensive timber production (Strategy for sustainable... 1994, see also Grayson 1993).

Agger and Koch (1989) elaborate the future needs and possibilities of increasing the variation and ecological stability of Danish forests. First of all, the very few still remaining original natural forests should be protected. Other forests should be managed so that the conditions resemble as much as possible the structure and functions of the original, untouched natural forests. Generally, a high percentage of deciduous forests should be preferred with tree species of local origin, and the continuity of the stands should be high, with a number of old and dead trees. Hübertz and Ovesen (1989) give further advice for forest managers. For example, monocultures should be converted to mixed forests and the underbrush should be retained, especially in forest edges.

In 1991, the Forest and Nature Agency published a handbook on multiple–use forestry prepared by Koch and Kristiansen. This booklet includes detailed recommendations on how to integrate timber production and the other forest benefits.

Finland

Selection cuttings were the main method for timber harvesting before the 1940s in Finland. They resulted in the disappearance of large diameter trees and in poor quality and slow growth of the remaining trees. From the 1950s, new silvicultural methods were increasingly utilized. They included regenerating pine by the seed tree method and spruce with the help of shelterwood. When using these methods, the forest was grown even–aged until maturity and regeneration felling. Thinning was done from below (Parviainen & Seppänen 1994). However, selection cuttings were still a common practice in private forests until the 1960s (Leikola 1987).

After the Second World War, the timber–processing industry was developed to earn export income and to create capital for the building of the industrial sector. Consequently, the utilization of timber resources increased and in the 1950s cuttings exceeded growth. This led to programs aiming at the intensification of timber production (Seppälä 1994). The period 1965–1974 was the most intensive time for Finnish silviculture; investments in clearcutting, artificial regeneration, fertilization, drainage and soil preparation were increased substantially (Leikola 1987).

Since the 1970s, environmental organizations and forest owners have been arguing for more ecological and diverse silvicultural methods. However, the Forest 2000 –Programme (a long–term plan for forestry and forest industry in Finland) from the year 1985 still recommended clearcuttings, planting and fertilization (Metsä 2000... 1985, Leikola 1987). The Programme was revised in 1992. The new version included information on the environmental impacts of clearcutting, soil preparation, fertilization and drainage but explicit new silvicultural recommendations were not outlined (Metsä 2000... 1992, The representation... 1992).

Private forest owners, forestry companies and state foresters started to diversify their silvicultural methods little by little already in the 1980s. For example, the Forestry Centre Tapio, an organization for guiding private forestry, published new environmental guidelines in 1989 (Metsäluonnon hoito... 1989). Really radical changes have come about as a consequence of the United Nations Conference on Environment and Development (UNCED) in Rio de Janeiro in June 1992 and the second Ministerial Conference on the Protection of Forests in Europe which was held in Helsinki in June 1993. The aim of the Helsinki conference was to promote the implementation of the forest–related UNCED agreements in Europe (Ministerial Conference... 1993, UNCED... 1993).

Simultaneously with active participation in international processes, Finland has started to reform its own silvicultural practices. In May 1993, a working group consisting of representatives from the most prominent interest groups in the forestry sector was set up to outline new environmental guidelines for forestry in Finland. The work was concluded in July 1994. The report of the working group, which has been confirmed both by the Ministry of Agriculture and Forestry and the Ministry of the Environment, can be regarded as the plan for ecologically sustainable forestry in Finland. It will form part of a more comprehensive national forest plan, which will be prepared in the future (Metsätalous ja ympäristö 1994, New environmental... 1994). The Ministry of the Environment has produced a separate plan for the protection of biodiversity in forests (Suomen metsäluonnon... 1994). The need for reforms has also led to the rewriting of Finnish forestry legislation. The new laws are expected to be finalized in 1995.

The new environmental guidelines give explicit silvicultural advice concerning regeneration, final fellings, prescribed burning, soil preparation, field afforestation, stand management, use of herbicides, fertilization, drainage, building of forest roads, harvesting, and management of fragile forests, for example, in high elevations and on islands (New environmental... 1994). In addition to the official process, Finnish forestry organizations and enterprises have produced their own silvicultural guidelines (e.g. Finnish Forest and Park Service: Forestry; environment guide 1994, National Forest Extension Service: Luonnonläheinen metsänhoito... 1994, Finnish Forest Industries Federation: Kalland & Pätilä 1993).

The Finnish Forest Research Institute has also published a report on ecologically sustainable silviculture (Parviainen & Seppänen 1994). According to Parviainen's summary in the Journal of Forestry (1994), Finland's new silvicultural principles include small–scale forest regeneration, prescribed burning, favoring mixed forest stands, avoidance of over–mature and excessive stand densities, preservation of mosaic–like variability, and maintenance of ecologically valuable corridors and protective zones.

Iceland

Accurate figures for the plantations in Iceland are not available, but the area of planted woodland is estimated to be about 15,000 ha. The majority of the plantations consist of exotic species, most of which have been established in the birch woodland (Blöndal 1993).

Native birch has been planted increasingly in the 1990s. This is done by planting small scattered islands of birch and allowing the land between to regenerate by seeding from the mother trees (Blöndal 1993). Private landowners favour the local birch over the exotics (Loftsson 1993).

The use of soil preparation has increased lately in Icelandic forestry. From 1960, several small areas of marshland have been drained for planting by ploughing. Only the Christmas tree plantations are fertilized. *Lupinus nootkatensis* is used for site amelioration in eroded areas. It rapidly colonizes areas and through symbiotic bacteria fixes nitrogen. Some of the nitrogen becomes available for other species which invade or are planted among the lupines (Blöndal et al. 1986). Icelandic silvicultural specialities also include fencing against sheep and horses (Blöndal et al. 1986) and the production of seedlings in greenhouses which are sometimes heated by geothermal spring water (Lines 1990).

A long–term strategy for the protection of Iceland's birch forests will be set up in the near future. It will be based on the information obtained from an inventory of native woodlands which was carried out in 1985–1990 (Naturskogar i Norden 1994).

Norway

Like its eastern neighbours, Norway entered the era of clearcuttings followed by artificial or natural regeneration after the Second World War. Natural regeneration has been mainly practised in pine forests. Timberline forests have been harvested with selection cuttings. The development of mechanization in the forest started also in the mid–1950s (Forestry in Norway 1991).

The rationalization, use of large machines and building of forest roads was accelerated in the 1960s and 1970s. Clearcut areas and plantations became prominent components of the forest landscape. This led to dissatisfaction and demands to increase the amount of protected forest areas and to modify forestry practices to be more acceptable from the viewpoint of recreation and nature protection (Flersidig skogbruk... 1990). According to Grayson (1993), the highly mechanized methods of the 1970s also led to growing emphasis on favouring smaller–scale labour–intensive technologies. Silviculture was further disintensified in 1989 when it was decided that no more state subsidies be given for the use of herbicides, strong soil preparation and fertilization on peatlands (Flersidig skogbruk... 1990).

The Norwegians have produced a considerable amount of environmental guidelines for carrying out silvicultural activities. The leading forestry journal Norsk Skogbruk has published three special issues containing advice for practical foresters (Landskapsvern... 1978, Flerbruk... 1986, Flersidig skogbruk... 1990). Also the Royal Norwegian Ministry of Agriculture has contributed to the discussion (e.g. Flersidig skogbruk... 1989). Researchers have been involved in developing alternative methods as well (e.g. Petersen 1989, Solbraa 1989, Solbraa & Grønvold 1992).

As a follow–up to the UNCED Conference, Norway is preparing a national strategy for the preservation and sustainable utilization of biological diversity (Landbruksdepartementet's... 1994). Measures for the forestry sector will include "amendments to regulations and administrative rules to give more emphasis to biodiversity, more active use of legal powers against environmentally unsound operations or practices, further development of planning tools and development of strategies for management of ecologically important areas" (Norwegian forest... 1994).

Sweden

Selection cutting was also the dominant harvesting method in Sweden until the Second World War, after which management by compartments (trakthyggesbruk) became the most common method. This meant that large areas of selectively thinned forests were clearfelled and planted or sown. The method led to the development of other silvicultural technologies like soil preparation practices, establishment of seed tree orchards, seedling production and forest tree breeding (Andersson 1992).

According to Andersson (1992), Sweden outlined an economic policy in the 1940s, which clearly chose the forest and metal industry as the means of financing for the building of the desired welfare state. Consequently, the main objective of silvicultural methods became the production of timber for industry.

In the 1960s and 1970s, the biological and environmental results of the rationalization and adoption of large–scale methods were found unsatisfactory. This led to public debate on clearcutting and the use of herbicides (Remröd 1991, Andersson 1992). Still, the annual amount of clearcutting kept on increasing until 1983, after which it started to decrease slowly. The use of herbicides for weeding was abolished by law in 1980. Since 1983, there has been the possibility of exemptions, but this has been sparsely used. No spraying of herbicides has occurred since 1986 (Skogsstatistisk årsbok 1994). Generally, silviculture was most intensive in the first half of the 1980s in Sweden.

The Swedish forest legislation is being liberalized and subsidies other than those for environmental purposes are disappearing (Outlook for the European... 1994). The Swedish parliament approved the new forest policy in May 1993. The previous policy was found to be unsuccessful in the fields of nature conservation and production of high quality timber. The new environmental and production goals of Swedish forestry are the following (Sweden's new... 1994):

1) The productivity of forest land shall be preserved. Biodiversity and genetic variation in the forests shall be secured. Forest must be managed so that plant and animal species which exist naturally in the forest ecosystems, can survive under natural conditions and in vigorous populations. Endangered species and vegetation types shall be protected. The forests' historical, aesthetic and social values must be defended.
2) Forests and forest land shall be utilized efficiently aiming at a sustainable and valuable yield. The composition of forest production must be such that it has the potential of satisfying different human needs in the future.

Swedes have a good knowledge basis for developing ecologically and culturally sound forestry practices. Many handbooks on fauna and flora management in forestry were published already in the 1980s (e.g. Ahlén et al. 1984, Ingelög et al. 1984, Ehnström & Waldén 1986). Practical silvicultural guidelines for private forest owners became available in 1990 (Rikare skog... 1990). Recently, comprehensive books have been published on special issues such as cultural heritage and landscape (e.g. Kulturmiljövård... 1992, Gustavsson & Ingelög 1994). Also the big forest companies have started to experiment with new methods and to document their results (A search... 1994).

5.4 Forest industry

Until the 20th century, timber was mainly used as a raw material in the production of tar and potash, for construction and as fuelwood for household needs and industry, for example in metal processing. The first wood–processing enterprises, waterpowered sawmills, were established in the 16th century. The present–day forest industry started to evolve in the latter half of the 19th century, after the invention of the technology necessary for making paper from wood fibre (Seppälä 1994).

Up until at present, it has been customary to divide the forest industry into mechanical and chemical wood processing. The main products of mechanical wood processing are sawnwood, plywood, particle boards, fibreboards and more elaborated products such as furniture, houses and building components. The chemical forest industry produces pulp, paper and paperboard. Today, a new terminology is being developed. For example, in Finland the two main sectors of forest industry are increasingly called wood products industry and pulp and paper industry.

Recently, new industrial ways of using timber and wood have started to evolve. They include large–scale Christmas tree production and trade, the livelihoods connected to the growing of biomass for energy production and the so called tourism industry, which is partly based on culturally ingenious carpentry products and landscapes created by standing timber. The processing of non–wood forest–based raw materials has traditionally not been included in the scope of the forest industry concept in the Nordic countries.

The wood–processing forest industry is a prominent sector in the national economies of Finland, Norway and Sweden. In these countries, the total production of forest industries is still increasing, although the share of income generated is decreasing compared to other branches of economy. In Denmark, and to some extent also in Iceland, wood processing is recognized as a significant local livelihood.

Denmark

The industrial sector of Denmark is mainly based on the processing of imported raw materials and the country is dependent on a high level of exports. The principal export articles are agricultural products, machinery, chemicals and furniture (Scandinavia... 1993). Most of the wood–processing enterprises are situated in rural areas. Consequently, although forestry and forest industry do not form a major branch of the economy in Denmark, they may be crucially important employers locally (*Table 5.9*) (En fremtidig skovpolitik 1986).

Table 5.9 Forest products in the economy of Denmark, million US $ in 1991 (Forestry statistics...1993).

Production	% of GDP	Imports	Exports	% of trade	Consumption
584	0	1 766	417	1	1 933

In the Danish forest industry, the emphasis is on mechanical wood processing. Because of this the industry is very dependent on building activity and the demand for furniture. About three quarters of the furniture manufactured in Denmark is exported, making the country the fourth largest European furniture exporter (Scandinavia... 1993). Other important wood–processing branches include sawmills, particle–board factories and value–added production including, for example, manufacturing of wooden floors. The production structure in the mechanical forest industry sector is changing, as in the other Nordic countries, towards less but bigger enterprises (*Table 5.10*) (Brief glimpses... 1982, En fremtidig skovpolitik 1986).

Table 5.10 Production and trade of roundwood and forest industry products in Denmark in 1992 (Metsätilastollinen vuosikirja... 1994).

	Production	Imports	Exports
Roundwood, 1000 m^3	2 000	553	576
Sawnwood and wood–based panels, 1000 m^3	867	2 246	189
Pulp, 1000 MT	100	96	88
Paper and paperboard, 1000 MT	341	1 198	216

The production and export of Christmas trees and greenery has become an increasingly important income source for Danish forest owners. Though covering only about 5 % of total forest area, they produce 20–25 % of total income from forests (Linddal 1993). Most of the current exports go to Germany, but

Hong Kong and Japan are emerging as new markets (Scandinavia... 1993). Christmas trees and decoration greenery may be cultivated in short rotation cycles in up to 10 % of the forest area (Act of forests... 1989).

A plan of the Ministry of Agriculture for the development of wood and wood–based products (1993/94) intends to make use of the non–utilized resources of the forestry sector. The plan proposes, among other things, increased production of particle boards and other wood–based panels and improvement of technologies to reduce the amount of waste in industrial processing. Another plan, the Biomass Agreement (1993) aims at increasing the utilization of biomass for energy production in big power stations. The agreement is expected to lead to the development of markets for wood of small dimensions and poor quality (Strategy for sustainable forest... 1994).

Finland

In Finland, the forest industry plays a greater role in the national economy than in any other industrial country of the world and a significant proportion of manufacturing in the metal industry is connected to the forest sector (e.g. paper machines, pulp boilers and forest harvesting machines). The majority of forest industry products are exported. The share of imported raw materials is smaller than in other export sectors, on the average only 16 %. This makes the forest industry the most important net export sector in Finland (*Tables 5.11 and 5.12*) (Forest industry... 1993).

The diversification of the structure of exports to decrease dependence on one sector of industry was set as a goal of national policy in the 1970s. Since then, the share of forest industry of the total value of exports has decreased, although slowly, being 56 % in 1970, 44 % in 1980 and 38 % in 1990 (Metsätilastollinen vuosikirja 1994, Seppälä 1994).

Table 5.11 Forest products in the economy of Finland, million US $ in 1991 (Forestry statistics... 1993).

Production	% of GDP	Imports	Exports	% of trade	Consumption
8 959	7	512	8 238	36	1 233

The most important production sector within mechanical wood processing is the sawmill industry. Most of its products (99 %) are made from conifers (Gullichsen et al. 1994). In the 1980s, the production and ownership structure of the sawmill industry changed; small independent sawmills were increasingly replaced by large enterprises. In the 1990s, about 50 % of the total production capacity of sawmilling was covered by 5 big companies. Some 120 of the 4,000 sawmills in Finland export their products. The rest are small enterprises acting mainly in local markets (Gullichsen et al. 1994).

The main products of the plywood industry include birch, spruce and mixed plywoods as well as special and processed plywood. The bulk of plywood and more than half of fibreboards are exported. Particle boards for building and the furniture industry are mainly used in the domestic market. The production of plywood is hampered by the lack of suitable birch raw material (Forest industry... 1993).

The 1980s witnessed a general decrease in production in the mechanical wood–processing industry. In the 1990s, the market share of Finland has increased, but structural problems make the industry unstable and vulnerable to international market fluctuations. There is pressure to increase value added, to modernize the equipment and to develop special products for specific customer groups (Seppälä 1993).

The production of pulp and paper grew strongly in the 1980s. Today, most investments are made in the production of high quality printing and writing papers. The production of newsprint has not grown significantly since the 1970s.

The share of wood in the final paper products has decreased as the degree of processing has risen. In addition to woodpulp, the paper industry nowadays consumes considerable quantities of pigments either as fillers mixed with the fibre or as paper coating pigments. They improve printing properties (Forest industry... 1993). Recycled fibre forms only about 5 % of the raw material, although 45 % of domestic waste paper is collected for recycling. Lack of domestic raw material limits the possibility of increasing the share of recycled fibre in Finnish paper products. Transporting waste paper from abroad would be environmentally unfeasible (Seppälä 1993, Gullichsen 1994). The combined use of fillers and recycled fibre in paper and paperboard is expected to increase from 17 % in 1990 to 25 % in 2010, the rest being woodpulp (Seppälä 1993).

Table 5.12 Production and trade of roundwood and forest industry products in Finland in 1992 (Metsätilastollinen vuosikirja... 1994).

	Production	Imports	Exports
Roundwood, 1000 m^3	39 000	6 075	572
Sawnwood and wood–based panels, 1000 m^3	7 922	198	5 156
Pulp, 1000 MT	8 617	199	1 291
Paper and paperboard, 1000 MT	8 823	144	7 882

The structure of pulp and paper industry is developing towards large companies with part of their activities abroad. Production costs, especially timber and labour, are relatively high in Finland. Consequently, the strategy of the industry is to invest in innovative products and technologies to be able to compete with enterprises situated in countries with lower costs (Seppälä 1993).

A period of intensive forestry and forest industry policy planning started in Finland in the 1960s as a consequence of the shortage of wood for industrial purposes. The latest plan, the Forest 2000 Programme, extends to the year 2010 and it is the first one to begin with a timber surplus situation (Saastamoinen 1987). The recommendations concerning the forest industry of the revised Forest 2000 Programme are the following (The representation... 1992):

1) Improvement of competitiveness by lowering the labour cost per unit produced and increasing productivity. The costs of roundwood, harvesting and transporting should also be decreased.
2) Pointing research and product development activities towards new and environmentally friendly products and technologies.
3) Continued foreign investments, especially in marketing and distribution.
4) The sawmill industry and the forest owners should come to a consensus on how to improve the competitiveness of sawmills and how to increase their production.

The Forest 2000 Programme has been criticized over inadequate integration of non–wood and multiple–use aspects with timber production (Saastamoinen 1987). Furthermore, the industrial policy planning tradition has recently started to change in Finland. According to the National Industrial Strategy, control and subsidies should be replaced by long–term promotion of the overall preconditions. The aim of the Strategy is to serve as a basis for the future industrial programs and measures (Kansallinen teollisuusstrategia 1993).

Iceland

About 3 % of Iceland's land area is capable of growing production forests but less than 1 % of the total land area would be needed to supply the domestic demand for wood raw material (Gunnarsson et al. 1987). However, practically all sawnwood, wood–based panels and paper products are imported to the island. Local wood–processing enterprises are likely to develop when the plantations grow older.

In 1987, the main wood–based forest products of Iceland were Christmas trees (10,765 trees), firewood (141 tons), turnery wood (12 m^3) and fencing stakes (3,000 stakes) (Snorsson 1992). The native birch is used for fencing stakes, turnery wood and firewood. The latter is used both for home fires and for smoking food (mutton, herring). Larch from thinnings is good for fencing stakes and posts. Norway spruce and Lodgepole pine have been popular as Christmas trees since 1970. The price of Christmas trees has been decided on the basis of the price of trees imported from Denmark (Blöndal et al. 1986).

Forestry has recently been accepted as a part of agriculture and as a means of offering the rural districts new livelihood opportunities. The reason for this is that Icelandic farming has suffered a marketing crisis and the number of sheep has had to be reduced. However, production forestry in Iceland can be achieved only with the help of financial support from the central government (Blöndal et al. 1986).

Norway

Forestry products account for the fifth largest export commodity group from Norway, after oil and gas, metals, shipping and fish. About 80 % of the pulp and paper and 40 % of the sawnwood production are exported. Three quarters of the pulp and paper exports and the greater part of sawnwood exports go to central Europe (*Tables 5.13 and 5.14*) (Agriculture in Norway 1993).

Table 5.13 Forest products in the economy of Norway, million US $ in 1991 (Forestry statistics... 1993).

Production	% of GDP	Imports	Exports	% of trade	Consumption
2 400	2	708	1 517	4	1 592

The Norwegian sawnwood industry consists of small and medium–size enterprises, mainly catering to home markets and often also to local markets. In 1991 there were 225 sawmills and planing mills, 7 particle–board mills located throughout the country in districts with easy access to mountain birch and sawmill residues, and 4 fibreboard mills, two of which were integrated with sawmills (Forestry in Norway 1991). In recent years, the size of sawmills has increased and their number has decreased. At the same time the amount of employees has decreased (Økt verdiskapning... 1994).

Norway is the leading country in Scandinavia in using wood in buildings. This is a result of conscious efforts including a regularly awarded prize to remarkable achievements in wooden architecture (Puun käytön... 1994). The overall strategy of the Norwegian sawnwood industry is to specialize in high quality products. This requires improved selection of raw material in an early stage of production and increased cooperation between the various actors involved in the harvesting and processing chain (Økt verdiskapning... 1994).

The structure of the Norwegian pulp and paper industry has changed during the past 10 years. The branch has become dominated by a few big companies. There has been "continuous concentration in fewer and larger mills, an extensive consolidation into stronger groups through take–over bids and share capital transfer and large capital investment in modern production equipment" (Forestry in Norway 1991).

Newsprint is still the main product of the Norwegian chemical forest industry (Key figures 1994). However, Norske Skogindustrier, which is the biggest forest industry enterprise in Norway, aims at developing "new grades of wood–based paper to meet the demands of an ever more complex and quality–conscious market" (Norske Skogindustrier... 1993).

Table 5.14 Production and trade of roundwood and forest industry products in Norway in 1992 (Metsätilastollinen vuosikirja... 1994).

	Production	Imports	Exports
Roundwood, 1000 m^3	11 000	1 667	880
Sawnwood and wood–based panels, 1000 m^3	2 833	572	970
Pulp, 1000 MT	2 006	45	533
Paper and paperboard, 1000 MT	1 683	302	1 383

According to a working group consisting of representatives of forest industry, local administration, forest owners and forest research, the most important objectives of the Norwegian mechanical and chemical forest industry during the next 10–15 years are: 1) cost efficient, environmentally sound production of real and even quality products, and 2) production of products with high value added and creation of wood products which can compete succesfully with other materials (Økt verdiskapning... 1994).

Sweden

In 1951, the share of forest industry products of the total value of exports was 42.5 % in Sweden. The forest industry is still the most important net export sector although its share of exports has decreased. The foreign currency earning capacity of the industry can be explained by the low demand for imported raw materials. Since 1951, the engineering industry has grown to be clearly the biggest industry sector in the Swedish economy (*Tables 5.15 and 5.16*) (Skogspolitiken... 1992, A search... 1994, Nordisk statistisk... 1994).

Table 5.15 Forest products in the economy of Sweden, million US $ in 1991 (Forestry statistics... 1993).

Production	% of GDP	Imports	Exports	% of trade	Consumption
10 756	4	1 168	9 873	18	2 051

In 1990, there were 2,423 sawmills in Sweden. Mills with an annual production exceeding 5,000 m^3 account for 94 % of the total production. The number of mills has declined steadily from the 1960s but production and productivity have increased (Skogsstatistisk årsbok 1994). The production of boards has been decreasing since the late 1980s. In the near future only plywood production is expected to increase (Scandinavia 1993). As in the other Scandinavian countries, the sawmilling industry of Sweden is in need of reorganization to be able to compete internationally (Lönner & Libäck 1992).

The development in the pulp and paper industry has been towards producing different types of paper instead of pulp for sale. During the past 30 years, the production of newsprint, paperboard and writing and printing papers has expanded significantly. The reason behind the increase in newsprint production is the suitable long–fibre raw material from spruce and relatively cheap electricity. Production of special papers has also increased (Lönner & Libäck 1992).

The structure of the Swedish forest industry has changed during the past few years through fusions and alliances. In 1950 there were 130 pulp and paper mills, by the beginning of 1970s the amount had fallen to 80, and by the turn to the 1990s only 20 were left. Today, three enterprises dominate the field and are responsible for 75 % of the turnover. These companies also have activities abroad, mainly in western Europe (Skogspolitiken... 1992).

Table 5.16 Production and trade of roundwood and forest industry products in Sweden in 1992 (Metsätilastollinen vuosikirja 1994).

	Production	Imports	Exports
Roundwood, 1000 m^3	54 000	6 335	1 008
Sawnwood and wood–based panels, 1000 m^3	12 931	528	8 485
Pulp, 1000 MT	9 908	202	2 756
Paper and paperboard, 1000 MT	8 376	318	6 624

Swedish forest policy was re–evaluated in 1992. The emphasis of the extensive research and planning effort was on silviculture and its environmental consequences (Skogspolitiken... 1992). The forest industry has also been emphasizing the ecological aspects of timber production and harvesting in its public relations activities (e.g. Remröd 1991, The forest cycle... 1993, A search... 1994).

5.5 Future challenges

Growing forest resources in Europe

Timber resources are growing in all the Nordic countries and the annual cut is also expected to be smaller than the annual growth in the future. The same is happening in most European countries (Kuusela 1994, Outlook on the European... 1994).

Central Europe has traditionally been the main market of Nordic forestry products, more than half of the exports go to member states of the European Union. In the long run, the change from shortage to abundance of timber resources is likely to affect the structure of the forest industry and trade in Europe. Because of the increasing resource base and the huge supply of waste paper raw material for paper making, many of the present importing countries will have the possibility of developing their own forest industries to cover domestic demand (e.g. Lönner & Libäck 1992, Thörnqvist 1994). This new situation forces the Nordic countries to look for a new role and strategies in the European context. The process of adaptation and reorganization will be speeded up by Finland's and Sweden's joining the European Union in 1995.

Despite the increase of forest area and the amount of timber, the quality of the forests is unsatisfactory in Europe. A lot of the original biodiversity has been lost. There is very little old–growth forests left and it will take a long time to rehabilitate the man–made forests to resemble natural ecosystems (Dudley 1992). From the wood utilization point of view, the quality of forests is getting worse because of the lack of silvicultural treatment. The forests are becoming thick, which leads to reduced growth and poor quality of timber (Kuusela 1994). In addition, European forests are weakened by air pollution and the global climate change may cause unexpected changes in ecosystems (Forest condition... 1993, Decline... 1994, Kellomäki 1994a, 1994b).

The above mentioned phenomena are also affecting the Nordic forests. If these problems can be solved both internationally and locally, increasing forest resources will provide good conditions for future timber utilization as well as for other forest uses, and for the restoration and protection of forest ecosystems.

Global competition

The Nordic countries are nowadays more dependent on the trends of global trade of timber and forest products than ever before. The prices of most forestry products have been decreasing in global market during the past 30 years (Enroth 1992) and the industry is characterized by vulnerability to price fluctuations in markets. The latest recession in the branch was around the turn of the 1990s. At the moment, the Nordic forest industry is thriving and expanding again.

The global sawn timber markets are changing. The availability of good quality timber is likely to decrease in the future (Økt verdiskapning... 1994). This concerns especially the supply of tropical timber. Environmental conservation measures have decreased the supply of raw material from the large coniferous forests of western North America. Political instability of the former Soviet Union has kept the vast Russian timber resources away from the world markets (Svanborg 1994). These trends have created at least temporarily new markets for Nordic sawnwood products, especially in the fast growing Asian economies.

The rapidly growing use of recycled fibre is the most important single factor which has affected the pulp and paper markets in recent years. The amount of recycled fibre in paper products in the world is expected to increase to 36 % by the year 2000, in western Europe the corresponding figure is 45 % (Seppälä 1993). This will change the competition situation for newsprint and packaging materials. The production of these products will move close to cities where the raw material and markets are. The advantage of closeness to traditional wood raw material will decrease.

Another factor affecting the trade of timber and forest products is the increased availability of short–rotation wood in the markets. Pulp production based on plantations in South America and in southeast Asia has increased significantly (Arjas & Häggblom 1993). Plantations are also increasing wood production in Europe, especially in France, Great Britain, Spain and Ireland (Enroth 1992). Short–rotation wood production has increased competition and lowered the prices in pulp markets (Kajaste 1994).

Timber for sawmilling can also be produced in the plantations. This will increase the availability of low quality sawnwood and reduce the prices. The prices of high quality slowly grown northern sawn timber are, however, likely to stay on a rather high level because of the scarcity of raw material (Arjas & Häggblom 1993).

To be able to act profitably in world markets, the Nordic countries have to specialize in products which are not affected by the competition created by recycled fibre and short–rotation wood production. Such products include high quality writing and printing papers, special papers and paperboard for food packages (Arjas & Häggblom 1993).

The global demand for paper products has been estimated to grow by 2.5 % annually in the 1990s. The technological readiness of the Nordic chemical forest industry to respond to the growing markets in paper products is among the best in the world. The international competitiveness of the Nordic sawnwood industry has not developed on a par with the chemical wood processing. About 70 % of the products of the mechanical forest industry are used for building. The skill to use wood in houses and in other constructions has badly deteriorated in Scandinavia during the past 50 years. One reason for this has been strict fire regulations. Now attitudes are changing and efforts are being made to develop, increase and diversify the product assortment (e.g. Økt verdiskapning... 1994, Puun käytön... 1994, Thörnqvist 1994).

To improve international competitiveness, the Nordic forest industries have been reorganizing themselves from the 1980s through company fusions. This has been done in order to rationalize production structure, to lower production costs, to increase the ability to invest and to increase financial stability. Scandinavian forest companies have also been buying and building enterprises abroad to be closer to the markets. In 1990, 35 % of Swedish and 25 % of Finnish forest industry capacity was situated abroad (Seppälä 1993). Other ways to increase competitiveness include developing new products and technologies, increasing the efficiency of raw material utilization and manufacturing products in which the special properties of the Nordic tree species are important to the quality of the product (e.g. Økt verdiskapning... 1994).

The benefits gained by large–scale production are most significant in the chemical forest industry. In the mechanical forest industry, also small and medium–size enterprises can be competitive because of customer specific products and their flexibility in adapting to changes in markets (Enroth 1992). The governments of Finland, Norway and Sweden are nowadays increasing support to small–scale mechanical wood processing for trade policy and employment reasons (e.g. Puun käytön... 1994). The new fast–evolving industry livelihoods having forest related applications include environmental technology, data management systems and gene technology (Arjas & Häggblom 1993).

The national assets of the Nordic countries include the relatively low price of energy, good transportation systems in and out of country, and state policies supporting the forest industry. Finland and Sweden are now facing new challenges because of the common monetary policy in the European Union which reduces the possibility of using monetary operations, for example devaluations, to support the domestic economy (Heikka 1992).

Environmental awareness

Because of the increased availability of information on problems like air and water pollution, extinction of species and destruction of landscapes, the environmental awareness of ordinary people, public administrators and industries is increasing. Consumers and retail customers of forestry enterprises are nowadays probably the most powerful force guiding development in forestry. The strong multinational environmental organizations, for example World Wide Fund for Nature and Greenpeace, have acquired an active and influential role in global and national forestry debates (e.g. WWF submission... 1994, Karjalainen 1994, Pennanen 1994). Furthermore, the development of environmental awareness is speeded up by the fast improving provision of information through the media and worldwide computer–based electronic information networks.

Buyers are increasingly requiring affirmation of the harmlessness of the utilization of forest resources and the impacts of industrial production. Consequently, various ecolabelling and timber certification systems affecting Nordic forestry are being developed (e.g. Forest Stewardship Council... 1994, The forest conservation... 1994). The public sector is participating in defining the ground rules for forest utilization by developing criteria and indicators for sustainable forestry as an aftermath of the United Nations Conference of Environment and Development (UNCED) in Rio de Janeiro in 1992 (e.g. Ministerial Conference... 1994). The official and voluntary Environmental and Social Impact Assessments (EIAs and SIAs) also serve the purpose of increasing awareness of the various consequences of resource use. Industries are utilizing lifecycle analyses (LCAs) to analyze their activities and to facilitate environmentally conscious planning (Johdatus elinkaarianalyysin... 1994). The above–mentioned systematic analyzing methods and procedures will be increasingly used by various interest groups acting in the fields of timber production and wood processing.

Commercial forestry companies have responded to public pressure by starting to pay more attention to the origins of raw materials and by improving their production processes. They have modified silvicultural practices in their own forests to be more ecologically sound and started to avoid buying timber from producers who are known to use unacceptable cutting methods. In processing, the "trend is towards tightly closed mills operated with the goal of eliminating all hazardous waste emissions" (Scandinavian... 1994). The Nordic forest industries have already succeeded in cutting down many harmful emissions and specific plans have also been made to invest in environmentally sound technology in the future (e.g. Forest industry... 1993).

Environmental problems related to the forest industry have exceptionally complicated societal connections. For example, in Finland the wood processing industries account for about a quarter of the total consumption of primary energy and roughly 40 % of electricity consumption. This dependence connects the forestry branch to energy policy decision making, including the problems created by fossil fuels and nuclear power. Future energy production decisions may lead to considerable restructuring in timber utilization and industrial processes. For example, chemical pulp production produces more energy than it consumes (cf. mechanical pulp production consumes a lot of energy) and efficient power stations can be built in pulp mills. If the price of energy rises in the future, the amount of chemical pulp production is likely to increase. Consequently, the demand for timber will also increase because chemical pulping consumes more wood raw material than mechanical pulping (Wahlström et al. 1994).

Generally, recycling of waste paper is considered an environmentally desirable goal, not least to reduce the waste management problems in densely inhabited areas. However, the fibres can not be recycled forever and it may not be reasonable to transport the waste long distances. That is why it has been proposed that a balanced mix of recycling and burning of the waste should be sought (e.g Arjas & Häggblom 1993). Burning of excessive waste paper to produce energy could reduce the harmful effects of using non–renewable fossil fuels in energy production and transport (The forest cycle... 1993).

Despite the complicated environmental challenges, the Nordic forest industries are optimistic concerning their future. The forest–based industrial system with its renewable raw materials and recyclable products has every possibility of working in accordance with the natural cycles that control climate and the environment (The forest cycle... 1993, Forest industry... 1993).

Employment

Employment in the forestry sector has been decreasing steadily over the past 30 years because of increasing productivity in timber production and wood processing. For example, in Finland the amount of workers in forestry was 63,000 in 1982 and 28,000 in 1993. The corresponding figures for the forest industry are 120,000 and 73,000 (*Table 5.17*) (Skogspolitiken inför... 1992, Metsätilastollinen vuosikirja 1994, Strategy for sustainable forest... 1994). The strongest declining trend can be seen in forest work, due to the increased share of mechanized harvesting.

Table 5.17 Employment in timber production and wood processing in Denmark, Finland, Norway and Sweden.

	Timber production	Wood processing	Total
Denmark[a]	5 000	30 000	35 000
Finland[b]	28 000	73 000	101 000
Norway[c]	8 000	29 000	37 000
Sweden[d]	30 000	84 000	114 000

Sources: a) En fremtidig skovpolitik 1986.
b) Metsätilastollinen vuosikirja 1994.
c) Agriculture in Norway 1993.
d) A search... 1993.

Fundamental changes are taking place in the rural areas of the Nordic countries. The profitability of farming is decreasing and the share of forestry in the economy of farmers is likely to increase as a consequence of European integration. Because of the small average size of private forest properties, most farms will have to reorganize their activities in order to survive (e.g. Hyttinen 1994, Selby 1994). New supplementary sources of income are needed. They include public welfare services, tourism, professions based on information technology, production of raw material for bioenergy, and small and middle–sized manufacturing, for example processing non–wood forest–based raw materials (Idékatalog... 1988, Oksa 1994). A succesful and balanced creation of additional earning possibilities requires increased cooperation between the administrations involved in rural policy.

Most of productive forest land is owned by private people in Scandinavia. However, an increasing amount of forest owners live far away from their properties. New types of professions and private enterpreneurship could be created to provide silvicultural and timber harvesting services to the increasing amount of the absentee forest owners. These activities could provide working opportunities and supplementary income for the farmers and forest owners living in the countryside and could also be combined with efforts to produce high quality timber of various species for various purposes.

To reverse the trend of decreasing employment in wood processing and to contribute to the economic structure of rural areas, it has been recommended that the amount of small and middle–size mechanical wood–processing enterprises should be increased (Toropainen 1994). A vital sawmill industry combined with competitive wood–processing enterprises is also important for the forest owners, because 60–70 % of their stumpage earnings come from sawlog sales (Lönner & Libäck 1992, Økt verdiskapning... 1994, Palo 1994). Networking and cooperation are seen as essential means when organizing the new small–scale production activities. The creation of markets for the products requires investments in the promotion of wood as a building material and as a replacement for the materials based on non–renewable resources such as plastic, aluminium, concrete and metals (e.g. The language of wood... 1987, Nordic wood... 1994, Thörnqvist 1994).

The main objective of regional policy in the Nordic countries has been to promote balanced development in all parts of the country (Grayson 1993, Regional utveckling... 1994). The policy goals have included securing the rural population the same standard of living, both economically and socially, as in urban areas and narrowing the gap between incomes and employment in various parts of the country (Skogspolitiken... 1992, Suomi... 1993, Selby 1994, Økt verdiskapning... 1994). Grants and other economic support to agriculture and forestry have been an essential part of this policy. Today, the grants to timber production are being cut back and support to nature conservation, landscape management and other new branches of rural activity is increasing. The new forms of support to rural areas are being developed by the European Union and the national governments.

References

Act of forests: Denmark. 1989. Translation 7 June 1989. Miljøministeriet/Skov– og Naturstyrelsen, Hørsholm. 18 pp.

Agger, P. & Koch, N.E. 1989. Sammenfatning og hovedkonklusioner. Summary: Conclusions. In: Naturen i Skoven. (Nature in forest.) Miljøministeriet/Skov– og Naturstyrelsen, Hørsholm. p. 6–10. (In Danish.)

Agriculture in Norway. 1993. Royal Ministry of Agriculture, Oslo. 14 pp.

Ahlén, I. et al. 1984. Faunavård i skogsbruket; allmän del. (Fauna management in forestry: general part.) Skogsstyrelsen, Jönköping. 60 pp. (In Swedish.)

Andersson, S. 1992. Dagens skogsbruk växer fram, 1950–1991. (The development of today's forestry, 1950–1991.) In: Elmberg, J., Bäckström, P.–O. & Lestander, T. (eds.). Vår skog: vägvalet. (Our forest: choosing the way.) LTs förlag, Stockholm. p. 39–87. (In Swedish.)

Arjas, A. & Häggblom, R. 1993. Ajatuksia Suomen metsäteollisuuden tulevaisuudesta. (Thoughts about the future of Finnish forest industry.) In: Suomi 2020: visioita kansakunnan tulevaisuudesta. (Finland 2020: visioning the future of the nation.) Valtioneuvoston kanslia, Helsinki. p. 19–31. (In Finnish.)

Blöndal, S. 1991. Socioeconomic importance of forests in Iceland. In: Alden, J., Mastrantonio, J.L. & Odum, S. (eds.). Forest development in cold climates. Plenum Press, New York. 13 pp.

Blöndal, S., Benedikz, P. & Ottósson, J. 1986. Forestry in Iceland: a brief description of its history and present status. Iceland Forestry Service, Reykjavik. 32 pp.

Brief glimpses of Danish forestry. 1982. Danish Forestry Society, Copenhagen. 52 pp.

Contortatallen i Sverige: en lägesrapport. (Lodgepole pine in Sweden: a status report.) 1992. Skogsstyrelsens contortautredning. SLU Reprocentralen, Umeå. 226 pp. (In Swedish.)

Decline and dieback of trees and forests: a global overview. 1994. FAO Forestry Paper 120. FAO, Rome. 90 pp.

Dudley, N. 1992. Forests in trouble: a review of the status of temperate forests worldwide. WWF, Gland. 260 pp.

Ehnström, B. & Waldén, H. 1986. Faunavård i skogsbruket. Del 2. Den lägre faunan. (Fauna management in forestry. Part 2. The invertebrate fauna.) Skogsstyrelsen, Jönköping. 351 pp. (In Swedish.)

En fremtidig skovpolitik. (Future forest policy.) 1986. Dansk Skovforenings Tidsskrift, Særnummer april 1987. 125 pp. (In Danish.)

Enroth, R.–R. 1992. Euroopan integraatio ja metsäsektori. (European integration and forestry sector.) Metsäntutkimuslaitoksen tiedonantoja 345. 95 pp. (In Finnish.)

Flersidig skogbruk. (Multiple–use forestry.) 1990. Norsk skogbruk 11: 4–42. (In Norwegian.)

Flersidig skogbruk: skogbrukets forhold til naturmiljø og friluftsliv. (Multiple–use Forestry: the relationship of forestry with natural environment and recreation.) 1989. Norges offentlige utredninger, NOU 1989:10. Forvaltningstjeneste, Oslo. 139 pp. (In Norwegian.)

Flersidig skogbruk: veiledende retningslinjer for det praktiske skogbruk. (Multiple–use forestry: guidelines for practical forestry.) 1986. Norsk Skogbruk 5: 1–40. (In Norwegian.)

Forest condition in Europe: results of the 1992 survey. 1993. CEC–UN/ECE, Brussels/Geneva. 37 pp.

The forest conservation program: program description and 1994 operations manual. 1994. Scientific Certification Systems, Oakland.

The forest cycle: piece by piece. 1993. Swedish Pulp and Paper Association, Stockholm. 48 pp.

Forest industry, environment, nature. 1993. Finnish Forest Industries Federation, Helsinki. 67 pp.

The forest resources of the temperate zones: the UN–ECE/FAO 1990 forest resource assessment. Volume I. General forest resource information. 1992. United Nations, New York. 348 pp.

Forestry: environment guide. 1994. Finnish Forest and Park Service, Vantaa. 101 pp.

Forestry in Norway. 1991. The Norwegian Forestry Society, Oslo. 23 pp.

Forestry in Sweden. 1994. Skogsstyrelsen, Prognosavdelningen. Update 6 May 1994. 1 p.

Forestry statistics today and tomorrow 1961–1991...2010. 1993. FAO, Rome. 46 pp.

Forest Stewardship Council. 1994. Principles and criteria for natural forest management. Oaxaca, Mexico.

Grayson, A.J. 1993. Private forestry policy in western Europe. CAB International, Wallingford. 329 pp.

Gullichsen, J. Paulapuro, H. & Kohtala, J. 1994. Metsäteollisuus. (Forest industry.) In: Kurki–Suonio, I. & Heikkilä, M. (eds.). Kestävän kehityksen edellytykset Suomessa. (The prerequisites for sustainable development in Finland.) Tammi, Rauma. p. 141–193. (In Finnish.)

Gunnarsson, E., Gudmundsson, E., & Arnason, R. 1987. Hagkvæmni nytjaskógræktar. (Feasibility of production forestry.) Skógræktarrit 10, Sérprent úr ritinu "Audlindir um aldamót", Sérrit 3 úr Framtídarkönnun ríkisstjórnar. (Iceland Forestry Service, Forestry Report 10, Reprint from Government Paper No. 3, "Resources for the Future".) Reykjavík. 50 pp. (In Icelandic.)

Gustavsson, R. & Ingelög, T. 1994. Det nya landskapet: kunskaper och idéer om naturvård, skogsodling och planering i kulturbygd. (The new landscape: knowledge and ideas of nature management, afforestation and planning in cultural areas.) Skogsstyrelsen, Jönköping. 360 pp. (In Swedish.)

Heikka, T. 1992. Länsi–Euroopan integraatio ja metsäteollisuus. (Integration in West Europe and the forest industry.) In: Enroth, R.–R. Euroopan integraatio ja metsäsektori. (European integration and forestry sector.) Metsäntutkimuslaitoksen tiedonantoja 345: 82–89. (In Finnish.)

Hofstätter, M. 1993. Danmark er et skovland: hvis vi self vil. (Denmark is a forest land: if we want it to be.) Jord og viden 8: 3–5. (In Danish.)

Honoré, S. & Jespersen, C. 1989. Maskiner og pesticider i skovbruget. (Forest machines and pesticides.) In: Naturen i Skoven. (Nature in forest.) Miljøministeriet/Skov– og Naturstyrelsen, Hørsholm. p. 54–56. (In Danish.)

Hübertz, H. & Ovesen, C.H. 1989. Særlige fredningsinteresser i skov. Summary: Conservation interests in Danish forests. In: Naturen i Skoven. (Nature in forest.) Miljøministeriet/Skov– og Naturstyrelsen, Hørsholm. p. 57–59.

Hyttinen, P. 1994. The effects of the potential EC –membership on the role of forestry on Finnish farms. Scandinavian Forest Economics 35: 38–49.

Idékatalog för skogsbrukare. (Handbook of ideas for forest owners.) 1988. LTs förlag, Stockholm. 181 pp. (In Swedish.)

Ingelög, T. 1981. Floravård i skogsbruket: allmän del. (Flora management in forestry: general part.) Skogsstyrelsen, Jönköping. 153 pp. (In Swedish.)

Ingelög, T., Thor, G. & Gustafsson, L. (eds.). 1984. Floravård i skogsbruket. Del 2. Artdel. (Flora management in forestry. Part 2. Species.) Skogsstyrelsen, Jönköping. 488 pp. (In Swedish.)

Jespersen, C. 1989. Skovdyrkning og skovdyrkningsmetoder. Summary: Forestry and forestry methods. In: Naturen i Skoven. (Nature in forest.) Miljøministeriet/Skov– og Naturstyrelsen, Hørsholm. p. 52–53.

Johdatus elinkaarianalyysin käyttöön yrityksessä. (Introduction to the use of lifecycle analysis in an enterprise.) 1993. Teollisuuden työnantajain keskusliitto, Helsinki. 20 pp. (In Finnish.)

Kajaste, I. 1994. The Finnish economy and the prospects of forest industries: trade, competition and environment. Ministry of Finance, Economics Department, Discussion Paper no. 43. 23 pp.

Kalland, F. & Pätilä, A. 1993. The green change. Finnish Forest Industries Federation, Helsinki. 29 pp.

Kansallinen teollisuusstrategia. Summary: National industrial strategy for Finland. 1993. Kauppa– ja teollisuusministeriö, Helsinki. 124+54 pp.

Karjalainen, H. 1994. Suomen WWF:n metsäohjelma. (Forestry program by WWF Finland.) Suomen WWF, Helsinki. 32 pp. (In Finnish.)

Kellomäki, S. 1994a. Metsätalous. (Forestry.) In: Kurki–Suonio, I. & Heikkilä, M. (eds.). Kestävän kehityksen edellytykset Suomessa. (The prerequisites for sustainable development in Finland.) Tammi, Rauma. p. 417–482. (In Finnish.)

Kellomäki, S. 1994b. Response of the boreal forest ecosystem to climatic change and its silvicultural implications. In: The Finnish research programme on climate change: second progress report. Publications of the Academy of Finland 1/94. p. 209–210.

Key figures 1993. 1994. The Norwegian Pulp and Paper Association, Oslo. 9 pp.

Klingberg, T. 1994. Det nya träoldern. (The new wood era.) In: Skogsåret 1993–94. (Forest year 1993–94.) 1994. Skogsägarnas Riksförbund, Stockholm. p. 6–7. (In Swedish.)

Koch, N.E. & Kristiansen, L. 1991. Flersidigt skovbrug: et idékatalog. (Multiple–use forestry: a handbook of ideas.) Skov– og Naturstyrelsen, Hørsholm. 39 pp. (In Danish.)

Kulturmiljövård i skogen. (Protection of cultural heritage in forests.) 1992. Skogsstyrelsen, Jönköping. 259 pp. (In Swedish.)

Kuusela, K. 1985. Euroopan puuntuotanto ja teollisuuspuun ulkomaankauppa 1950–2000. (Timber production and trade of industrial roundwood in Europe 1950–2000.) Sitra, Sarja B, Nro 79. 184 pp. (In Finnish.)

Kuusela, K. 1990. The dynamics of boreal coniferous forests. Sitra, Helsinki. 172 pp.

Kuusela, K. 1994. Forest resources in Europe. European Forest Institute, Research report 1. 154 pp.

Landbruksdepartementet's handlingsplan for bevaring og bærekraftig bruk av biologisk mangfold. (Action plan of the Department of Agriculture for protection and sustainable use of biological diversity.) 1994. Høringsutkast. Landbruksdepartementet, Oslo. 67 pp. (In Norwegian.)

Landskapsvern og naturvern i skogen: veiledende retningslinjer i skogen. (Landscape and nature protection in forest: guidelines for forestry.) 1978. Norsk Skogbruk 2: 3–43. (In Norwegian.)

The language of wood: wood in Finnish sculpture, design and architecture. 1987. Museum of Finnish Architecture/Finnish Society of Crafts and Design/ Museum of Contemporary Art/Association for Contemporary Art, Helsinki. 230 pp.

Leikola, M. 1987. Metsien hoidon aatehistoriaa. Summary: Leading ideas in Finnish silviculture. Silva Fennica 21(4): 332–341.

Linddal, M. 1993. Denmark. In: The Forest Resources of the Temperate Zones. The UN–ECE/FAO 1990 Forest Resource Assessment. Volume II. Benefits and functions of the forest. United Nations, New York. p. 88–94.

Lines, R. 1990. Forestry in Iceland. Scottish Forestry 44(2): 85–93.

Loftsson, J. 1993. Forest development in Iceland. In: Alden, J., Mastrantonio, J.L. & Odum, S. (eds.) Forest development in cold climates. Plenum Press, New York. p. 453–461.

Lönner, G. & Libäck, K. 1992. Omvärlden och svensk skogsnäring. (Swedish forestry and the rest of the world.) In: Elmberg, J., Bäckström, P.–O. & Lestander, T. (eds.). Vår skog: vägvalet. (Our forest: choosing the way.) LTs förlag, Stockholm. p. 149–165. (In Swedish.)

Luonnonläheinen metsänhoito: metsänhoitosuositukset. (Nature–like silviculture: silvicultural guidelines.) 1994. Metsäkeskus Tapion julkaisuja 6. 71 pp. (In Finnish, a shortened version available in English: Good forests for the good of all. 1994. Forestry Centre Tapio, Helsinki. 17 pp.)

Mattsson, L. & Östlund, L. 1992. Människan och skogen: en tillbakablick. (Man and forest: historical overview.) In: Elmberg, J., Bäckström, P.–O. & Lestander, T. (eds.). Vår skog: vägvalet. (Our forest: choosing the way.) LTs förlag, Stockholm. p. 11–38. (In Swedish.)

Metsä 2000 –ohjelman pääraportti. (Main report of the Forest 2000 Programme.) 1985. Talousneuvosto, Helsinki. 189 pp. (In Finnish.)

Metsä 2000 –ohjelman tarkistustoimikunnan mietintö. (Report of the revising committee of Forest 2000 Programme.) 1992. Komiteamietintö 1992:5. Valtion painatuskeskus, Helsinki. 112 pp. (In Finnish.)

Metsätalouden ympäristöopas. 1993. Metsähallitus, Vantaa. 112 pp.

Metsätalous ja ympäristö. (Forestry and environment.) 1994. Maa– ja metsätalousministeriö, Metsätalouden ympäristöohjelmatyöryhmän mietintö 1994:3. 100 pp. (In Finnish.)

Metsäteollisuuden vuosikirja 94. (Annual report of forest industry 94.) 1994. Metsäteollisuus ry, Helsinki. 68 pp. (In Finnish.)

Metsätilastollinen vuosikirja 1993–94. (Yearbook of forest statistics 1993–94.) 1994. The Finnish Forest Research Institute, Helsinki. 348 pp. (Charts and summary in English.)

Ministerial Conference on the Protection of Forests in Europe, 16–17 June 1993 in Helsinki: documents. 1993. Ministry of Agriculture and Forestry, Helsinki. 55 pp.

Ministerial Conference on the Protection of Forests in Europe, 16–17 June 1993 in Helsinki: European list of criteria and most suitable quantitative indicators. 1994. Ministry of Agriculture and Forestry, Helsinki. 20 pp.

Naturskogar i Norden. (Natural forests in the Nordic countries.) 1994. Nordisk Ministerråd, Nord 1994:7. 109 pp. (In Swedish.)

New environmental programme for forestry in Finland. 1994. Ministry of Agriculture and Forestry/Ministry of the Environment, Helsinki. 63 pp.

The Nordic Environment: present state, trends and threats. Nord 1993:12. 211 pp.

Nordic Wood: Pohjoismainen puuteollisuuden tutkimus- ja tuotekehitysohjelma 1994–1996. 1994. (Nordic Wood: Nordic research and product development plan for wood industry 1994–1996.) 1994. TEKES, Helsinki. 4 pp.

Norske Skogindustrier A.S: annual report 1993. 1993. Norske Skogindustrier A.S, Skogn. 49 pp.

Norwegian Forest Management: information. 1994. The Royal Norwegian Ministry of Agriculture, Oslo. 6 pp.

Oksa, J. (ed.). 1994. Syrjäisen seudun uudet kerrostumat. (New activities in remote rural areas.) Joensuun yliopisto, Karjalan tutkimuslaitoksen julkaisuja 110. 123 pp. (In Finnish.)

Økt verdiskapning fra skog. Skogbruk og skogindustri: perspektiver, utfordninger og områder for forskning. (Increased value from the forest. Forestry and forest industry: perspectives, challenges and topics for research.) 1994. Norges Forskningsråd, Oslo. 40 pp. (In Norwegian.)

Outlook for the European forest resource and roundwood supply. 1994. FAO/ECE working paper. Draft.

Palo, M. 1994. Ympäristötietoisen metsäpolitiikan strategia. (A strategy for environmentally conscious forest policy.) In: Palo, M. & Hellström, E. 1994. Metsäpolitiikka valinkauhassa. (Forest policy in a melting pot.) Metsäntutkimuslaitoksen tiedonanoja 471: 307–467. (In Finnish.)

Parviainen, J. 1994. Finnish silviculture: managing for timber production and conservation. Journal of Forestry, September 1994: 33–36.

Parviainen, J. & Seppänen, P. 1994. Metsien ekologinen kestävyys ja metsänkäsittelyvaihtoehdot. (Ecological sustainability of forests and silvicultural alternatives.) Metsäntutkimuslaitoksen tiedonantoja 511. 110 pp. (In Finnish.)

Pennanen, J. 1994. Villi ja vapaa metsä: raportti Suomen metsien luontaisesta rakenteesta ja dynamiikasta sekä mahdollisuuksista kehittää luonnondynamiikkaan perustuva metsänhoitomalli. Summary: Wild and free forest: a report on the natural structure and dynamics of Finland's forests and the possibilities of developing a forest management model based on the natural forest dynamics. Greenpeace, Helsinki. 61 pp.

Petersen, K. 1989. Skoglig flerbruk: en litteraturstudie. Norsk institutt for skogforskning, Ås. 80 pp.

Plum, P.M., Billeschou, A. & Thomsen J. 1990. Skovrejsning i Danmark. Summary: Increasing forest area in Denmark. Sveriges Skogsvårdsförbunds Tidskrift 3: 28–33. (In Danish.)

Puun käytön edistäminen rakentamisessa: puun rakennuskäytön edistämistyöryhmän muistio. (Increasing the use of wood in construction: a report by the committee for the promotion of the use of wood in construction.) 1994. Maa– ja metsätalousministeriö, Työryhmämuistio 1994:21. 26 pp. (In Finnish.)

Regional utveckling i Norden: basprojektets årsrapport 1993/94. (Development of regions in the Nordic countries: annual report of a base project 1993/94). Nord 1993:30. 148 pp. (In Swedish.)

Remröd, J. 1991. The forest of opportunity. Skogsindustrierna, Stockholm. 102 pp.

The representation of the revised Forest 2000 Programme: Finnish forest policy in the 1990s. 1992. Finnish Forestry Association, Helsinki. 26 pp.

Rikare skog: 90 –talets kunskaper om naturvård och ekologi. (Richer forest: the knowledge of nature protection and ecology in the 90s.) 1990. Skogsstyrelsen, Jönköping. 133 pp. (In Swedish.)

Saastamoinen, O. 1987. Multiple use and the Forest 2000 Programme. Scandinavian Forest Economics 29: 39–47.

Scandinavia: forestry and the wood production industry. 1993. World Forest Center, Portland. 38 pp.

Scandinavian forestry: the forest should be used wisely – not used up. 1994. Swedish Pulp and Paper Association/Svensk Skog/Norwegian Pulp and Paper Association/The Norwegian Forest Owners' Federation/Finnish Forestry Industries Federation/Finnish Forestry Association/Nordic Timber Council. 27 pp.

A search for sustainable forestry: the Swedish view. 1994. Swedish Pulp and Paper Association, Stockholm. 59 pp.

Selby, A. 1994. Primary sector policies and rural development in Finland. 1994. In: Gilg, A.W. (ed.). Progress in rural policy and planning. John Wiley & Sons, Chichester. p. 157–176.

Seppälä, H. 1993. Metsäteollisuus 2010: arvio Suomen metsäteollisuudesta ja sen puunkäytöstä. (Forest industry 2010: a scenario for Finnish forest industry and utilization of timber resources.) Metsäntutkimuslaitoksen tiedonantoja 454. 54 pp. (In Finnish.)

Seppälä, R. 1994. Metsäsektoriko uuden nousun veturi? (Forest sector: an engine for a new economic boom?) In: Snellman, V. (ed.). Tutkimus metsien kestävyyden ja käytön perustana. (Research as a basis for sustainability and utilization of forests.) Metsäntutkimuslaitoksen tiedonantoja 523: 97–106. (In Finnish.)

Skogspolitiken inför 2000 –talet. (Forest policy for the 21st century.) 1992. Jordbruksdepartementet, Statens offentliga utredningar, SOU 1992:79. 343 pp. (In Swedish.)

Skogsstatistisk årsbok 1994. (Statistical Yearbook of Forestry 1994.) 1994. National Board of Forestry, Jönköping. 349 pp. (Charts and summary in English.)

Snorsson, A. 1992. Iceland. In: The forest resources of the temperate zones: the UN–ECE/FAO 1990 forest resource assessment. Volume I. General forest resource information. United Nations, New York. p. 184–187.

Solbraa, K. 1989. Flersidig skogbruk. Summary: Multiple use in forestry. Norsk institutt for skogforskning, Rapport 7. 35 pp.

Solbraa, K. & Grønvold, S. 1989. Skogøkologi og flersidig skogbruk III. Del. B. Skoglige tilpasninger, friluftsliv og konsekvensvurderinger. Summary: Silvicultural adaptations, open air recreation, and consequence evaluation. Norsk institutt for skogforskning, Rapport 14. 54 pp.

Strategy for sustainable development: proposals for a Swedish programme. 1993. Swedish Environmental Protection Agency, Solna. 280 pp.

Strategy for sustainable forest management. 1994. Ministry of the Environment/The National Forest and Nature Agency, Copenhagen. 65 pp.

Suomen metsäluonnon monimuotoisuuden turvaaminen. (Protecting the diversity of Finnish forest nature.) 1994. Ympäristöministeriö, Alueiden käytön osasto. 83 pp. (In Finnish.)

Suomi: eurooppalainen maaseutumaa. (Finland: a rural country in Europe). 1993. Valtioneuvoston maaseutupoliittinen selonteko eduskunnalle, Helsinki. 41 pp. (In Finnish.)

Svanborg, A. 1994. Äntligen har skogkonjunkturen vänt. (The forest trade conditions have finally changed.) In: Skogsåret 1993–94. (Forest year 1993–94.) 1994. Skogsägarnas Riksförbund, Stockholm. p. 38–42. (In Swedish.)

Sweden's new forest policy. 1994. The National Board of Forestry, Jönköping. 13 pp.

Thörnqvist, T. 1994. Mer trä i byggandet. (More wood in building.) In: Skogsåret 1993–94. (Forest year 1993–94.) 1994. Skogsägarnas Riksförbund, Stockholm. p. 52–56. (In Swedish.)

Toropainen, M. 1994. Voiko metsäsektori nostaa kansantalouden lamasta. (Can the forestry sector help the national economy out of recession?) In: Toropainen, M. & Mäkkeli, P. (eds.). Metsäsektori myllerryksessä. (Forestry sector in turmoil.) Metsäntutkimuslaitoksen tiedonantoja 500: 45–54. (In Finnish.)

UNCED: YK:n ympäristö– ja kehityskonferenssi, Rio de Janeiro 3.–14.6.1992. (UNCED: United Nations Conference on Environment and Development, Rio de Janeiro 3.–14.6.1992.) Ympäristöministeriö/Ulkoasiainministeriö, Helsinki. 239 pp. (In Finnish.)

Wibe, S. 1992. Sweden. In: Wibe, S. & Jones, T. (eds.). Forests: market & intervention failures. Five case studies. Earthscan Publications Ltd., London. p. 58–89.

Wahlström, E., Hallanaro, E.–L. & Reinikainen, T. 1993. The state of the Finnish environment. Ministry of the Environment, Helsinki. 163 pp.

WWF Submission to the Commission on Sustainable Development. 1994. WWF, Gland. 34 pp.

Yearbook of Nordic statistics 1994. 1994. Nordic Council of Ministers, Nord 1994:1. 431 pp.

6 Non–timber forest products and their utilization

Kauko Salo[1]

Abstract

The economically most significant and most popular berry species in Finland, Sweden and Norway are cowberry (*Vaccinium vitis–idaea*), bilberry (*V. myrtillus*), cloudberry (*Rubus chamaemorus*) and raspberry (*R. idaeus*). They are primarily forest and mire species inhabiting various forest and peatland site types. Cowberry is the most important household berry in Finland and it also provides the most abundant crop. Bilberry takes first place in Sweden and Norway. The majority of the best edible mushrooms occur on dry, dryish or moist upland forests. Most of them are mycorrhizal species that accompany pine, spruce and birches. Three mushroom species: *Boletus edulis*, *Suillus luteus* and *Cantharellus cibarius* are considered to be the best edible mushrooms in all the Nordic countries. Also, other common wild berry species and edible mushroom species are popular. The reindeer lichen species *Cladonia alpestris* forms extensive carpets in the driest woodlands of the northern part of the boreal coniferous zone. The lichen is used for ornamental purposes. Lichen is mostly exported to central Europe, where the leading importers are Germany, Austria and Italy. Norway spruce (*Picea abies*) has been the traditional choice as a Christmas tree. The estimated annual consumption of Christmas trees in the Nordic countries is 6.5–7.5 million trees. Christmas trees and the ornamental foliage of numerous exotics (e.g. *Abies nordmanniana*, *A. procera*) have a significant role in Danish forestry. Ornamental foliage consists of evergreen branchlets used for decorative purposes. In this area of forestry, Denmark is the leader in Europe.

Keywords: forest products, berries, mushrooms, lichen, Christmas trees, ornamental foliage.

---

[1] Finnish Forest Research Institute
Joensuu Research Station
P.O.Box 68, FIN–80101 Joensuu, Finland

6.1 An aspect of multiple–use forestry

In addition to timber production, forest uses include berry picking, mushrooming, other forms of gathering, Christmas tree harvesting, gamekeeping and hunting, reindeer husbandry and cattle grazing, recreation and outdoor activities, landscape management and conservation. Forests and peatlands embody a vast amount of intangible values. A fine example of such a value is the experience of encountering a rare woodpecker next to its nesting tree or the enthralling spectacle of a whooper swan winging over a vast mire.

Our ancestors knew what multiple use of forests was all about. Grazing of cattle in the woods is no longer practised. Charcoal, resin, tar and birch bark are examples of wood products no longer used very much. It was timber that people used to make their dwellings out of, and moss gathered from the forest floor was used to keep out the cold that would otherwise have crept in through the gaps in the walls. Wood was the source of heating energy as well. Two traditional forms of forest use still remain: picking of berries and mushrooms, and hunting.

Generally speaking, berries and mushrooms are considered to be parallel or by–products provided by forests. In years when the berry and mushroom crops are plentiful, cowberry, sometimes called lingonberry (*Vaccinium vitis–idaea*) and ceps (*Boletus pinophilus*) may be considered to be parallel forest products on dryish Scots pine sites in Finland. In times of poor berry or mushroom crops on the same sites they are forest by–products. The monetary value per hectare of the berries and mushrooms waiting to be collected on certain sites can exceed the income to be obtained from the final cutting of the tree crop. Bounteous communities of cloudberry (*Rubus chamaemorus*) and cranberry (*Vaccinium oxycoccos*) have been obliterated by draining of vast areas of peatland in southern and central Finland. Instead of being reworked, failed drainage and planting areas should be allowed to return to their former paludified state. With time, the water and nutrient status of such peatlands will once again provide conditions suitable for cloudberry and cranberry crops to flourish. Remedial draining and thinning of peatland sites should be directed at sites where the tree crops show good growth and where the benefits to be gained are in proportion to the expenses involved.

6.2 Rights of public access

Finland and Sweden are the Nordic countries where rights of public access (in Finland the term is translated as *Everyman's rights*) are recognized. Denmark, Iceland and Norway have more restrictions. It means that citizens of Finland and Sweden have the right of free access to outdoor areas, where they can pick wild berries, mushrooms, wild herbs, wild flowers, fallen dry branches and cones lying on the ground. However, you are not allowed to gather logging waste nor the branches and cones of windthrown trees without the landowner's permission. Income earned from the sales of wild berries, herbs and mushrooms is tax–free in Finland, whereas income earned from the collecting of lichen is taxable. In the case of the latter, the landowner collects him/herself or rents the collecting right to an outsider. In Sweden, people have to pay taxes for all kinds of income including producing and selling honey, and picking berries and mushrooms. However, it is possible to sell up to 5,000 SEK/person without paying taxes. A new "hobby–tax" will be introduced during spring 1993 in Sweden. The collecting of cones without a permit is punishable under the Criminal Act either as theft or damage to property.

Rights of public access have also been limited by the nature conservation legislation and by certain regulations pertaining to conservation areas. In addition, the Ministry of Agriculture and Forestry in Finland is entitled to prohibit the picking of cloudberries by non–residents and foreigners in some districts in the Province of Lapland. Also in Norway, the picking of cloudberries by non–residents is restricted in the Nordland, Troms and Finnmark counties.

Rights of public access is a tradition which has developed over the years. It is a kind of common usage that has its roots in the early stages of the settlement of Finland, Norway and Sweden. They are based, on the one hand, on usufruct from time immemorial and, on the other hand, on the written law. Regulations pertaining to the rights in Sweden are set out in a law from the time of King Christian in 1442, and they were put into writing in 1608 when Karl IX ratified the law.

Regulations connected to rights of public access were included in the Law of Moses over 5,000 years ago. One of these laws states: "When thou comest into thy neighbour's vineyard, then thou mayest eat grapes thy fill at thine own pleasure; but thou shalt not put any in thy vessel. When thou comest into the standing corn of thy neighbour, then thou mayest pluck the ears with thine hand; but thou shalt not move a sickle unto thy neighbour's standing corn" (Raamattu 1954).

The application of rights of public access in Denmark and Iceland is very much in the spirit of the Law of Moses. In Iceland, a person may collect wild berries for eating on the spot. If you intend to pick berries to take home, you need to have the landowner's or tenant's permission. In Denmark, there are restrictions on the recreational use of forests owned by private persons. Citizens are allowed to walk only along paths and the landowner is entitled to prohibit the picking of wild berries and mushrooms – and yet landowners there do not themselves organize the picking of berries and mushrooms. State–owned forests allow the collecting of mushrooms for private consumption but not for sale (Hallgrimsson 1990).

In many other countries in Europe, the picking of berries and mushrooms is subject to license. In France, for example, the state forestry authority sells bilberry picking licenses in the mountain regions for those state forests where sufficient amounts of berries have been observed. According to calculations made in France, the amount of money collected from the selling of bilberry picking licenses corresponds to almost a half of the total income obtained from bilberry forests.

Rights of public access in the Nordic countries also apply to foreigners. It is common that people from Poland and the Baltic states go to northern Sweden and Finland for a month or two to pick berries for sale. The constantly increasing volume of tourism and outdoor activities will eventually force the Nordic countries to review their policies. Some of the fundamental issues will be the following: Will free access to move about outdoors and pick berries and mushrooms in forests and on mires without a need for the landowner's permission be retained? Will legislative measures be undertaken or will separate picking licenses be issued by the Ministry of Agriculture and Forestry to reg-

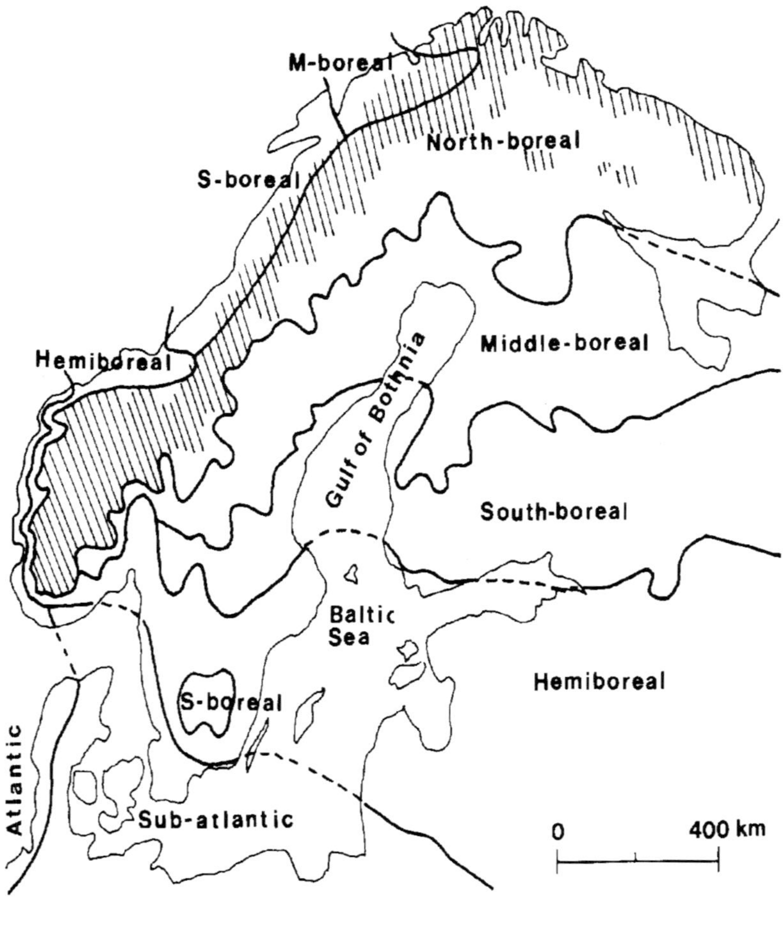

Figure 6.1 Vegetation zones of Fennoscandia. The Arctic–Alpine vegetation zone has been shadowed, mainly according to Ahti et al. (1964, 1968).

ulate entry to private forest estates? Will we find ourselves purchasing picking licenses for state–owned forests and peatlands?

6.3 Occurrence of wild berries in the Nordic countries

Forest berries and site types

The species of wild berries presented in this publication are primarily forest and mire species inhabiting various forest and peatland site types, fells,

Table 6.1 European, North American and scientific names of some edible wild berries. *Economically most important berries in the Nordic countries. **Here the English name is a common name of numerous species and hybrids of *Rubus fruticosus.*

Europe	North America (when different)	Scientific name
Alpine bearberry		*Arctostaphylos alpinus*
Arctic bramble		*Rubus arcticus*
Bearberry		*Arctostaphylos uva-ursi*
"Bear bramble"**		*Rubus fruticosus*
Bilberry*	Blueberry	*Vaccinium myrtillus*
Bird cherry		*Prunus padus*
Blackthorn		*P. spinosa*
Bog bilberry	Bog blueberry	*Vaccinium uliginosum*
Cloudberry*		*Rubus chamaemorus*
Cowberry*	Lingonberry	*Vaccinium vitis-idaea*
Granberry		*V. oxycoccos*
Crownberry		*Empetrum nigrum*
Juniper		*Juniperus communis*
Mountain crowberry		*Empetrum hermaphoroditum*
Raspberry*		*Rubus idaeus*
Roses		*Rosa canina, R. tomentosa, R. dumalis, R. rugosa, R. majalis*
Rowanberry		*Sorbus aucuparia*
Sea buckthorn		*Hippophaë rhamnoides*
Small cranberry		*Vaccinium microcarpum*
Stone bramble		*Rubus saxatilis*
Wild cherry		*Prunus avium*
Wild strawberry		*Fragaria vesca*

peripheral areas around cultivations, and shorelines within the boreal vegetation zone of the Fennoscandia region (Finland, Sweden, Norway) (*Figure 6.1*).

The names given to some berry bearing dwarf shrubs differ in Europe and North America (*Table 6.1*). This is partly because they refer to different species. For instance, the European bilberry is *Vaccinium myrtillus*, whereas the North American blueberry is *V. corymbosus* and *V. angustifolium*. North American names for the European bilberry and cowberry are widely used in the trade in Nordic regions.

Finland has some fifty indigenous berry species and of these 37 are edible. Sixteen species are picked for human consumption (*Table 6.2*); these include a few (e.g. bog bilberry (*Vaccinium uliginosum*), crowberry (*Empetrum nigrum*), mountain crowberry (*E. hermaphroditum*), stone bramble (*Rubus saxatilis*) and alpine bearberry (*Arctostaphylos alpinus*)) that have found very little use in the recent past. According to Eriksson et al. (1979), Sweden has at least 35 edible berry species and Norway has 30 indigenous berry species whose berries and fruits are fit for human consumption (Valset et al. 1976).

The economically most significant and most popular berry species in Finland, Sweden and Norway are cowberry, bilberry, cloudberry and raspberry. Cowberry (*Figures 6.2* and *6.3*) is the most important household berry in Finland and it also provides the most abundant crop. Bilberry takes first place in Sweden and Norway; according to Kardell (1980), the average annual bilberry crop in Sweden is larger than that of cowberry. In Norway, cloudberry comes second after bilberry with cowberry taking third place (Valset et al. 1976). The aforementioned berry species are common, but there is large variation between their occurrence in the different parts of the countries.

The southern, central and northern boreal zones of Finland and Sweden are typified by the presence of a contiguous belt of coniferous forest dominated by Norway spruce (*Picea abies*) and Scots pine (*Pinus sylvestris*). The common and economically most significant wild berries are to be found in these coniferous forests interspersed with two birches (*Betula pendula* and *B. pubescens*). Cowberry (*Vaccinium vitis–idaea*) grows in pine dominated dry and dryish upland forests and occasionally in upland forests on moist soils. Bilberry (*V. myrtillus*) is at its most typical in spruce dominated moist upland forests; it is also to be found in dryish upland forests composed of pine and birch in addition to spruce. Grovelike upland forests may also contain bilberry. According to the results of the third national forest inventory conducted in Finland (1951–1953), half of Finland's forest area was covered by cowberry and bilberry at the field layer level.

Crowberries (*Empetrum nigrum* and *E. hermaphroditum*) are a common field layer species in dry and dryish upland pine forests and in the fells of the coniferous zone. Moist and grovelike upland sites are typically populated by raspberry (*Rubus idaeus*), rowan (*Sorbus aucuparia*) and stone bramble (*Rubus saxatilis*). Along with crowberries, these have not been very popular in the

Table 6.2 The distribution of edible berries on different site types and in different parts of Finland (Luonnonmarjaopas 1988).

Berry species	**Site type**									**Distribution**			
	Shoreline vegetation	CT	VT	MT	OMT	Spruce mires	Pine mires	Treeless bogs	Mountain vegetation	Southern Finland	Central Finland	Kainuu and Southern Lapland	Lapland
Cowberry [1] *Vaccinium vitis-idaea*	○	◒	●	◒	○	◒	◒		◒	C	C	C	O
Bilberry [1] *Vaccinium myrtillus*		○	◒	●	◒	◒	○		◒	C	C	C	U
Cloudberry [1] *Rubus chamaemorus*						◒	●		○	U	O	C	C
Raspberry [2] *Rubus idaeus*	●			○	●	○				C	C	U	-
Cranberries [2] *Vaccinium oxycoccus*						○	◒	●		C	C	O	-
V. microcarpum							●			O	C	C	C
Arctic bramble [2] *Rubus arcticus*	●			○	◒	○				U	C	U	-
Crowberries *Empetrum nigrum* *E. hermaphroditum*	●	●	●	○		○	●		●	U	O	C	C
Rowan *Sorbus aucuparia*	◒		○	●	●	○				C	C	C	U
Sea buckthorn *Hippophaë rhamnoides* [3]	●									C	C	-	-
Bog bilberry *Vaccinium ugilinosum*				○		◒	●		●	U	O	C	C
Wild strawberry *Fragaria vesca*	●				◒					C	C	U	-
Juniper *Juniperus communis*	◒	◒	◒	●	◒	○			○	C	C	C	C
Stone bramble *Rubus saxatilis*	●			◒	●	○				C	C	C	U
Black bearberry *Arctostaphylos alpina*							○		●	-	-	U	C

1 Economically most important berries in Finland.

2 Second economically most important groups of berries. Site types have changed and yields have been decreased since the early 1930s.

3 Sea buckthorn occurs only on the islands and the shoreline of Gulf of Bothnia and on the Åland Islands.

● Common
◒ Less Common
○ Rare

C = Common
O = Occasional
U = Uncommon

CT = Calluna Type
VT = Vaccinium Type
MT = Myrtillus Type
OMT = Oxalis-Myrtillus Type

Nordic countries in the recent past. There are years when some tens of tons of rowan berries and crowberries will be bought for industrial consumption in Finland. Juniper (*Juniperus communis*) is common in moist and grovelike upland forests and along the coast. Juniper berries are mainly used to spice food and beverages.

Extensive mountain areas in eastern Norway, western Sweden and in Finnish Lapland belong to the Arctic–Alpine cold vegetation zone that encompasses the tundra north of the timber line and the alpine parts above the timber line in mountainous areas (*Figure 6.1*). The edible wild berries of the fell country are bilberry, crowberry, bog bilberry (*Vaccinium uliginosum*) and alpine bearberry (*Arctostaphylos alpinus*), but only minor amounts of these are picked in the fells of Finland, Sweden and Norway.

Boreal zones with their oceanic birch forests are to be found along the coast of Norway north of Trondheim. The hemiboreal zone of Fennoscandia is limited to the coastal region of southern Norway, most of southern Sweden and only a narrow strip of southern Finland. This zone includes pine and spruce forests and forests composed of broadleaved species; e.g. oak–hazel and elm–ash forests (Kalliola 1973). In these forests, cowberry occurs only in the more barren parts and even bilberry is limited to patches in moist upland forests.

The western part of Denmark and the western coast of southern Norway belong to the central European Atlantic zone while eastern Denmark and the southernmost western coast of Sweden belong to the sub–Atlantic zone. Both of these zones are outside the natural area of distribution of spruce and, as an indigenous part of the flora, even pine occurs only rarely. This being the case, the wild berries of the Atlantic and sub–Atlantic zones are rare, and it may be that they are to be encountered only in plantations of conifers.

The other common berry species of Sweden and Norway include cranberry, small cranberry, crowberry, bog bilberry, rowan, juniper, wild strawberry, and arctic bramble. In Sweden, and especially in Norway, local popularity is enjoyed by the species *Rubus plicatus*, *R. nessensis* and *R. fissus* (Valset et al. 1976). This group of berries includes several perennial, dense, sharp–thorned bush species and hybrids. The only species indigenous to Finland, *R. plicatus* and *R. pruinosus*, are rare and the latter is protected in the Åland Islands (Ingmanson & Holmberg 1988).

In some parts of southern Sweden, southern Norway and Denmark, blackthorn (*Prunus spinosa*) and wild cherry (*P. avium*) are picked as well as cultivated in gardens. Blackthorn occurs as a rare species in southwestern Finland and in the Åland Islands (Hämet–Ahti et al. 1986). The berries of bird cherry (*P. padus*), common in Sweden, Norway and Denmark but rare in the southern parts of Finland, have been traditionally used to make fruit drinks, wines and liqueurs (Ingmanson & Holmberg 1988). Due to the rarity of the species in Finland, its berries have not found household use. In the Nordic countries there are several species of roses (*Rosa canina*, *R. tomentosa*, *R. dumalis*, *R. majalis*,

Figures 6.2 and *6.3* Cowberry is in bloom and makes berries in the dry and dryish heath forest sites. The best yields of cowberry can be 100–200 kg/ha in very good sites but normally the average yields are 5–20 kg/ha in large heath forest areas. Photos: Kauko Salo (flowers) and Erkki Oksanen (berries).

R. rugosa) that find a lot of local use. Rose berries have especially been used in the form of purées and soups (Valset et al. 1976, Ingmanson & Holmberg 1988).

The Iceland of today is a barren land. Only 1 % of the country's land area, 1,250 km^2, is covered by woodland with downy birch (*Betula pubescens*) as the dominant species. The mean height of the trees (80.8 % of all) is under 2 metres. The native Icelandic woodlands are shrubby (Blöndal 1993). The other native tree species are willows (*Salix phylicifolia*, *S. lanata*), rowan (*Sorbus aucuparia*) and aspen (*Populus tremula*).

In Iceland there are 10 indigenous plant species that produce berries. Most of the species are found in woodlands but none of them is restricted to such sites. The most common berries are crowberry (*Empetrum nigrum*), mountain crowberry (*E. hermaphroditum*), bog bilberry (*Vaccinium uliginosum*), bilberry (*V. myrtillus*) and stone bramble (*Rubus saxatilis*). Other less abundant species are wild strawberry (*Fragaria vesca*), rowan (*Sorbus aucuparia*), one of the few native tree species, juniper (*Juniperus communis*) and bearberry (*Arctostaphylos uva–ursi*) (Hallgrimsson 1990).

Peatland berries

In terms of its area of peatland, Finland is the world leader. The total peatland area was originally about 10.4 million hectares (or one third of the country's entire land area). At present the total area drained for forestry and agricultural purposes is about 6 million hectares with part of this area classified as being paludified mineral soil forest. The drained area corresponds to one quarter of Finland's forest area (Päivänen 1990).

Sweden's area of peatland (peat depth 30 cm or more) amounts to 6.3 million hectares (Hånell 1990). The peatland area in Norway is 3 million hectares, of which 1.1 million hectares are either cultivated for agricultural purposes or lie above the timberline (Brække 1990). Iceland's peatland area is 1 million hectares, and that of Denmark is 0.6 million hectares (Kivinen & Pakarinen 1980).

The two wild berry species typical to virgin pine mires and bogs in Finland, Sweden and Norway are cloudberry (*Rubus chamaemorus*) (*Figure 6.4*) and bog bilberry (*Vaccinium uliginosum*); both of these also grow well on spruce mires. Small cranberry (*V. microcarpum*), crowberry (*Empetrum nigrum*) and mountain crowberry (*E. hermaphroditum*) are typical of pine bogs.

Cranberry (*Vaccinium oxycoccos*) grows on treeless bogs. Cowberry (*Vaccinium vitis–idaea*) and bilberry (*V. myrtillus*) are typical forest berries, but they also grow and produce good yields on transitional drained peatlands and in old peat–based forests. A peat substrate is also suitable for berry–producing individuals of raspberry (*Rubus idaeus*) on logged spruce mire sites and for arctic bramble (*R. arcticus*), an increasingly rare species, on meadows and in spruce bogs that are regularly flooded (Salo 1988a). In Denmark and Iceland, these mire berry species are rare and have no economic significance.

6.4 Yields of the most important wild berries

The overall crops of wild forest and peatland berries in Sweden have been studied by Eriksson et al. (1979) and Kardell and Carlsson (1982). Corresponding calculations for Finland have been made by Raatikainen et al. (1988) and Salo (1991a). The average yield of bilberry in Sweden for the years 1975–1977 was 255 million kilos and it varied within the range of 219–307 million kilos. The normal bilberry crop in Finland has been calculated to be 150–200 million kilos. Sweden's average cowberry crop for the above years was calculated to have been 155 million kilos with a variation of 142–168 million kilos. Finland's cowberry crop for a normal year is about 200 million kilos; in exceptionally good years, as in 1988 and 1989, the crop can rise to 500 million kilos.

The cloudberry crop in Sweden for the years 1978–1980 was reckoned to average 75.7 million kilos. In 1979, the crop was 93 million kilos. Finland's cloudberry crop is estimated to vary within the range of 25–50 million kilos. The main part of the annual crop is collected from the mires of Lapland and occasionally from southeast Finland. Southern and central Finland have an average of only three good cloudberry crops per decade. Cloudberry crop failures are known to be less frequent in northern Finland (Salo 1991a).

The main crop of Sweden's foremost household berries (cowberry, bilberry, cloudberry) has been calculated to be 485 million kilos. The corresponding figure for Finland for an average year is 450 million kilos. Thus, the two countries do not greatly differ from one another on the average. The biggest difference appears to be in the size of the cowberry crops; Finland's overall crop can occasionally rise to a level which is double (500 million kilos) the normal year's crop. The biological crop of cowberry, bilberry and cloudberry was about 600 million kilos in 1988. This means that the cowberry and cloudberry crops were abundant and bilberry crop above average (Salo 1991a).

The annual crops of cranberry have declined as a consequence of draining of peatlands during the past couple of decades; in Finland they are estimated to be on average 20–30 million kilos. According to Kardell and Carlsson (1982), Sweden's cranberry crops in 1978–1980 averaged 20.8 million kilos.

The annual crops of crowberries, raspberry, rowan and bog bilberry amount to some tens of millions of kilos. Sweden's overall raspberry crop varied between 15–20 million kilos per year during the period 1974–1977 (Eriksson et al. 1979). The combined crop of all edible wild berries in Finland is estimated to amount to 500 million kilos in poor years and to 1,000 million kilos in a good year such as 1988 (Salo 1991a). In Sweden, the overall crop of wild berries is probably about 500 million kilos per year (Eriksson et al. 1979).

Figure 6.4 Cloudberry (*Rubus chamaemorus*) is economically most valuable berry species to people in the northern parts of Finland, Sweden and Norway. Photo: Juha–Pekka Hotanen.

The coverage of bilberry plants on the sample plots in Sweden during the period 1974–1977 was on average 25 % in stands older than 70 years of age. In younger stands the coverage varied within the range of 5–20 %. During the said period, the average coverage of cowberry on sample plots representing forests of varying ages from 10–160 years was 4–8 % (Eriksson et al. 1979). In Swedish forests, bilberry forms contiguous communities while cowberry occurs in a more patchlike manner on the forest floor.

According to Eriksson et al. (1979), the average crops of bilberry per hectare of forest land in 1975–1977 fell within the range of 9.3–13.2 kilos. The study also indicates that forests in northern Sweden produce 70–75 % of the country's entire bilberry crop. The cowberry crops for the above period averaged 6.1–7.1 kg/ha. Northern Sweden accounts for 70–75 % of Sweden's cowberry crop. The average cloudberry yield for the period 1975–1977 was almost 20 million kilos.

The biggest yields and highest mean coverages of cowberry on mineral soils are to be found in pine forests and the lowest in deciduous forests. The same relationship applies to mires, except for a lower coverage on treeless bogs. Clearcut areas on mineral soils show a somewhat higher relative coverage (Kardell & Carlsson 1982).

Sweden's cloudberry production during the period 1978–1980 was 75.7 million kilos (fresh weight). The mean yield for forest land was 13.5 kg/ha, and 18.4 kg/ha for mires. In the same period, the cranberry crop was 20.8 million kilos; of this, 90 % originated from mires. The average annual production was 5.0 kg/ha. The corresponding figure for forest land (paludified forest land) was 2.2 kg/ha (Kardell & Carlsson 1982).

The average cowberry crop was 8.0 kilos per hectare of forest land in central Finland. Berry vegetation coverage on forest land averaged 5.8 %. The biggest crops were recorded on VT (*Vaccinium* Type) and CT (*Calluna* Type) site types, on which the average mean crops were 24–27 kg/ha. About three quarters of the cowberry crop was estimated to occur in amounts worth picking. The differences between districts were considerable. The average bilberry coverage on forestry land was 11.3 % and the average crop was 4.3 kg/ha; the years included in the study were poorer than average in terms of the amount of bilberry. According to the same study, crowberry yields on mineral soil forests averaged 1.5 kg/ha, bog bilberry 0,7 kg/ha (2.0 kg on mires), raspberry 0.2 kg/ha, and stone bramble 0.017 kg/ha. Calculations for different mire and bog types gave a cranberry average of 1.3 kg/ha and 0.9 kg/ha for cloudberry (Raatikainen et al. 1984).

The highest average cowberry crop for the years 1981–1984 in the districts of Nurmes and Lieksa (both with vast areas of forest and peatland), as recorded on the permanent sample plots (berry sample plot size 10 m^2) laid out for National Forest Inventory (NFI) purposes, was 5.4 kg/ha on dryish upland sites (EVT: *Empetrum–Vaccinium* Type) while the average bilberry crop on moist upland sites (VMT: *Vaccinium–Myrtillus* Type) was 23.4 kg/ha. Systematic random sampling on forest land always includes sample plots with no cowberry present or present but sterile. This is why the average crops remain low when such crops are compared to crops produced by cowberry sites with good yields (Salo 1991a).

The bilberry crops in a few districts in eastern Finland were inventoried in 1983 by means of 120 sample plots located on moist and dryish upland sites. The sample plots were 10 m^2 in size and they were located on sites known to be good for bilberry. The average bilberry crop for the year was found to be 94 kg/ha. On the best sites, the crop had been over 160 kg/ha. The same sample plots also yielded an average of 46 kg/ha of cowberries. The results obtained showed that average cowberry and bilberry crops can be fairly high if the sample plots are placed on sites where there is an abundance of fertile berry vegetation (Salo 1991a).

Even in the best years, the crops of cloudberry in Finland are only a few tens of kilos for vast contiguous areas. On more compact areas, however, populated by female plants, the crop per hectare can rise to several hundreds of kilos (Huttunen 1978, Kortesharju et al. 1978, Kortesharju 1984, 1988, Veijalainen 1979).

According to Ruuhijärvi (1974), the cranberry crop in southern Finland's mires and bogs varies within the range of 50–150 kg/ha, whereas Huttunen (1978) reports the cranberry crops on northern Finland mire and bog types to be in the region of 13–42 kg/ha. On vast mire areas, the average crops per hectare in good years are probably in the region of 20–40 kilos.

Factors influencing berry crops

The following is a list of the foremost factors influencing berry crops in certain forested areas of Finland (e.g. Nousiainen 1983, Salo 1982, 1988a, 1991a):

1) The preceding growing season's conditions influence the physiological state of berry shrubs. Temperature and moisture influence the nutritional state of the shoots in autumn when the flower primordia are differentiated.
2) The number of flower primordia increases at the beginning of the growing season.
3) Thickness of snow blanket; if it is thin, the shoots stick out above the snow and the flower buds freeze.
4) Spring frosts at a time when the berry shrubs are at the bud and flowering stage; summer frosts when the berries are still green.
5) Successful pollination. All economically important berries are insect–pollinated. Thus, the weather conditions should be congenial for insects during the flowering stage.
6) There should be an abundance of insect pollinators during the flowering stage; e.g. the duration of bilberry flowering is of significance for the success of pollination.
7) Heavy rains and hail showers during the flowering stage can result in reduced crops, for instance of cloudberry.
8) A lengthy period of no rain while the berries are still green can result in reduced berry size or shedding of immature berries by the plant.
9) Local reductions in crops can be caused by certain insects (e.g. cloudberry weevil, certain mushroom diseases, grouse).
10) The final crop is influenced by the density of berries on the shoots and the size of the berries.
11) Forest site types and tree species composition, bushes and the field layer.
12) Stand age (development class) influences the occurrence of berry shrubs.
13) The canopy cover of trees influences berry fertility.

Silvicultural measures constitute an important category of their own: 14) clearcutting, 15) ploughing and 16) scarifying significantly reduce berry crops. Bilberry, a shade seeking species, does not tolerate clearcutting, nor the desiccating impact of direct sunlight. Its thin, wax–free leaves dry up and berries disappear as of the year after clearcutting. Cowberry often flourishes on clearcut sites as a consequence of increased light and temperature; on dry and dryish upland sites it can produce good crops as long as the cowberry vegetation is not broken up in connection with logging. Cowberry shrubs on clearcut moist upland sites are invigorated. In addition to wavy hair–grass (*Deschampsia flexuosa*), such sites are taken over by dense growths of sprouts of hardwood species; this puts an end to the supply of light to the berry shrubs, and berry production falls. When the hardwood sprouts are cleared away, the berry shrubs become covered by the cut–down trees, and picking of berries becomes difficult (Salo 1991a).

17) Draining and 18) fertilization have a debilitating effect on cloudberry and cranberry crops. Drainage has often encouraged the growth of cloudberry on the upturned piles of peat and along the edges of ditches. The prerequisite to this is that there are communities of female plants within the drainage area as they propagate via underground rhizomes. Cloudberry on drained and fertilized peatland can produce higher crop volumes than before draining. This situation often lasts for several years on peatlands in southern and central Finland. In northern Finland, a drained peatland site can be productive even 40 years after being drained (Numminen 1979).

Drainage of peatlands has led to a situation in which productive strains of cranberry disappear as early as the first and second post–drainage year following a fall in the groundwater level. The peak drainage years are now behind us, but the challenges that await peatland forestry during the next few decades lie in the thinning of peatland forests and remedial drainage. Failed drainage areas are not worth the trouble of being redrained; they should be left to revert to their former state. With the passing of time, the original mire water state will re–establish itself and cloudberry and cranberry will become productive again (Salo 1991a).

What proportion of the wild berries do we actually pick?

The picking of wild berries is of great economic significance to the regions of Kainuu and Lapland in Finland. It has been estimated that the value of all edible wild berries in a normal crop year amounts to about 400 million FIM (80 million US $). In a normal crop year, the yield of wild berries (cowberry, bilberry, cloudberry) is 450 million kilos. In recent years, Finns have actually picked 40 million kilos; i.e. 9–10 % of the total amount. The year 1991 was a good crop year, and the amount of economically significant berries picked was estimated to have amounted to 50 million kilos (Salo 1991b).

According to the results of a questionnaire–study conducted in Sweden (1,000 questionnaire sheets were sent to persons 16–74 years of age), the Swedes picked 78 million litres (about 47 million kilos) of bilberries, cowberries, cloudberries and raspberries in 1977; this corresponded to 7 % of the year's overall crop. On average, Swedes picked 13 litres (or 7.8 kilos) per person (*Table 6.3*). Bilberry is Sweden's most popular berry (picked by 47 % of the population); cowberries were picked by 38 % of the population (Hultman 1983). According to the results of studies conducted at district level in Finland (the districts of Suomussalmi, Lieksa, Nurmes and Valtimo), cowberries and bilberries were picked by 70–90 % of the local people. The differences in the popularity of cowberry and bilberry were small (Salo 1985). Women were slightly more active than men. In Sweden, the central and northern part of the country (Norrland, Svealand) were the leading areas; the lowest picking figures were obtained for densely populated areas in Götaland around Gothenburg and Malmö in southern Sweden (Hultman 1983).

According to a paper published by the Department for the Environment, about 56 % of the population of Norway picked berries for their own use, of those 44 % picked bilberries, and 43 % picked lingonberries. About half as many picked raspberries and cloudberries. It is assumed that, in 1979, the picking of wild berries amounted to (in total) between 20 and 35 million kilos. Of these, about 40 % were lingonberries, 30 % bilberries and 25 % raspberries and cloudberries. Based on the actual prices in 1984 in Norway, the value of wild berries picked amounted to at least 900 million NOK (Friluftsliv 1985).

Several district–level investigations have been conducted in Finland to find out the berry picking activeness of people and the amounts of wild berries picked per person. Kainuu (in eastern central Finland) is renowned for its abundance of wild berries, and the per capita figures there are very high. The highest average wild berry amount for 1983 was 56.4 kg/person and it was achieved in the district of Suomussalmi (Salo 1985). The corresponding figure for 1988 was 61.9 kg/person, this time in the district of Kuhmo (Kujala et al. 1989a).

The wild berries that are picked are cowberry, bilberry, cloudberry, raspberry, rowan berry and cranberries. Cowberry is the most important of these; it accounted for 60 % of the total amount picked at Suomussalmi in 1983 (Salo 1985). *Table 6.3* shows the amounts of berries picked in different years in Finland and Sweden. The sampling applied in these investigations was directed at the adult population – persons 16(18)–70(74) years of age. According to the figures presented in the table, the amounts of berries picked per person varied a lot depending on the year. There does not appear to be great differences between Finland and Sweden in terms of berry picking activeness.

Most people pick berries for private consumption. It is a highly appreciated activity for the whole family for a few days a year. In Sweden and Finland, different berry products are common on the food menu, and are still an important source of vitamins during the hard, dark winter period. Jam cooking is a traditional activity, too. There is a market for a range of tools for picking and processing different wild berries. Berry picking for sale is organized in a more commercial way, with a chain of buyers and sellers in the countryside set up especially during the berry picking period.

According to Salo (1985), the people of Suomussalmi are the best at making use of their berry crops in regard to the amounts of berries sold; the average amount of cowberries sold per inhabitant in 1983 was 24.6 kilos. This amounted to 73.7 % of the amount picked. In 1983, the adult population of Suomussalmi picked, on the average, 22.2 kilos of cowberries (and 10 kilos of bilberries) per person per trip, and those under 18 years of age picked an average of 7.8 kilos of cowberries and a few kilos of bilberries. The average time spent picking berries on a cowberry trip, including getting there and back and

Table 6.3 Amounts of wild berries picked in some towns and districts in Finland and Sweden. The amounts of berries reported in litres in Raatikainen's and in the Swedish studies have been converted into kilos by multiplying with 0.6.

Picking year	**Study object**	**Picking activity, %**	**Amount picked per person, kg**	**Source**
1977	Swedish population	54	7.8	Hultman (1983)
1977	Åtvidaberg	73	13.8	Kardell (1979)
1977	Pihtipudas	94	30.6[1]	Raatikainen (1978)
1977	Pihtipudas	92	6.7[2]	Raatikainen and Raatikainen (1983)
1978	Lavia	85[1] 75[2]	20,7	Rossi et al. (1984)
1978	Mänttä	87[1] 74[2]	7.3	Rossi et al. (1984)
1979	Enonkoski	89[1] 61[2]	22.0	Rossi et al. (1984)
1979	Konnevesi	80[1] 77[2]	9.8	Rossi et al. (1984)
1079	Gislaved	79	15.0	Kardell and Johansson (1982)
1981	Ilomantsi	72[1] 86[2]	22.8	Rossi et al. (1984)
1982	Joensuu	73	3.2	Salo (1984)
1982	Seinäjoki	64	2.6	Salo (1984)
1982	Nurmes		19.9	Salo (1985)
1983	Nurmes	76[3]	31.8	Salo (1985)
1982	Valtimo		14.5	Salo (1985)
1983	Valtimo	86[3]	23.2	Salo (1985)
1982	Suomussalmi		49.9	Salo (1985)
1983	Suomussalmi	92[3]	56.4	Salo (1985)
1987	Suomussalmi	87	29.8	Kujala et al (1989)
1988	Suomussalmi	93	54.2	Kujala et al. (1989)
1987	Kuhmo	90	33.0	Kujala et al. (1989)
1988	Kuhmo	81	61.9	Kujala et al. (1989)

[1] cowberries, [2] bilberries, [3] picking activity in 1983

breaks, was 5 hours and 45 minutes. Those that made 9 hour picking trips clearly picked more cowberries than the others.

Commercial picking of wild berries fits best in with the daily work routine of those working in agriculture and forestry. They can pick berries in between their morning and evening chores, and men have more time in the autumn once hay making and other harvesting jobs are done. The amount of picking by men appears to increase along with an increase in the distance to the picking site and an increase in the inaccessibility of the site. Especially in Suomussalmi, forest workers pick a lot of cowberries and bilberries during breaks on long trips to work and they also know the best places to pick. These people were also the most prominent in selling cowberries and bilberries to shopkeepers (Salo 1985).

The picking of wild berries and edible mushrooms for sale is more like work to the rural population than it is for townspeople. The tax–free income obtained from selling what one has picked is of considerable importance in the purchasing of services and goods. Picking berries and mushrooms is an enjoyable pastime for people from urban areas. It provides them with exercise and the opportunity to pick berries and mushrooms for their personal consumption and even for relatives and friends (Salo 1984).

Recent years have seen the introduction of national berry picking competitions in Finland; 1,869 participants were involved in 1989. The minimum amount of picked wild berries in the competition was 100 kilos. Most of the leading competitors were from Kainuu. The winner reached a total of 4,743 kilos of wild berries and fourth place went to a person who had collected 3,622 kilos. The tax–free income per year from picking can rise to 30,000–40,000 FIM. Picking berries can be hard work, fuel is expensive and there are plenty of indirect costs involved before the buckets of berries have been delivered to the customer (Salo 1992).

In Finland, the value of the wild berry and edible mushroom harvest has corresponded to 3–4 % of the value of timber production during the past few years. It was at its highest in 1976 (4.5 %) and at its lowest in 1974 when it was 2.2 % (Saastamoinen 1983). In the years 1982–1983, the relative value of the wild berry harvest in the districts of Suomussalmi and Nurmes corresponded to 10 % of the value of wood production. This is a high figure (for mushrooms the corresponding figure was 1 %). In the districts of Lieksa and Valtimo, the value of the berry harvest is above average according to earlier studies covering the whole of Finland (Salo 1985).

6.5 Edible mushroom species in the Nordic countries

History

Of the Nordic countries, Finland and Sweden have a tradition of mushrooms being included in the staple diet of the population. Norway, Denmark and Iceland are lesser users of mushrooms. This is supported by the observation by Høeg (1974) that there is no evidence to indicate that mushrooms might have been used in Norway as food by people in older times, not even in times of famine. As a typical example of the traditional attitude in the matter, he quotes a man in Sandsvär (near Kongsberg) as having said on seeing a summer tourist preparing and eating a meal of *Cantharellus cibarius* in 1935: "Isn't that just what I've always said; eating mushrooms is only for upper class people and cows" (translated from a Norwegian dialect by Lars Helge Frivold). Høeg also mentions that *Elaphomyces* spp. was used for strengthening the sexual desire of cows; it was also sometimes eaten by people for the same purpose. But times have changed in Norway as well, and today's Norwegians are showing increasing interest in mushrooming – and not only in the species *Cantharellus cibarius* and *Hydnum repandum* (Høiland & Ryvarden 1984).

One of the oldest stories of edible mushrooms was written by Pontoppidan (1752) in his book on the natural history of Norway. He tells that morels (*Morchellaceae*) are dried, and sold, and that they mainly occur in certain counties in southeast Norway (Buskerud, Hedmark) where they are sought after by buyers and sent to other places. He also mentions that *Amanita muscaria* is boiled in milk and used to catch flies, because it is very poisonous.

No statistics are available on the Danes' consumption of forest mushrooms (and the same applies to Iceland). Nevertheless, Danes pick less wild mushrooms than their northern neighbours. The primary reason for this is that only 9.5 % of Denmark is forested and most of this area is privately owned. Iceland's forested area is a mere 1 %. Denmark and Iceland lack the policy of rights of public access which are customary in the other Nordic countries.

It is known that the ancient Romans and Slavic nations of Europe were great appreciators of mushrooms. The Germanic peoples, on the other hand, hardly included mushrooms in their diet at all. Mushrooms were introduced to the royal court of Sweden and the homes of the aristocracy during the 16th and

Figure 6.5 Delicious cep (*Boletus edulis*) is one of the most desired edible mushrooms in the Nordic countries. Very near by, there is growing another mycorrhizal species, poisonous *Cortinarius* sp. Photo: Kauko Salo.

17th centuries from France. However, it is only since the beginning of this century that the rural population has begun to use mushrooms. A campaign for promoting the use of mushrooms was conducted in Sweden during the difficult famine years of the last century, but this was of no avail (Kardell 1980). Finns, too, failed to realize the value of mushrooms during the famine years of the 1860s. Thousands of people would have survived the famine if they had used protein–rich mushrooms. Old books contain descriptions of domestic animals and reindeer eating mushrooms (e.g. *Leccinum scabrum* and *Boletus* and *Suillus* species) in the woods, but people themselves had no knowledge of mushrooms and how they might be used.

Apart from the western and southwestern coastal areas and eastern and southeastern parts of the country, use of mushrooms within the borders of present–day Finland was very moderate right up to the end of the past century. The gentry in western and southwestern Finland had taken a fancy to chanterelles in particular after being introduced to these mushrooms by their Swedish acquaintances. Finns in eastern and southeast Finland and the orthodox population of the territory ceded to the Soviet Union in the aftermath of World War II were accustomed to eating ceps (*Boletus* spp. (*Figure 6.5*), *Leccinum* spp., *Suillus* spp.) as part of their diet (the latter particularly when fasting). Early last century, the use of mushrooms in Finnish territories in the proximity of St. Petersburg (former Leningrad) was promoted by the closeness of the city and the numerous summer cottages of the aristocracy along the coast of the Gulf of Finland, as these also ensured a constant demand for mushrooms (Vuorela 1975).

Mushrooms were already a part of the staple diet of the population in northern Karelia in the 18th century. Due to the simplicity of the use of salted mushrooms, this form of consuming mushrooms became widespread even in the impoverished backwoods of eastern Finland (Jäppinen et al. 1985). The species favoured for salting were milk caps (*Lactarius torminosus*, *L. trivialis*, *L. utilis*) and occasionally species of the genus *Russula* (Mäyränpää 1981).

Occurrence and use of the foremost edible mushrooms

Finland has 2,000 mushroom species, microfungi excluded. Two hundred of these are edible mushrooms, but only some 30 species are used as food. There are 23 really poisonous mushroom species and about 30 other mushrooms suspected to be poisonous (Harmaja et al. 1978). In july 1994, 22 mushrooms and mushroom groups were accepted as commercial edible mushrooms in Finland.

A commercial edible mushroom means a mushroom species that is of value for the mushroom trade and is approved for trade by the National Board of Trade and Industry. Commercial mushroom species approved for trade are normally easy to recognize. They must be species not to be confused with poisonous or inedible mushroom species (Kauppasieniopas 1988). Most of the foremost commercial mushrooms are common species with high yields which are found growing on mineral soils and favoured by Finns on autumn mushroom picking trips (Salo 1984, Salo 1985).

A list of the best edible mushroom species (commercial mushrooms) has also been drawn up in Sweden. Mushrooms on this list may be sold in the market places, for instance. In addition, the Swedes have a list of mushrooms that can be included in mixed mushroom products (von Hofsten & Holmberg 1980). Corresponding commercial mushroom guides and lists of commercial mushrooms are lacking in the other Nordic countries. However, a range of other types of books on mushrooms is available in each of the Nordic countries. These books explain the distinguishing characteristics of the best mushrooms, where they grow, their distribution, and how to prepare food from them. *Table 6.4* lists the Nordic countries' best edible mushrooms and their scientific names.

Most of the best edible mushrooms in Finland, Sweden and Norway occur in dry, dryish or moist upland forests. Most of them are mycorrhizal species that accompany pine, spruce and the birches. Many are capable of having a mycorrhizal relationship with two or three species and are therefore to be found commonly on several kinds of site types. Examples of these include *Lactarius trivialis* (spruce and birches), *Russula paludosa* (spruce and pine) and *Cantharellus cibarius* (birches and occasionally spruce and pine). There are only a few saprophytic species included in the list of edible mushrooms: *Scutiger ovinus*, *Gyromitra esculenta*, *Armillaria mellea* coll.; these may also happen to be mycorrhizal species. Purely saprophytic species are *Craterellus cornucopioides*, *Morchella* spp., *Macrolepiota procera*, *Lepista nebularis* and *L. nuda*.

The most common and most popular edible mushrooms are to be found among ceps, milk caps, *Russula* species and *Cantharellus cibarius* and groups close to them. Three mushroom species (*Boletus edulis*, *Suillus luteus* and *Cantharellus cibarius*) are considered to be the best edible mushrooms in all of the Nordic countries. Good edible mushrooms that inhabit conifer and mixed forests include *Lactarius deterrimus*, *Cantharellus tubaeformis*, *Craterellus cornucopioides*, *Hydnum repandum* and *Rozites caperatus*; these occur in all the other Nordic countries except in Iceland (*Table 6.4*). The choicest edible mushrooms are served fried, stewed or as soups. Drying and/or deep–freezing are the usual way to preserve them.

While the most common edible mushrooms in Iceland occur in virgin birch woods, the species *Suillus grevillei* and *S. luteus* are common in conifer plantations. Common edible birch wood species are *Tricholoma fulvum* (*T. flavobrunneum*) and *T. album*; these are not used in the other Nordic countries (Hallgrimsson 1990). Denmark has three ceps: *Boletus badius*, *B. erythropus* and *B. luridus*, and these occur in hardwood (often beech) forests. Many mushroom devotees like to enjoy them boiled (Albertsen et al. 1981).

Table 6.4 The most important edible mushroom species in the Nordic countries and their scientific names (Hansen & Knudsen 1992). Data from Finland according to Kauppasieniopas (1988) and Korhonen (1986), Sweden (von Hofsten & Holmberg 1980), Norway (Høiland & Ryvarden 1984, Egeland & Myhr 1988), Denmark (Knudsen 1983) and from Iceland (Hallgrimsson 1990).

C = Commercial edible mushroom species, E = Good edible mushrooms. Other good edible mushrooms in Finland (EF), in Sweden (ES) and in Norway (EN).

Mushroom species	Finland	Sweden	Norway	Denmark	Iceland
Ceps:					
Boletus edulis	C	C	E	E	E
B. pinophilus	C				
B. badius		ES		E	
Suillus variegatus	C	C	E		
S. granulatus		C	E		
S. grevillei		C	EN		E
S. luteus	C	C	E	E	E
Leccinum versipelle	C	ES	E		
L. vulpinum	C				
L. aurantiacum	C	ES	EN	E	
L. scabrum			E		E
L. holopus			E		
Milk caps:					
Lactarius trivialis and L. utilis	C				
L. rufus	C				
L. torminosus	C				E
L. deterrimus	C	C	E	E	
L. deliciosus	C	C	E		
L. volemus		C	E		
Russulas:					
Russula paludosa	C			E	
R. aeruginea	EF	S			E
R. claroflava	C				E
R. decolorans	C	ES	E		
R. integra		C			
R. vinosa	C				
R. xerampelina	EF	ES		E	E
Champignons:					
Agaricus silvicola	EF	ES	E	E	
A. arvensis	EF	C	E	E	
A. campestris	EF	C	E	E	
A. bitorquis	EF	C	E	E	
A. silvaticus	EF	ES	E		
Hygrophorus camarophyllus	C	C			
Hygrocybe punicea		C			
Camarophyllus pratensis		C		E	
Rozites caperatus	C	ES	E	E	
Armillaria mellea coll.	C	ES		E	E
Macroleopiota procera and M. rhacodes	EF	C		E	
Lepista nebularis	EF				
L. nuda	EF		EN	E	
Hypholoma capnoides	EF				
Cantharellus cibarius	C	C	E	E	E
C. tubaeformis	C	C	E	E	
C. lutescens	C	C			
C. xanthopus			E		
Craterellus cornucopioides	C	C	E	E	
Scutiger ovinus	C	ES	E		
Hydnum repandum	C	C	E	E	
H. rufencens	C	C	E		
Gyromitra esculenta	C	C			
Morchella esculenta and M. conica	C	C	EN	E	
Tricholoma flavorirens	EF	C	EF	E	
T. fulvum					E
T. album					E
Calocybe gambosa		C	EN	E	
Coprinus comatus		ES	E	E	
Kuehneromyces mutabilis	EF	ES		E	E
Pleurotus ostreatus				E	
Total	41	39	31	26	13

Albertsen et al. (1981) have made the observation that Danish mushroom devotees (Foreningen til Svampekundskabens Fremme) tend to favour *Cantharellus cibarius*, *Boletus edulis*, *Coprinus comatus*, *Lepista nuda* and *Kuehneromyces mutabilis*. For them, *Russula* species and milk caps are the least preferred. The use of milk caps in Denmark, Sweden and Norway is rare, and they are not salted as in Finland. Alongside chanterelles and ceps, Finland's most popular edible mushrooms include the milk caps, especially *Lactarius trivialis*, *L. utilis*, *L. rufus* and *L. torminosus* are favoured as salted mushrooms.

Good wild champignons (*Table 6.4*) are to be found in Finland, Sweden, Norway and Denmark. In Finland, only cultivated champignons (*Agaricus bisporus*, *A. hortensis*) may be sold as commercial mushrooms. Finland's indigenous champignons are not common, and they are collected by persons who definitely recognize them. The novice mushroom collector can easily confuse them with the fatally poisonous destroying angel (*Amanita virosa*) whose gills are permanently white. The gills of champignons are reddish in juvenile individuals and finally they turn to blackish brown. Due to emissions of heavy metals *Agaricus bitorquis* and *A. arvensis* should not be collected from urban parks and roadsides.

6.6 Mushroom yields

The first survey of mushroom yields in Finland was carried out on mineral soils and peatlands by Rautavaara (1947). A large, scientifically implemented mushroom study was carried out in Sweden together with Sweden's National Forest Survey (NFS) in 1974–1977 when 29,000 sample plots were investigated (Kardell et al. 1980).

It was not until the late 1970s and early 1980s that research directed at mushroom yields became more widespread in Finland (e.g. Ohenoja 1978a, 1978b, 1984, 1993, Ohenoja & Koistinen 1984). Mushroom yields have been investigated over smaller areas; e.g. Sjöblom et al. (1979), Vauras and Huhtinen (1980), Ohenoja and Metsänheimo (1982) and Salo (1983). After the work of Kardell et al. (1980), the practice in Sweden has been to concentrate on studying the influence of silvicultural measures on mushroom yields (e.g. Wästerlund & Ingelög 1981, Wästerlund 1982, Kardell 1984, Kardell & Eriksson 1987). The yields of mushrooms and mushroom communities on peatland sites have been studied by Lange (1948) in Denmark and Salo (1979), Saari and Salonen (1983), and Salonen and Saari (1990) in Finland.

Many recent studies dealing with the structure of mushroom communities have also looked into the yields of mushrooms on different forest site types; e.g. Mehus (1986) in Norway, Jäppinen et al. (1986), Hintikka (1988) and Salo (1988b, 1993a) in Finland and Dahlberg (1991) in Sweden.

According to the results of the mushroom study included in the NFS conducted in Sweden in 1974–1977, the total mushroom crop was estimated to have been 2,400 million kilos; of this amount, 20 % was thought to be made up of edible mushroom species. There seems to be great variation in annual crops; 89.6 % of the sample plots were void of mushrooms in 1975 while in 1974 the figure was 63.4 %. In this study, good mushroom sites were those sample plots (78 m^2) that contained 11 or more fruit bodies of mushrooms; i.e. 1,400 fruit bodies per hectare. In a good mushroom year in Sweden, this applied to 7–8 % of the forests. The average weight of an individual mushroom fruit body was estimated to be 50 g. It was estimated that forest land would carry an average of 2,000 fruit bodies per hectare; i.e. one mushroom per 5 m^2. This corresponds to about 100 kilos fresh weight per hectare (Kardell et al. 1980).

In the mushroom studies conducted using the permanent sample plots (100 m^2 in size) laid out in the course of the NFI (National Forest Inventory) in Finland (for different forest and peatland site types, uneven–aged forests and drained and virgin state–owned peatlands), it has been possible to determine the mushroom species present, their ecological groups, the structure of mushroom communities and the yields per hectare (Salo 1988b, 1993b).

The biggest average mushroom yields were collected in the study years 1981–1984 from mineral soil sample plots on VMT (*Vaccinium–Myrtillus* Type) site types. In 1981, the average mushroom yield (of mycorrhizal and saprophytic species) on VMT site types was 129.0 kg/hectare (fresh weight) while on EVT (*Empetrum–Vaccinium Type*) site types it was 53.4 kg/hectare (fresh weight). The distribution of mushroom species and yields were influenced by weather factors (rainfall and temperature) as well as by tree species relationships, field and ground layer vegetation, and the humidity and temperature of the top layer of the soil. On drained sites, the quantity of mushroom species and biomass increased more in the pine mire group than in the spruce mire group. Sites drained a long time ago (i.e. now in the category of transitional drained peatlands) and mature forests established on peatland carried higher amounts of mushrooms than recently drained peatlands or peatlands still in their virgin state (Salo 1988b).

Commercial edible mushrooms and other good edible species formed 30 % of the total mushroom crop of upland forests and drained peatland sites in Finland. In the good crop year of 1981, the theoretical overall mushroom crop was estimated to be 2,145 million kilos (Salo 1988b).

The yields (kg/ha, fresh weight) of fifteen commercial edible mushroom species and groups were established on the permanent sample plots in 1985–1986 in connection with the 8th NFI. Forested mineral soil was represented by 2,859 of the plots (300 m^2) or segments (< 300 m^2). Moist (MT: *Myrtillus* Type) and dryish site (VT: *Vaccinium* Type) types accounted for 69.5 % of the segments on mineral soil, the grovelike site type (OMT: *Oxalis–Myrtillus*

Type) amounted to 16.9 % while the dry site type (CT: *Calluna* Type) represented 6.2 % (Salo 1993a).

The numbers of commercial edible mushrooms on the plots and segments were counted, and the yields for each species were then expressed in terms of fresh weight/ha. The average fresh weight (g) of a single fruit body of the various commercial edible mushroom species was obtained from Salo's (1988b) investigation carried out on permanent sample plots of the 7th NFI in the province of North Karelia during 1981–1984.

The yields of mushrooms were calculated only for segments with some mushrooms growing on them. The most common commercial edible mushroom was *Lactarius rufus* with a mean yield of 19.3 kg/ha for 329 segments. Other common commercial species with significant yields were *Lactarius utilis* (with an average yield for all segments of 12.2 kg/ha, where *L. utilis* has been found), *Suillus variegatus* (mean yield 11.4 kg/ha), *Russula paludosa* (mean yield 11.0 kg/ha) and *Cantharellus tubaeformis* (mean yield 8.3 kg/ha). *C. tubaeformis* was collected from 27 segments, but its fruit body grew with an average density of 2,938/ha in mature stands on moist sites (MT). However, its yield was 14.7 kg/ha, since the average weight of a single fruit body was 5 g. *Scutiger ovinus* also occurred in large groups on moist sites in coniferous and mixed woods, and hence its average yield on the segments was 36.7 kg/ha. The most common cep species was *Leccinum versipelle* with an average yield of 11.0 kg/ha on moist sites. The average yield of *L. aurantiacum* on the two grovelike sites was 34.7 kg/ha (Salo 1993a).

In the studies of both Ohenoja and Koistinen (1984) and Jäppinen et al. (1986), the highest yields of commercial edible mushrooms occurred on dryish (VT) sites. The reason for this is that high–yielding species growing on infertile sites have been chosen as commercial edible mushrooms. According to Salo (1993a), *Lactarius rufus*, *Suillus variegatus* and *Russula paludosa*, three high–yielding commercial edible mushrooms, and known to be mycorrhizal fungi of pine, grow on moist MT site type as well as on dry CT site type and dryish VT site type. This is assuming that pines are mixed in with spruce and birch on moist sites. The plots in the Pohjanmaa–Kainuu region on VMT (*Vaccinium–Myrtillus* Type) site type and Peräpohjola on HMT (*Hylocomium–Myrtillus* Type) site type, and on the so called forest–Lapland LMT (*Ledum–Myrtillus* Type) site type vegetation zones are characterized by the crowns of spruces having a columnar form, an abundance of birch, and with pine at times being the dominant species on both HMT and LMT sites.

How much mushrooms do we actually collect and use?

One out of every three Swedes picked an average of 2.4 kg of mushrooms in 1977; this means a total of 13 million kilos of mushrooms for Sweden as a whole. According to Salo (1991b), the amount of wild mushrooms collected in Finland in recent years has been 5 million kilos per year (i.e. 0.3 % of the biological crop of 2,000 million kilos and 3 % of the collectable crop of edible mushrooms of 170 million kilos). In Norway, mushrooms were picked by 12 % of the population. Those who picked berries or mushrooms or both make, on an average, 5 field trips a year. In 1979, Norwegians picked 0.5–0.9 million kilos of mushrooms (Friluftsliv 1985).

Finland's annual wild mushroom harvest during the period 1900–1965 averages out at about 1 million kilos; the war years of 1940–1944 are exceptions as the annual harvests then amounted to nearly 17 million kilos (Kunnas 1973). The population of southern Sweden is more active in collecting mushrooms than the northerners, according to Hultman (1983). Kardell's (1979) investigation reveals that Åtvidaberg's adult population (16–70 years of age) collected an average of 3.6 kilos of mushrooms per person in 1977. A similar sampling study conducted in the Gislaved district gave an average figure of 2.4 kilos per person of fresh edible mushrooms (Kardell & Johansson 1982).

Results similar to those of Sweden have been recorded also in Finland. According to the results of a study by Pekkarinen et al. (1980) and household questionnaires, Finns collected an average of 2.4 kilos of edible mushrooms in 1977. The consumption of mushrooms was at its highest in households in eastern Finland (the provinces of North Karelia, Kuopio and Mikkeli) and at its lowest in southern Finland. Salo (1984) observed distinct differences in the mushroom consumption of people from western and eastern parts of Finland. In the autumn of 1982 (a poor mushroom year), 28 % of the households of Seinäjoki (in western Finland) reported having collected mushrooms; the corresponding figure for the households in Joensuu (in eastern Finland) was 68 %. The average amount collected per person in Joensuu was 2.0 kilos, while that in Seinäjoki was 0.5 kilos.

The per person figures of edible mushrooms collected for the rural areas of eastern Finland during the years 1982–1983 varied within the range of 2.5–4.5 kilos. Families in eastern Finland and Kainuu preferred milk caps and mixed mushrooms, which they preserved by salting. One mushroom–collecting family out of three deep–froze their ceps and chanterelles; desiccated mushrooms were consumed in only one household out of ten (Salo 1985).

Milk caps were more popular in rural households, but chanterelles and ceps less popular, than in urban households. Households of better educated mothers collected and consumed more mushrooms and more species than the households of mothers with only basic educational backgrounds. Mushroom consumption was at its highest in families belonging to the higher social strata; this segment of the population collected ceps and chanterelles more than the others and consumed a lot of edible mushrooms either fresh or after having been deep–frozen (Pekkarinen et al. 1980). Also the familiarity with mushrooms was more common among people belonging to the higher social strata; this is exemplified by the observance that 45 % of the persons holding managerial or senior positions in the district of Suomussalmi and in the province of North Karelia could recognize 6–10 mushroom species and 23 % of them rec-

Figures 6.6 and *6.7* Edible mushrooms are healthy food in unpolluted natural forest habitats. Teaching of edible and poisonous mushrooms could start at young age by mother and father. Poisonous red fly agaric (*Amanita muscaria*) and a cep (*Leccinum scabrum*) in the hands. Photo: Pekka Vuojärvi. Poisonous red fly agaric (*Amanita muscaria*) in the right hand and edible milk cap (*Lactarius utilis*) in the left hand (right). Photo: Kauko Salo.

ognized more than 10 species. Only one out of two persons from the other professional groups could recognize 2–5 mushroom species (Salo 1985) (*Figures 6.6* and *6.7*).

6.7 Other forest products

Lichens

The genus of reindeer lichens *Cladonia* has had the greatest economic role in the Nordic countries in the lichen category. In times of famine, people have eaten certain lichen species (e.g. Iceland moss (*Cetraria islandica*)); in the case of Norway, it has been known to be a shop commodity (Kauppi 1990). Lichen have been used in folk medicine through the ages, even if this use has been based more on beliefs than on actual proof of the medicinal effect of lichens. Reindeer lichens and bearded lichens (*Alectoria* spp.) have been made use of by extracting usnic acids from them, as these are useful as ointments in the treatment of fungal infections.

Over twenty years ago in Finland, reindeer lichens were still used as sources of usnic acid, but since then their production has ceased for economic reasons (Kauppi 1990). In the reindeer husbandry regions of Finland, Sweden and Norway, reindeer lichens form the staple diet of reindeer. Lichen communities have been depleted in recent years because of overgrazing and their species compositions have also undergone changes. Especially the genus *Stereocaulon* and some other *Cladonia* species have increased in number in areas trampled by reindeer (Salo 1991a).

The reindeer lichen species *Cladonia alpestris* is boreal and tends to avoid the most oceanic regions, hence its absence from Iceland and the outer coastal areas of Norway (Ahti 1977). It forms extensive carpets in the driest woodlands of the northern parts of the boreal coniferous zone, where the tree stands are generally composed of pine (*Pinus sylvestris*). *Cladonia alpestris* is a typical climax species in such woodlands, attaining dominance in the ground layer as long as the plant community is allowed to develop without external disturbance (Kauppi 1979) (*Figures 6.8* and *6.9*).

Large amounts of *Cladonia alpestris* have been collected outside the reindeer husbandry region of Finland along the river Oulunjoki, in the island of Hailuoto in the Gulf of Bothnia, and in upland forests of the *Cladonia* site type. The practice is to collect the lichen into wooden crates after rain or after the lichen patches have been irrigated. The income obtained from collecting lichen is taxable. Landowners collect the lichen themselves or sell collecting licenses to outsiders. The lichen *Cladonia alpestris* is used for ornamental purposes, especially in making wreaths. Smaller amounts are used to make lichen bands for flower arrangements and in architects' miniature models (Kauppi 1979).

Figures 6.8 and *6.9* The reindeer lichen species (*Cladonia alpestris*) is growing in dry tree stand composed of Scots pine. The practice is to collect the lichen into wooden crates after rain or after the lichen patches have been irrigated. The income is taxable. Landowners collect the lichen themselves or sell collecting licenses to outsiders. Lichen is mainly used for ornamental purposes. Photos: Kauko Salo.

Table 6.5 The weight (1,000 kg) and value (£ 1,000) of *Cladonia stellaris* exports from Fennoscandia in 1970–1975 (Kauppi 1979).

	1970	1971	1972	1973	1974	1975
Finland	1907 (733)	1476 (536)	1560 (699)	1537 (975)	837 (701)	746 (665)
Norway	1302 (274)	980 (258)	942 (305)	911 (373)	1100 (518)	980 (550)
Sweden	533 (128)	459 (116)	585 (195)	623 (236)	772 (400)	671 (374)
Total	3742 (1135)	2915 (910)	3087 (1199)	3071 (1584)	2709 (1619)	2397 (1589)

The international trade of this species is largely carried on by Finland, Sweden and Norway, whose combined lichen exports in 1970–1975 were in the region of 3,000 tons a year, representing an annual value of 12 million FIM (Kauppi 1979) (*Table 6.5*). Lichen is mostly exported to central Europe where the leading importers nowadays are Germany (78 % of exported volume), Austria, Italy and Netherlands (Yearbook of forest statistics... 1990). In 1979, the number one importer of lichen from Fennoscandia was Germany (82 %) and second place (10 %) went to Denmark (Kauppi 1979). As of 1990, Norway has begun to tax incomes obtained from collecting lichen. Sweden has begun to subsidise exports of *Cladonia alpestris* as a means of improving its competitiveness (Salo 1991a).

Christmas trees and ornamental foliage

A custom in Finland has been to take one's Christmas tree from the forest, with or without the permission of the landowner. Some have it given to them by friends. The third alternative is to buy it from the market place just before Christmas. Norway spruce (*Picea abies*) has been the traditional choice (*Figure 6.10*). Landowners themselves have begun to sell Christmas trees at markets and other such places. Logging operations are the main source of Christmas trees; the tops of felled trees are trimmed to suitable size. Imports of Christmas trees to Finland have been minimal as exotic spruces and certain pines (e.g. *Pinus contorta*) have not yet become popular.

There are some tens of Christmas tree farms in Finland; these are estimated to produce roughly 40,000 trees. This includes a few farms that raise the popular Serbian spruce (*Picea omorica*). Although the population of Finland is made up of 1.3 million families and an estimated 95 % of them acquire a Christmas tree each year (and then there are the trees for schools, kindergartens, business enterprises and government offices), it is still reckoned that there is a need for raising about 250,000 trees. The overall worth of the Christmas tree trade is about 10 million FIM, based on the average price of 40 FIM per tree. The use of green branches has not become popular in Finland; the little demand there is has been met with branches of the indigenous Norway spruce and Scots pine (Lassheikki & Holmberg 1989).

In Sweden, people normally buy a Christmas tree if they live in urban areas, and most of the Christmas trees are Norway spruces. The estimated annual consumption of Christmas trees in Sweden is in the region of 2.5–3.0 million trees. In 1989, Swedish tree farmers had about 2 million exotic spruces under cultivation. The most common exotic was blue spruce *(Picea pungens)* of which there were 1.3 million plants. Second place (0.4 million plants) went to Caucasian silver fir *(Abies nordmanniana)*. According to one study, winter damage and damage caused by browsing elk, etc., means that only one out of four of these firs reach full size as Christmas trees. The corresponding figure for *Picea pungens* is 40 % (Normann 1989).

Denmark, Norway and Iceland are characterized by their numerous exotics arranged into different provenance trials. One of the purposes of the trials is eventually to provide Christmas trees and branches for ornamental purposes. Along with *Picea abies* and *Pinus sylvestris*, there are plantations of *Picea omorica*, *P. engelmannii* and *Abies procera*, *A. nordmanniana*, *A. lasiocarpa* and *A. amabilis* (Froland et al. 1989).

The original Christmas trees used by the people of Iceland were birches with juniper branches hung on them. Lately, plantations of *Picea abies* have been established in significant amounts as a supply of Christmas trees (Arnason 1989). Today, the market for Christmas trees in Iceland is estimated to be 30,000–40,000 trees. It is reckoned that voluntary forestry associations supply another 2,000 trees (mainly *Picea abies* and *Pinus contorta*). Iceland's home–grown trees probably satisfy 25–30 % of the demand. All the imported trees come from Denmark. These have gradually gained in popularity as the supply of cheap *Abies nordmanniana* has increased. Nowadays, Danish fir and cypress greenery dominate the market. Icelanders have nothing with which to challenge these imports (Blöndal 1993).

Christmas trees and ornamental foliage have a significant role in Danish forestry. Danes use annually some 1.3 million Christmas trees in their homes and another 0.2 million in public facilities. The leading Christmas tree species is Norway spruce (consumption 0.85 million trees), and *Abies nordmanniana* (0.5 million) comes second. *Abies procera* (0.15 million) is an example of another species used for this purpose (Østergård 1989).

Caucasian silver fir is Denmark's economically most significant Christmas tree species while noble fir (*Abies procera*) is mainly used as a raw material for ornamental works. The first cuttings are made when the tree is 8–10 years old (Jensen 1989). Christmas tree cultivations cover a total area of 5,217 hectares in Denmark. Caucasian silver fir accounts for 73 % of the cultivated area, noble fir for 6 % and the remaining species (incl. cypress, thuja, Douglas fir and Norway spruce) for 21 %.

Plantations aimed at producing ornamental foliage cover an area of 7,367 hectares. The leading tree species, Caucasian silver fir, covers 37 % of the cultivated area, next is noble fir with 34 %. The rest of the cultivated area is covered by various species (Christensen 1979). The cultivated area of Christmas

Figure 6.10 Christmas trees help to bring the joy of Christmas into the home. Norway spruce (*Picea abies*) has been the traditional choice. The estimated annual consumption of Christmas trees in the Nordic countries is 6.5–7.5 million trees. Photo: Kauko Salo.

trees and decorative greenery is currently believed to be twice the figures mentioned above. Ornamental foliage consists of evergreen branchlets used for decorative purposes mostly around Christmas, but smaller amounts of cypress, for instance, are used for wreaths the year round. This form of production expanded so much in the 1970s and 1980s that today Denmark is Europe's leader in the field.

Ornamental foliage plantations are cultivated very intensively, and large investments are necessary for the purchasing of first–rate plants, weed control equipment and fertilizers. Irrigation is often also an expense category. Plantations for the purpose of producing ornamental foliage make up 3.1 % (over 12,000 ha) of Denmark's forested area. Ten million Christmas trees are produced and about 27,000 tons of ornamental foliage were cut in 1991. The majority of this production is exported.

References

Ahti, T. 1977. Lichens of the boreal coniferous zone. In: Seaward, M.R.D. (ed.). Lichen Ecology. London, New York, San Francisco. p. 145-181.

Ahti, T., Hämet–Ahti, L. & Jalas, J. 1964. Luoteis–Euroopan kasvillisuusvyöhykkeistä ja kasvillisuusalueista. (Vegetation zones and vegetation regions in North–West Europe.) Luonnon Tutkija 68: 1–28. (In Finnish.)

Ahti, T., Hämet–Ahti, L. & Jalas, J. 1968. Vegetation zones and their sections in northwestern Europe. Annales Botanici Fennici 5: 169–211.

Albertsen, J.F., Knudsen, H. & Sorensen, P.G. 1981. Svampespisning i Danmark. (Eating of mushrooms in Denmark.) Svampe 4: 49–58. (In Danish.)

Arnason, A. 1989. Erfaringer med ulike treslag/provenienser på Island. (Experiences with different tree species/provenances in Iceland.) PS Nåledrys 10: 37–38. (In Norwegian.)

Bengtson, B. (ed.). 1976. Nordisk miljörätt: en översikt. (Nordic environmental law: an overview.) Nordisk utredningsserie 25: 1–117. Stockholm. (In Swedish.)

Blöndal, S. 1993. Socioeconomic importance of forests in Iceland. In: Alden, J., Mastrantonio, J.L. & Odum, S. (ed.). Forest development in cold climates. Plenum Press, New York. 13 pp.

Blöndal, S., Benedikz, P. & Ottósson, J.G. 1986. Forestry in Iceland: a brief description of its history and present status. Iceland Forestry Service, Forestry report 8. 32 pp.

Brække, F.H. 1990. Peatland and paludified forest on mineral soil in Norway: potentials for forest production. In: Hånell, B. (ed.). Biomass production and element fluxes in forested peatland ecosystems. SLU, Umeå. p. 29-32.

Christensen, P. 1979. Areal– og produktionsundersögelse av nobilis og nordmannsgran. Skovteknisk institut, Rapport 3: 1–76.

Dahlberg, A. 1991. Ectomycorrhiza in coniferous forest: structure and dynamics of populations and communities. Swedish University of Agricultural Sciences, Uppsala. 38 pp.

Egeland, I.L. & Myhr, S. 1988. Sikre sopper. (Safe mushrooms.) Århus. 104 pp. (In Danish.)

Eriksson, L. Ingelög, T. & Kardell, L. 1979. Blåbär, lingon, hallon: förekomst och bärproduktion i Sverige 1974–1977. Summary: Bilberry, lingonberry, raspberry: occurrence and production in Sweden 1974–1977. Sveriges lantbruksuniversitet, Avdelningen för landskapsvård, Rapport 16. 124 pp.

Friluftsliv. (Outdoor recreation.) 1985. Miljøverndepartementet, Oslo. 148 pp. (In Norwegian.)

Froland, Å., Gislerud, O., Haug, G. & Rønshoff, E. 1989. Treslag og plante materiale for juletrær og pyntegrønt i Norge. (Tree species and seedling material for Christmas trees and ornamental foliage in Norway.) PS Nåledrys 10: 24–30. (In Norwegian.)

Hallgrimsson, H. 1990. Ber og sveppir. Summary: Berries and mushrooms. In: Skógraktarbókin. Skógraktarfélag Islands, Reykjavik. p. 239–243.

Hämet–Ahti, L., Suominen, J., Ulvinen, T., Uotila, P. & Vuokko, S. 1986. Retkeilykasvio. (Flora handbook for hikers.) Suomen luonnonsuojelun tuki, Helsinki. 598 pp. (In Finnish.)

Hånell, B. 1990. Present situation and future possibilities of peatland forestry in Sweden. In: Hånell, B. (ed.). Biomass production and element fluxes in forested peatland ecosystems. Proceedings of a seminar in Umeå, Sweden, September 3–7, 1990. Swedish University of Agricultural Sciences, Umeå. p. 45–48.

Hansen, L. & Knudsen, H. (eds.). 1992. Nordic Macromycetes. Vol. 2. Polyporales, Boletales, Agaricales, Russulales. Copenhagen. 474 pp.

Harmaja, H., Korhonen, M. & Åkerblom, H. 1978. Myrkkysienet ja sienimyrkytykset. (Poisonous mushrooms and mushroom poisonings.) Weilin & Göös, Espoo. 55 pp. (In Finnish.)

Hintikka, V. 1988. On the macromycete flora in oligotrophic pine forests of different ages in South Finland. In: Vänninen, I. & Raatikainen, M. (eds.). Proceedings of the Finnish–Soviet symposium on non–timber forest resources in Jyväskylä, Finland, 25–29 August 1986. Acta Botanici Fennica 136: 89–94.

Høeg, O.A. 1974. Planter og tradisjon. (Plants and tradition.) Universitetsforlaget, Oslo. 751 pp. (In Norwegian.)

von Hofsten, B. & Holmberg, P. 1980. Svamp – gott? nyttig? giftig? (Mushrooms – tasty? useful? poisonous?) Konsumentverket, Helsinborg. (In Swedish.)

Høiland, K. & Ryvarden, L. 1984. Norsk matsopp. (Norwegian edible mushrooms.) Universitetsforlaget, Oslo. 123 pp. (In Norwegian.)

Hultman, S.–G. 1983. Hur mycket bär och svamp plockar vi egentligen? Summary: How much berries and mushrooms do we actually pick? Vår Föda 35: 284–297.

Huttunen, A. 1978. Hilla– ja karpalosadoista Siuran alueella. Summary: On the cloudberry and cranberry yields in Siura district, N–Finland. Suo 29(1): 17–21.

Ingmanson, I. & Holmberg, P. 1988. Suuri marjakirja. (The big berry book.) Gummerus Oy, Jyväskylä/Helsinki. 223 pp. (In Finnish.)

Jäppinen, J.–P., Kirsi, M. & Salo, K. 1985. Luonnonvaraisten sienten sadot ja kaupallinen poiminta Itä–Suomessa, ensisijaisesti Pohjois–Karjalan läänissä. (The yields of wild mushrooms and commercial picking in eastern Finland, especially in the province of Pohjois–Karjala.) Metsäntutkimuslaitoksen tiedonantoja 200. 103 pp. (In Finnish.)

Jäppinen, J.–P., Hotanen, J.–P. & Salo, K. 1986. Marja– ja sienisadot ja niiden suhde metsikkötunnuksiin mustikka– ja puolukkatyypin kankailla Ilomantsissa vuosina 1982–1984. Summary: Yields of wild berries and larger fungi and their relationship to stand characteristics on MT and VT –type mineral soil sites in Ilomantsi, eastern Finland, 1982–84. Folia Forestalia 670. 25 pp.

Jensen, J.H. 1989. Pyntegrøntproduktionen i Danmark: nuvärende stade. (The production of ornamental foliage in Denmark: present situation.) PS Nåledrys 10: 14–15. (In Danish.)

Kalliola, R. 1973. Suomen kasvimaantiede. (The botanical geography of Finland.) WSOY, Porvoo. 308 pp. (In Finnish.)

Kardell, L. 1979. Taltorpsmon: ett rekreationsområde i Åtvidaberg. (Taltorpsmon: a recreation area in Åtvidaberg.) Sveriges lantbruksuniversitet, Avdelningen för landskapsvård, Rapport 17. 92 pp. (In Swedish.)

Kardell, L. 1980. Skogsmarkens bär och svampar: en hotad resurs? (Forest berries and mushrooms: an endangered resource?) Sveriges Skogsvårdsförbunds Tidskrift 78: 5–19. (In Swedish.)

Kardell, L. 1984. Skogsgödlingens inverkan på bär och matsvampar. (Impacts of forest fertilization on berries and edible mushrooms.) Skogsfakta 5: 67–73. (In Swedish.)

Kardell, L. & Carlsson, E. 1982. Hjortron, tranbär, lingon: förekomst och bärproduktion i Sverige 1978–1980. Summary: Cloudberry, cranberry, cowberry: occurrence and berry production in Sweden 1978–1980. Sveriges lantbruksuniversitetet, Avdelningen för landskapsvård, Rapport 25. 139 pp.

Kardell, L. & Eriksson, C. 1987. Kremlor, riskor, soppar: skogsbruksmetodernas inverkan på produktionen av matsvampar. (Russulas, milk caps and boleti. The impacts of forestry practices on the production of edible mushrooms.) Sveriges Skogsvårdsförbunds Tidskrift 2: 3–23. (In Swedish.)

Kardell, L. & Johansson, M.–L. 1982. Gislavedsborna och torvmarksdikning: en attitydstudie. (The inhabitants of Gislaved and peatland drainage: a study of attitudes.) Sveriges lantbruksuniversitet, Avdelningen för landskapsvård, Rapport 26. 117 pp. (In Swedish.)

Kardell, L., Persson, O., Carlsson, E. & Eriksson, L. 1980. Skogsmarkens produktion av marksvampar. (The production of edible mushrooms on forest land.) Svensk Botanisk Tidskrift 74:91–102. (In Swedish.)

Kauppasieniopas. (A guidebook of commercial mushrooms.) 1988. Valtion painatuskeskus, Helsinki. 65 pp. (In Finnish.)

Kauppi, M. 1979. The exploitation of Cladonia stellaris in Finland. Lichenologist 11(1): 85–89.

Kauppi, M. 1990. Lectures about lichens. Picking forest products –course in Kullaa. 26.4.1990.

Kivinen, E. & Pakarinen, P. 1980. Peatland areas and the proportion of virgin peatlands in different countries. In: Proceedings of the 6th International Peat Congress, Duluth, Minnesota. p. 52–54.

Knudsen, H. 1983. Politikens svampebog. (Politiken's mushroom book.) Politikens Forlag A/S, Copenhagen. 192 pp. (In Danish.)

Korhonen, M. 1987. Uusi sienikirja. (New mushroom book.) 2. painos. Otava, Helsinki. 318 pp. (In Finnish.)

Kortesharju, J. 1984. Observation on cloudberry crops in Finland. In: Saastamoinen, O., Hultman, S.–G., Koch, N.E. & Mattsson, L. (eds.). Multiple–use forestry in the Scandinavian countries. Communicationes Instituti Forestalis Fenniae 120: 86–88.

Kortesharju, J. 1988. Cloudberry yields and factors affecting the yield in northern Finland. In: Vänninen, I. & Raatikainen, M. (eds.). Proceedings of the Finnish–Soviet symposium on non–timber forest resources in Jyväskylä, Finland, 25–29 August 1986. Acta Botanica Fennica 136: 77–80.

Kortesharju, J., Mäkinen, Y., Hippa, H. & Koponen S. 1978. Hilla Lapin luonnonvarana. (Cloudberry as a natural resource in Lapland.) Acta Lapponica Fenniae 10: 69–77. (In Finnish.)

Kujala, M., Malin, A., Ohenoja, E. & Sipola, K. 1989. Oulun läänin luonnonmarja– ja sienivarat, niiden satoarviot, hyödyntäminen ja sivutulollinen merkitys/ OUKA–projekti. (The wild berry and mushroom resources in the province of Oulu; estimates of the yields, utilization and significance as a source of additional income/OUKA –project.) Pellervo–Seuran markkinatutkimuslaitos, Helsinki. 74 pp. (In Finnish.)

Kunnas, H.J. 1973. Metsätaloustuotanto Suomessa 1860–1965. (The production of forestry in Finland 1860–1965.) Suomen Pankin julkaisuja, Helsinki. 193 pp. (In Finnish.)

Lange, M. 1948. The agarics of Maglemose: a study in the ecology of the agarics. Dansk Botanisk Arkiv 13(1): 1–141.

Lassheikki, M. & Holmberg, G. 1989. Produktion och hemmamarknad för julgranar och gröna kvistar i Finland. (Production and domestic markets for Christmas trees and green branches in Finland.) PS Nåledrys 10: 36–37. (In Swedish.)

Luonnonmarjaopas. (A guidebook of wild berries.) 1988. Valtion painatuskeskus, Helsinki. 58 pp. (In Finnish.)

Mäyränpää, P. 1981. Sienen silakkoo ja sirveliä – ruokaperinnettä pohjois–karjalaisittain. (...mushroom...herring... – food traditions in northern Karelia.) In: Pohjois–Karjala tutuksi. Opintotoiminnan keskusliitto ry, Joensuu. p. 41–47. (In Finnish.)

Mehus, H. 1986. Fruit body production of macrofungi in some North Norwegian forest types. Nordic Journal of Botany 6: 679–702.

Normann, C. 1989. Produktion och hemmamarknad för julgranar och pyntegrönt i Sverige. (Production and domestic markets of Christmas trees and ornamental foliage in Sweden.) PS Nåledrys 10: 40–43. (In Swedish.)

Nousiainen, H. 1983. Eräiden Vaccinium –lajien pölytysbiologiasta, kukinnasta ja marjonnasta. (Biology of pollination, flowering and berry production of a few Vaccinium –species.) Metsäntutkimuslaitoksen tiedonantoja 90: 66–86. (In Finnish.)

Numminen, E. 1979. Näkökohtia hillasta. (Viewpoints to cloudberry.) Rovaniemen tutkimusaseman tiedonantoja 21: 39–44. (In Finnish.)

Ohenoja, E. 1978a. Mushrooms and mushroom yields in fertilized forests. Annales Botanici Fennici 15: 38–46.

Ohenoja, E. 1978b. Lapin suursienistä ja sienisadosta. Summary: Aspects of the larger fungi and mushroom yields in Finnish Lapland. Acta Lapponica Fenniae 10: 84–88.

Ohenoja, E. 1984. Fruit body production of larger fungi in Finland. 1. Introduction to the study in 1976–1978. Annales Botanici Fennici 21: 349–355.

Ohenoja, E. 1993. Effect of weather conditions on the larger fungi at different forest sites in northern Finland in 1876–1978. Acta Universitatis Ouluensis. Series A 243: 1–69.

Ohenoja, E. & Metsänheimo, K. 1982. Phenology and fruit body production of macrofungi in subarctic Finnish Lapland. In: Larsen, G.A. & Ammirati, J.F. (eds.). Arctic and alpine mycology. p. 390–404.

Ohenoja, E. & Koistinen, R. 1984. Fruit body production of larger fungi in Finland. 2. Edible fungi in northern Finland 1976–1978. Annales Botanici Fennici 21: 357–366.

Östergård, K. 1989. Det danske hjemmemarked for juletræer og klippegrønt. (The domestic market for Christmas trees and ornamental foliage in Denmark.) PS Nåledrys 10: 4–5. (In Danish.)

Pekkarinen, M., Poikela, M. & Koskinen, E. 1980. Sienten käyttö kotitalouksissa. (The utilization of mushrooms in households.) Helsingin yliopisto, Elintarvikekemian ja –teknologian laitos, EKT–sarja 537: 1–159. (In Finnish.)

Pontoppidan, E. 1752. Det første forsøg på Norges naturlige historie, Bind 1. (The first effort on the natural history of Norway, Part. 1.) København. 338 pp. (In Danish.)

Päivänen, J. 1990. Peatland forestry in Finland: an overview. In: Hånell, B. (ed.). Biomass production and element fluxes in forested peatland ecosystems. Proceedings of a seminar in Umeå, Sweden, September 3–7, 1990. Swedish University of Agricultural Sciences, Umeå. p. 33–43.

Raamattu. (The Bible). 1954. Viides Mooseksen Kirja. Luku 23, jakeet 24–25.

Raatikainen, M. 1978. Puolukan sato, poiminta ja markkinointi Pihtiputaan kunnassa. Summary: The berry yield, picking and marketing of Vaccinium vitis–idaea L. in the commune of Pihtipudas. Silva Fennica 12(2): 126–139.

Raatikainen, M. & Raatikainen, T. 1983. Puolukan sato, poiminta ja markkinointi Pihtiputaalla. Summary: The berry yield, picking and marketing of Vaccinium myrtillus in the commune of Pihtipudas, northern central Finland. Silva Fennica 17(2): 113–123.

Raatikainen, M., Rossi, E., Huovinen, J., Koskela, M.–L., Niemelä, M. & Raatikainen, T. 1984. Metsä– ja suomarjasadot Väli–Suomessa. Summary: The yields of the edible wild berries in Central Finland. Silva Fennica 18(3): 199–219.

Rautavaara, T. 1947. Suomen sienisato. Summary: Studies on the mushroom crop in Finland and its utilization. WSOY, Porvoo. 534 pp.

Rossi, E., Raatikainen, M., Huovinen, J., Koskela, M.–L. & Niemelä, M. 1984. Luonnonmarjojen poiminta ja käyttö Väli–Suomessa. Summary: The picking and use of edible wild berries in Central Finland. Silva Fennica 18(3): 221–236.

Ruuhijärvi, R. 1974. Soiden karpalosadoista. Summary: On the cranberry yields on peatlands. Suo 25(2): 25–30.

Saari, V. & Salonen, V. 1983. Luonnontilaisten suotyyppien sienisadoista Korpilahden Ristisuolla vuosina 1981 ja 1982. (Mushroom yields on natural peatlands types in Ristisuo in Korpilahti, 1981–1982.) Metsäntutkimuslaitoksen tiedonantoja 91: 10–20. (In Finnish.)

Saastamoinen, O. 1983. Marjojen ja sienten talteenoton taloudesta. (Economics of gathering berries and mushrooms.) Metsäntutkimuslaitoksen tiedonantoja 91: 41–53. (In Finnish.)

Salo, K. 1979. Mushrooms and mushroom yield on transitional peatlands in Central Finland. Annales Botanici Fennici 16: 181–192.

Salo, K. 1982. Metsänhoitotoimenpiteet ja marja– ja sienisadot. (Forestry methods and the yields of berries and mushrooms.) Pohjois–Karjalan Luonto 12: 14–18. (In Finnish.)

Salo, K. 1983. Marja– ja sienisatojen seuranta VMI –tutkimuksen osana Pohjois–Karjalassa. (The survey of berry and mushroom yields as a part of a NFI (National Forest Inventory) –study in North Karelia.) Metsäntutkimuslaitoksen tiedonantoja 90: 122–134. (In Finnish.)

Salo, K. 1984. Joensuun ja Seinäjoen asukkaiden luonnonmarjojen ja sienten poiminta v. 1982. Summary: The picking of wild berries and mushrooms by the inhabitants of Joensuu and Seinäjoki in 1982. Folia Forestalia 598. 21 pp.

Salo, K. 1985. Luonnonmarjojen ja sienten poiminta Suomussalmella ja eräissä Pohjois–Karjalan kunnissa. Summary: Wild–berry and edible–mushroom picking in Suomussalmi and in some North Karelian communes, eastern Finland. Folia Forestalia 621. 30 pp.

Salo, K. 1988a. Soiden monikäyttö, marjat ja sienet. (Multiple use of peatlands, berries and mushrooms.) Metsäntutkimuslaitoksen tiedonantoja 308:187–198. (In Finnish.)

Salo, K. 1988b. Sienilajisto ja –sato Ylä–Karjalan metsä– ja suotyypeillä 1981–1984. (Mushroom species and yields on the different forest and peatland types in North Karelia, 1981–1984.) Lisensiaattityö. (Licentiate thesis.) Helsingin yliopisto. 213 pp. (In Finnish.)

Salo, K. 1991a. Marjat, sienet ja jäkälät: osa metsien monikäyttöä. (Berries, mushrooms and lichens: a component in multiple–use forestry.) In: Tapion Taskukirja. Gummerus Oy, Jyväskylä. p. 246–260. (In Finnish.)

Salo, K. 1991b. Metsien monikäyttö ja keräilytuotteet. (Multiple–use forestry and the by–products.) Karjalainen, yliö. 4.8.1991. (In Finnish.)

Salo, K. 1992. Regional significance of wild berry picking in Kainuu. In: Suomussalmi eco–municipality. Research. Results. Experimentation. Opinions. Suomussalmi Municipality. p. 52–57.

Salo, K. 1993a. Yields of commercial edible mushroom species in mineral soil forests in Finland, 1985–1986. Aquilo. Ser. Botanica 31: 115–121.

Salo, K. 1993b. The composition and structure of macrofungns communities in boreal upland type forests and peatlands in North Karelia, Finland. Karstenia 33(2): 61–99.

Salonen, V. & Saari, V. 1990. Generic composition of macrofungus communities on virgin mire site types in Central Finland. Acta Botanica Fennica 27: 33–38.

Sjöblom, M., Wessman, L., Albrecht, A. & Rancken, R. 1979. Svampproduktionen samt en jämförelse av virkes–, bär– och svampproduktionens värde i några skogar i Ekenästrakten 1976–78. (The production of mushrooms and a comparison of value of the production of timber, berries and mushrooms in a few forest areas in the Ekenäsregion, 1976–78.) Yrkesutbildningsstyrelsen, Avdelningen för forstundervisning, preliminär rapport. 51 pp. (In Swedish.)

Steindórsson, S. 1964. Gróður á Islandi. Summary: Vegetation in Iceland. Almenna bókafélagið, Reykjavik. 186 pp.

Valset, K. 1974. Naturbärens ekonomiska betydelse i Norden, diskussionsinlägg. (The economic importance of wild berries in the Nordic countries, a contribution to the discussion.) Lantbrukhögskolan, Konsulentavdelningens stencilserie, Trädgård 71: 8. (In Swedish.)

Valset, K., Gulden, G. & Lunder, R. 1976. Utmarksressurser i for– og matproduksjon. Bær, nøtter, andre nyttevekster. Sopp. Honning. (Wilderness resources for fodder and food production. Berries, nuts and other useful plants. Mushrooms and honey.) Statens Forskningsstasjon Kise, Nes Hedmark. Ås/Oslo/Asker. 61 pp. (In Norwegian.)

Vauras, J. & Huhtinen, S. 1980. Turun Ruissalon metsäsienisadosta vuosina 1977–78. (Mushroom yields in Ruissalo, Turku, 1977–78.) Sienilehti 32: 22–27. (In Finnish.)

Veijalainen, H. 1979. Luonnonvaraiset hillasadot. (Yields of wild cloudberry.) Rovaniemen tutkimusaseman tiedonantoja 21:10–13. (In Finnish.)

Vuorela, T. 1975. Marjat, sienet ja kasvikset. (Berries, mushrooms and vegetables.) In: Suomalainen kansankulttuuri. (Finnish culture.) Werner Söderström, Porvoo. 260–261 pp. (In Finnish.)

Wästerlund, I. 1982. Försvinner tallens mykorrhizasvampar vid gödsling? (Do the mycorrhiza fungi disappear after fertilization?) Svensk Botanisk Tidskrift 76: 411–417. (In Swedish.)

Wästerlund, I. & Ingelög, T. 1981. Fruit body production of larger fungi in some young Swedish forests with special reference to logging waste. Forest Ecology and Management 3(4): 269–294.

Yearbook of forest statistics 1989. (Metsätilastollinen vuosikirja 1989). 1990. Folia Forestalia 760: 1–246.

7 Reindeer husbandry and hunting

Timo Helle[1]

Abstract

Reindeer husbandry started in Scandinavia in the late medieval period. Most of the reindeer herders are Sámi people. The amount of reindeer is limited by winter food resources. Areas suffering from a range crisis are apparent in Finland, Norway and Sweden. Reindeer have been suspected of preventing or hindering natural regeneration of Scots pine, whereas forestry has been shown to reduce the carrying capacity of winter ranges. Reindeer husbandry is severely threatened by pollution; damage has been caused by radiocesium and sulphur dioxide. Tourism and outdoor recreation increase the demand for reindeer meat and souvenirs but they also have negative effects on this livelihood. Hunting is one of the most popular hobbies in Scandinavia. In terms of the amount of meat, moose is today the most important game species in Fennoscandia, but at the same time the greatest single damaging agent in forests. Moose benefit from the early stages of forest succession. The number of grouse has been decreasing in all the Scandinavian countries. The decline of capercaillie has been the most drastic. Capercaillie, the most preferred game bird, is adapted to mature coniferous forests, and thus suffers from forest regeneration. Experiments have recently been carried out in order to develop ways of including game management aspects into forestry planning.

Keywords: reindeer, game, hunting, carrying capacity, grazing, administration, planning.

---

[1] Finnish Forest Research Institute
Rovaniemi Research Station
P.O.Box 16, FIN–96301 Rovaniemi, Finland

7.1 Introduction

Definitions of multiple–use forestry in the Nordic countries include reindeer husbandry and hunting for game (Hytönen 1995). The reindeer husbandry area in Finland, Norway and Sweden covers about 30 % of the land area; about 15,000 people, most of them of Sámi origin, are involved in that livelihood. Hunting is a hobby for about one million persons. Professional hunters can be found only in the northernmost Fennoscandia.

Especially in Finland and Sweden, the most valuable reindeer ranges are located within the Eurasian taiga zone. Similarly, most game mammals and birds are adapted to forested habitats. This means that reindeer and game production commonly take place in the same areas which are also used for wood production. The relationships between wood production and these other products are highly variable. The ones most discussed are, understandably, competitive relationships. A typical example is capercaillie (*Tetrao urogallus* L.), the most preferred game bird, which is adapted to mature coniferous forests, and suffers thus from forest renovation (Järvinen et al. 1977, Rolstad 1989a). In the case of reindeer (*Rangifer tarandus* L.), the relationships are more complicated. Semi–domesticated reindeer have been suspected of preventing or hindering natural regeneration of Scots pine (*Pinus sylvestris* L.) (Wretlind 1931, Bergan 1962, Kollström 1988), whereas mechanical forestry has been shown to reduce the carrying capacity of winter ranges (Eriksson 1975, Mattsson 1981, Helle et al. 1990a). In contrast to capercaillie and reindeer, moose (*Alces alces* L.) is a species adapted to the early stages of forest succession and thus benefits from various kinds of silvicultural practices (Ahlén 1975, Lavsund 1987). As to the amount of meat, moose is nowadays the most important game species in the Fennoscandian taiga forests, but at the same time the greatest single damaging agent in forests.

This chapter gives an overall picture of reindeer husbandry and hunting for game in the Nordic countries nowadays. The main purpose is to describe and evaluate how forestry and, to a lesser extent, some other forms of forest utilization have influenced reindeer ranges and game habitats.

7.2 Reindeer husbandry

History

The Fennoscandian semi–domesticated reindeer is a descendant of the wild mountain reindeer (*Rangifer tarandus tarandus* L.) (Siivonen 1975). Tamed reindeer were used as decoy animals in prehistoric times in the hunting of wild reindeer; later on these reindeer were probably used as draft animals (e.g. Ruong 1982). In contrast, monocultural reindeer nomadism is of much younger origin. It is a commonly accepted tenet that in Fennoscandia it was "discovered" in the late medieval period in the Vefsen area of the Norwegian coast (Vørren 1973). From there it expanded its range over the centuries both to the south and northeast, along the Scandinavian mountain range. The reindeer nomads there were Sámi people. In the 17th century, semi–domesticated reindeer were introduced into northern Finland, where reindeer husbandry was also adopted by Finnish settlers. The present distribution of reindeer husbandry and wild reindeer in Fennoscandia is shown in *Figure 7.1*. Reindeer from Norwegian stock were introduced in the 18th century to Iceland, where at present there are a few thousand feral animals (Thorisson 1980). In southern Norway, there are about 40,000 wild mountain reindeer (Reimers 1986); their number in the Kola peninsula was about 20,000 in the late 1960s (Semenov–Tian–Shanskii 1975), but has since then strongly decreased. The only Fennoscandian wild forest reindeer (*Rangifer tarandus fennicus* Lönnberg) population lives in Russian Karelia and eastern Finland, where there is a total of 6,000–7,000 animals (Heikura et al. 1983).

Sámi reindeer nomadism was originally quite stationary (Ruong 1982), but later it has been characterized by long migrations between winter ranges located in coniferous forests, and summer ranges on fells or, as in northernmost Scandinavia, on the shore and islands of the Arctic Ocean. In contrast to what is commonly believed, migrations reaching the coast are not an adapta-

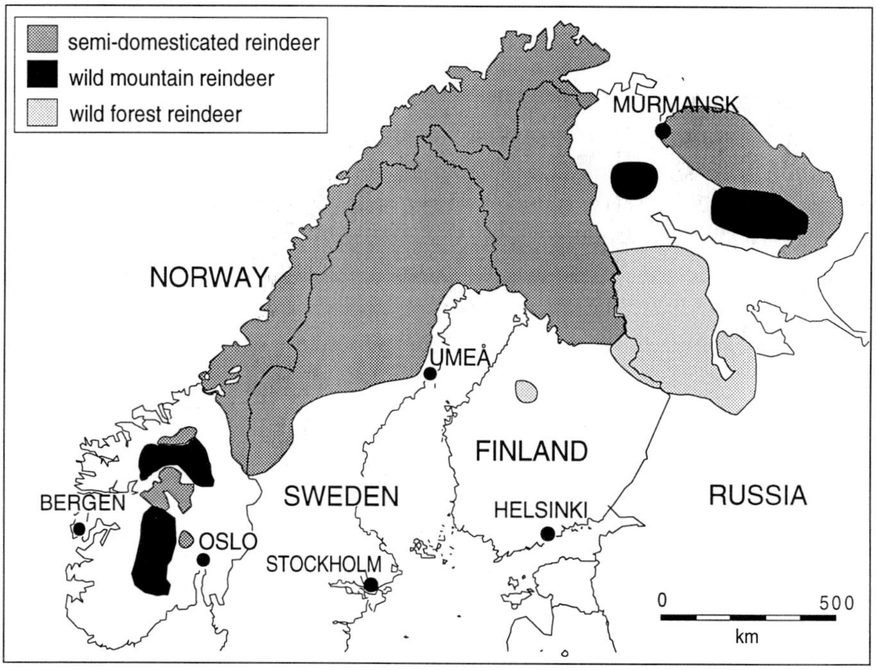

Figure 7.1 Distribution of reindeer husbandry and wild reindeer in Fennoscandia (Semenov-Tian–Shanskii 1975, Heikura et al. 1985, Skjenneberg 1989, Kojola 1993).

tion of the behavior of the wild ancestors of semi–domesticated reindeer (Tanner 1929). From the viewpoint of reindeer husbandry, migrations were reasonable because reindeer were easily kept under control on peninsulas and coastal islands, and at the same time lichen–rich winter ranges were saved from trampling (Oksanen 1978). Migration to the coast also served other purposes, such as fishing and trading, for instance. The right of free border crossing for reindeer–herding Sámi people was written into the Swedish–Danish/ Norwegian border treaty in 1776. The treaty is still partly in effect allowing some Swedish communities to graze during the summer in certain areas in Norway. The same treaty also applied to the border between Finland and Norway, because Finland belonged to Sweden until 1809. However, in 1852 the Finnish–Russian authorities unilaterally denied the rights of Sámi people to move with reindeer across the border. Thus, reindeer herders in Norway lost the opportunity of using winter ranges in Finland, and the Finnish herders, correspondingly, of migrating to the coast for the summer. The closing of the border confused reindeer management for several decades, and resulted in reorganization of reindeer husbandry in both countries. At present, the longest migrations, in Norway, are still about 350 km, and migrations are typical also to most Swedish herds. Finnish reindeer husbandry is, with few exceptions, quite stationary.

Legislation and organizations

Any EEA (European Economic Area) citizen living in the Finnish reindeer husbandry area has the right to own reindeer, while in Norway and Sweden ownership is restricted, almost exclusively, to Sámi people. The exceptions are reindeer companies in southern Norway and the "concession area" in Sweden beside the Finnish border, where owners can be non–Sámi residents and also, of course, partners or descendants of cross–cultural marriages.

The Finnish reindeer husbandry area is divided into 57 herding associations (paliskunta) with exact borders. The district is at the same time an administrative unit, which is responsible for the management of reindeer, although reindeer are owned by individual owners. In Norway, the Sámi reindeer area is divided into more than 100 seasonal or year round grazing districts. In Sweden, the basic unit is a grazing district referred to as a Sámi village ("sameby" in Swedish and "Sámi siita" in Samish), which is also an economic organization for the members. Their total number is 52, from which 33 are fell villages, 10 forest villages, and 9 concession villages. The number of families receiving their main livelihood from reindeer husbandry in Norway is about 650, in Sweden 800 (Skjenneberg 1989) and in Finland probably a little lower than in Norway because reindeer husbandry is, outside the Sámi area, commonly a subsidiary means of livelihood.

In each country, the Ministry of Agriculture is responsible for the public administration of reindeer husbandry. In Norway, Sámi herders have their own organization, The Norwegian Society of Sámi Reindeer Owners (Norges Reindrifsamers Landsforering). In Sweden, there are two organizations: the Swedish Sámi Society (Svenska Samernas Riksförbund) and the Reindeer Owners Union (Renägareförbundet). In Finland, herding associations belong to the Union of Reindeer Herding Associations (Paliskuntain Yhdistys).

Carrying capacity of winter ranges

The summer diet of reindeer consists of a great variety of graminids, herbs, and leaves of deciduous trees (see, for a review, Nieminen & Heiskari 1987). In winter, reindeer feed preferably upon reindeer lichens (*Cladina* sp.). If these are not available, they mostly rely on dwarf shrubs and grasses, and in forest areas particularly on arboreal lichens (*Alectoria* sp.and *Bryoria* sp.). Very lichen–poor winter diets have been reported, for instance, for semi–domesticated reindeer living in the southern part of the Finnish reindeer husbandry area (Sulkava & Helle 1975).

The effects of grazing and trampling are related to their intensity. On lichen heaths, the number of plant species as well as diversity are at a maximum on moderately grazed areas, whilst climax lichen associations are dominated by *Cladina stellaris* (Opiz) Pouzar & Venzda and, on the other hand, only few species tolerate extremely heavy grazing (Helle & Aspi 1983). Oksanen (1992) suggested that the intermediate disturbance caused by reindeer probably contributes to the maintenance of a high floristic diversity, and helps the rarities of Fennoscandian mountain flora to survive.

Most commonly, reindeer production is limited by food resources in winter. A central issue in reasonable herd management is the idea of carrying capacity. An equilibrium where the reproductive rate and mortality is equal is referred to as ecological carrying capacity (Caughley 1976). Low reproduction and high mortality are caused by the low food per capita ratio, which in turn is a result of overgrazing. By contrast, maximum sustained yield is obtained at economic carrying capacity. In reindeer husbandry, the aim is to adjust harvesting so that it allows maximum primary production of the feeding plants. The primary production is at a maximum when the biomass accounts for about 50 % of the ungrazed maximum biomass. Reindeer lichens are abundant only on dry and sub–dry sites, where competition from other vegetation is least (Mattila 1981). On such sites, the optimum biomass has a figure of about 1,000 kg/ha, and yearly production of 110 kg/ha (Kärenlampi 1973).

On the suggestion that reindeer can fulfil their energy requirement with lichens, both primary production of lichens as well as reindeer production peaked in wild mountain reindeer in Norway at a reindeer density of 1.8 individuals per km^2 in conditions where the proportion of lichen–rich vegetation types amounted to 12 % of the land area (Skogland 1986). Among semi–domesticated herds, the density is commonly higher, leading to a reduction in the lichen biomass. In northern Finland, the lichen biomass varies on dry sites between

50 and 400 kg/ha (Mattila 1981, 1988), depending on the districts and on animal density per lichen range (Helle et al. 1990b). In the districts with the lowest lichen biomasses, reindeer live largely on other food than reindeer lichen, including supplemental feeding in winter. A reduction of lichen biomass is reported both in Norway and Sweden (Tömmervik et al. 1991, Utsi 1992, Moxness et al. 1993).

In each country there are areas where range crisis is apparent. In Finland, reindeer owners have responded to this in two different ways. Intense supplemental feeding has increased meat production, but it is economically feasible only if the fodder can be obtained largely from one's own fields. Another method is calf harvest, which reduces the overall mortality rate, because calves are the first ones to starve when faced with a food shortage (Helle & Kojola 1993). This indicates that the detrimental effects of range depletion can be avoided to some extent by reasonable management practices. However, intense supplemental feeding seems inevitably to intensify overgrazing of lichen ranges, since reduced lichen availability does not have an effect on the population size; due to their high desirability, lichens are used already in the autumn and early winter (Helle & Kojola 1993). If intense supplemental feeding is rejected because of its high costs, the only way to maintain reindeer production at a stable level is by improvement of the condition of lichen ranges. Oksanen (1992) considered that in some areas in Sweden reindeer spoil lichen ranges by trampling them in summer. This problem may partly be solved by revising the border treaty allowing an increased number of Swedish reindeer to graze during the summer in the Norwegian mountains, which almost wholly lack lichen vegetation that would be sensitive to trampling. Above all, the number of reindeer has to be adjusted to overall range availability, since most Swedish Sámi communities have no legal right to graze in Norway. In Norway, range depletion is most obvious in Finnmark, where it has been planned to half the reindeer population, with the financial support of the state (see also Moxness et al. 1993).

Reindeer production

Criteria and accuracy of reindeer countings have varied at different times, and there are also apparent differences between countries. In general, the present data are more reliable than former ones. Finland is the only country whose law on reindeer husbandry categorically requires that reindeer must be tallied annually.

In 1990, the number of reindeer in the winter herd (i.e. after harvesting) amounted in Finland, Norway, and Sweden to 230,000, 220,000, and 300,000 animals, respectively. Density averages about 2 individuals per km^2. It varies considerably within each country, and is highest (up to 4 individuals per km^2) in northernmost Fennoscandia. In the long term, reindeer numbers have varied remarkably. Figures comparable to the present ones occurred in Sweden and

Finland in the 1930s. In each country, the reindeer population experienced a crash in the early 1970s, due to exceptionally difficult snow conditions, but has doubled since then (*Figure 7.2*).

According to a rule of thumb, meat production per reindeer (in the winter herd) is about 10 kg annually. However, the variation in production both per reindeer and per unit area is remarkable. In each country, the highest production per reindeer can be found in southern herds, whilst production per km^2 is highest in the northernmost herds, due to higher density (Lenvik 1988, Kojola & Helle 1991). In 1990, the total production in Finland, Norway, and Sweden was 3.5, 2.5 and 2.6 million kg, respectively. These figures exceed the average for the last 20 years. The value of production was in 1990 in Finland 120 million, in Norway 60 million, and in Sweden 80 million Swedish crowns. In each country, reindeer production is subsidized directly or indirectly by the state. Development of reindeer husbandry to its present state has been reviewed recently by Staaland and Nieminen (1993).

Anecdotes from the old days of reindeer husbandry tell that a rich reindeer owner could become poor in a few days or weeks due to wolves (*Canis lupus* L.), reindeer diseases, or starvation (e.g. Itkonen 1948, Ingold 1982, Ruong 1980). Deaths due to starvation are not unfamiliar to present reindeer husbandry, whilst epidemic diseases have almost wholly disappeared, and the numbers of large predators are controlled.

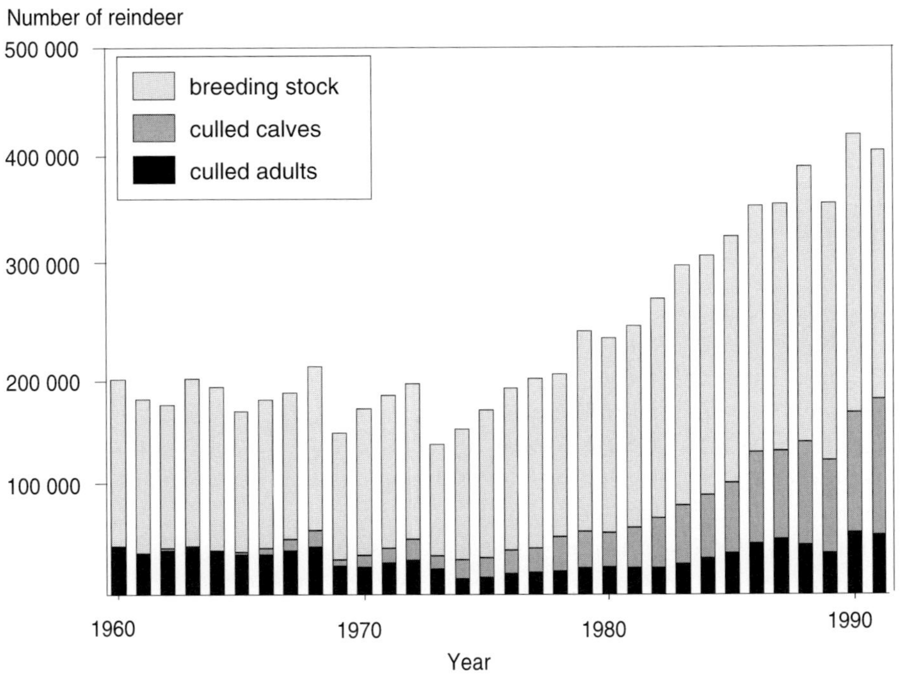

Figure 7.2 The number of reindeer (summer herd) and annual harvest in Finland 1960–91. Source: Official reindeer catalogues (Paliskuntain Yhdistys).

Most attention has been paid to the high mortality of calves during their first summer. In Finland, 45 % and 27 % of calves, on average, were lost between calf–marking and winter in 1981 and 1982, respectively (Nieminen 1993). It appeared that the calves which perished were in mid–summer somewhat lighter than those which survived. The same result was obtained by Haukioja and Salovaara (1978), suggesting that the high mortality was related to the poor condition of the reindeer.

However, the reasons for calf mortality may vary both annually and with area. Recently, Bjärvall et al. (1990) examined mortality and mortality factors in Jåkkakaska and Umby Sámi villages in northern Sweden. The study was carried out using ear–tagged and partly also radio–collared animals. Mortality averaged 17 % during the first year, and later about 3 % annually. Among calves, the most common mortality factor was large predators, which were responsible for 66 % of the losses; for diseases the figure amounted to 13 %. Wolverine (*Gulo gulo* L.) and Lynx (*Lynx lynx* L.) killed about 90 % of calves caught by predators. Also among adult reindeer, predators were a common mortality factor.

In 1990–91, the total number of reindeer killed by predators was in Finland, Norway, and Sweden, 900, 700, and 9,000, respectively (Bjärvall et al. 1991, Nieminen & Leppäluoto 1993). Those figures included only reindeer actually found and indemnified by the state. In Finland, losses due to predators are clearly smaller than in the 1970s, whilst in Sweden they have steadily increased.

Calf–marking takes place in Finland and Sweden generally in mid–summer, when reindeer gather into large herds in order to avoid the plague caused by blood–sucking and other parasitic insects. In Norway, calves of most of the herds are marked in autumn before the autumn migration to the winter grounds. Counting and harvesting of reindeer takes place in round–ups organized in autumn and winter. The adoption of snow–mobiles in the 1960s revolutionized gathering and herding routines; because reindeer were easy to get together, close herding was not needed (Pelto 1973, Müller–Wille 1975). Later, motor cycles and other terrain vehicles displaced a walking herder in summer, and nowadays helicopters are also used in placing and gathering reindeer. The possible stress effects of motorization on reindeer are still under discussion (Rehbinder et al. 1982).

Interestingly, totally new innovations are arising from the concern for the animals' well–being and various environmental issues, including, for exemple, ground deterioration caused by terrain vehicles, and the high costs of motorization that in turn forces the upkeep of maximal reindeer densities. For instance, the Sámi community of Grans in Sweden employs Icelandic ponies in gathering reindeer in summer and autumn.

Reindeer husbandry and forestry

In Finland, about 75 % of the reindeer population, and in Sweden almost all reindeer spend the winter in coniferous forests also commonly used for timber production. Confrontations between reindeer husbandry and forestry have been the "normal situation" since wood began to have commercial value in the later part of the 19th century. Foresters have suspected that reindeer may prevent natural regeneration of Scots pine, whilst reindeer owners have accused modern forestry of deterioration of winter ranges. The conflicts have been very similar in Finland and Sweden, partly also in Norway (Bergan 1962, Eriksson 1975, Mattsson 1981, Saastamoinen 1982).

The evaluation of the impacts of reindeer grazing on silviculture has appeared to be a difficult task, since it involves both apparent benefits and disadvantages. Brown and Mikola (1974) showed that, in laboratory conditions, lichen extracts inhibited the development of mycorrhiza in Scots pine seedlings. The experiment was then repeated in the field by surrounding the seedlings with a thick lichen mat. These seedlings grew more slowly than controls lacking lichen cover. Therefore, Brown and Mikola (1974) suggested that reindeer grazing may improve natural regeneration by reducing lichen extracts. However, the experimental design did not convince Kershaw (1985), who stated that a thick lichen mat also alters the growing conditions physically in several ways.

Another aspect to the problem is that reindeer can cause damage by rubbing their antlers against young Scots pines during the autumn and winter, or by breaking smaller plants in winter, when they dig for lichens beneath the snow. Mechanical damage can be quite abundant locally, but opinions on their silvicultural importance have been somewhat conflicting (Aaltonen 1919, Arnborg 1955, Bergan 1962, Kollström 1988).

Helle and Moilanen (1993) compared natural regeneration and growth of Scots pines on both sides of the southern border of the Finnish reindeer husbandry area. Thus, the field sample comprised both heavily grazed and ungrazed regeneration areas. Grazing did not affect the number of plants capable of developing. The death or poor condition of seedlings was in 80 % of the cases caused primarily by fungal diseases, and in 10 % by reindeer. Grazing slightly increased the frequency of Scleroderris canker (*Assocalys abietina* (Lagerb.) Schläpfer). On the other hand, there existed a strong negative correlation between mechanical damage caused by reindeer and the incidence of snow blight (*Phacidium infestans* Karst.). Feeding of large reindeer herds disturbs and hardens the snow cover thoroughly, and thus impedes the spreading of the fungus (see Roll–Hansen 1975).

Contrary to the results of Brown and Mikola (1974), removal of lichens did not improve the growth of seedlings which were less than 20 cm in height. On the contrary, seedlings between the age of 14–24 years grew faster on heavily grazed than on lightly grazed regeneration areas. Removal of lichens increases soil temperature in summer (Kershaw & Field 1975). Raitio (1990) showed that seedlings growing in warm soil (+20 °C) had more root tips and mycorrhizae than plants growing in cool soil (+10 °C); the difference was then also reflected in the growth of the shoot. One may assume that smaller seedlings, having a superficial root system, obviously suffered from lack of water, and lost the benefit of the improved soil temperature.

Another study examined how removal of lichen by reindeer affected the growth and condition of mature Scots pines (Helle & Nöjd 1992). In two of the three study areas, located in northern Finland, no difference existed in the radial growth or in the number of needle age classes between heavily grazed and ungrazed sample plots. In one area, Scots pines showed faster radial growth on grazed sample plots, and they had also more needle age classes. In the same area, the ungrazed lichen mat of the controls was significantly thicker (17 cm, including dead bases) than in the two other areas. It was suggested that the improved growth and better condition of trees on grazed sample plots were related to higher soil temperatures in summer. The other side of the confrontation, i.e. the accusation that forestry devastates winter ranges, is of younger origin and clearly associated with the adoption of clearcutting and soil preparation in forest renovation.

All the most important winter food resources are related to the age of the forest. Arboreal lichens occur in abundance only in forests more than 100 years old (Mattila 1979, Sparrevik 1983). On the other hand, clearcutting on fresh sites increases the biomass of hair grass (*Deschampsia flexuosa* L.) by 10–20 fold within a few years (Sulkava & Helle 1975, Mattila 1981). With regard to reindeer lichens, the relationship seems to be more complicated. Intensity of grazing in terms of pellet group density did not correlate with lichen biomass, but is strongly dependent on the age of the forest (*Figure 7.3*). The reason for the high preference of old forests remained partly unclear. On clearcut areas, heavy winds may harden the snow, restricting the access to lichens, and reindeer also seem to avoid cutting residue (Eriksson 1975). The avoidance of young sapling stands is probably associated with poor visibility, which may increase predation risk; similar behaviour is described in wild reindeer (Pruitt 1960).

Mattsson (1984) studied the opinions of reindeer owners in northern Sweden on acceptable and non–acceptable forestry, referred to as "harmony area" and "conflict area", respectively. The conflict area was characterized by a high proportion of clearcut areas and young Scots pine stands. The large size of a single regeneration area was also considered unacceptable. In the accepted harmony area, the age distribution of the forest stands was quite even compared to the conflict area. These two areas had quite a different forestry history. The harmony area had been subjected to frequent and probably selective cuttings between 1900 and 1940, whilst the forests of the conflict area were renovated more recently by clearcutting, and within a relatively short time span.

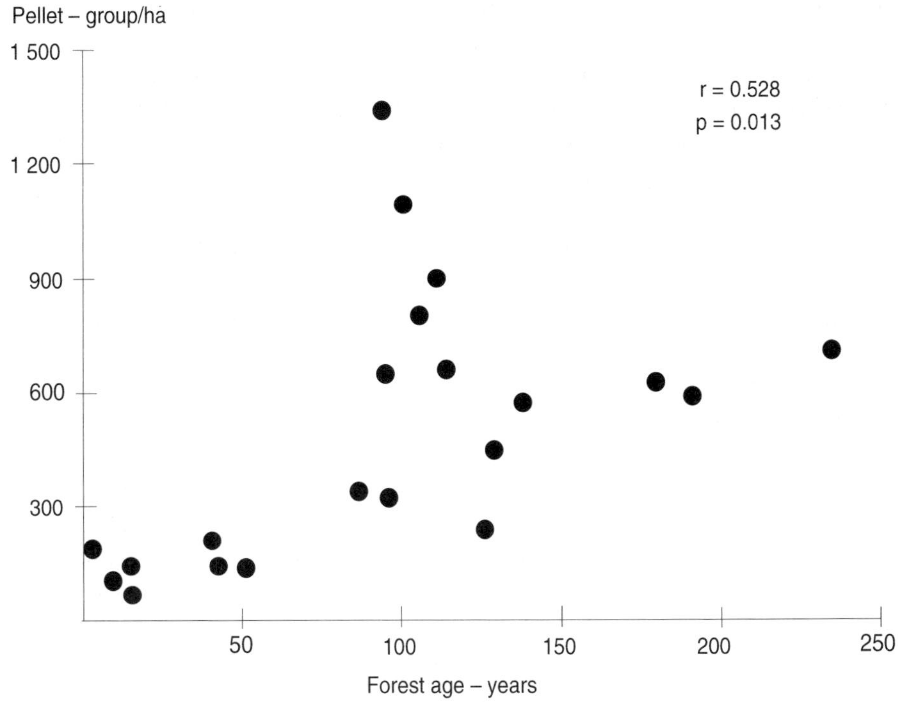

Figure 7.3 The relationship between the age of the stand and grazing pressure in terms of pellet–group density in Inari, Finnish Lapland (Helle et al. 1990a).

Eriksson et al. (1986) built up a model to evaluate the long term effects of timber production on reindeer winter ranges in Västerbotten and Norrbotten, Sweden. Availability of reindeer lichens and arboreal lichens was related to the stand age. It was assumed that clearcuttings and freshly thinned stands had no value as lichen ranges. The stands having arboreal lichens were at least 100 years old. Range data was then incorporated into a model calculating the potential cut up to the year 2080. In most areas, lichen availability was estimated to decline for some decades at the beginning of the next century, and the reduction of arboreal lichens is expected to continue through the whole period.

Forest fertilization was in common use especially in the 1970s. It was applied to increase the growth of middle age or older forests. The effects of fertilization on the behaviour of reindeer were studied in Västerbotten, Sweden, where about 41 % of pine forests with lichen were fertilized during 1960–74 (Eriksson 1980). Fertilization with ammonium nitrate reduced winter feeding activity of reindeer by 50 % during the winter after fertilization. To urea fertilization reindeer responded more strongly, and avoiding of the treated sites was still apparent the second winter after fertilization.

Forest fertilizers have also a direct impact on reindeer ranges (Eriksson & Raunistola 1993). Doses of around 150 kg of nitrogen per hectare had a positive effect on the growth of heather (*Calluna vulgaris* L.) and lingonberry

(*Vaccinium vitis–idaea* L.). By contrast, the standing crop of *Cladina* spp. decreased through the whole 11 year study period.

Among the silvicultural treatments, soil preparation using deep–ploughing has evoked the strongest criticism among reindeer owners (Eriksson 1975, Mattsson 1981). The method was developed in the 1960s, and was intended to be applied on grounds covered by a thick raw humus layer (e.g. Pohtila 1977). However, it almost totally replaced lighter soil preparation methods, and was used also on drier soils more than was originally planned. Deep–ploughing destroys a great proportion of the ground vegetation (Eriksson & Raunistola 1990), and when applied to large continuous areas, it may be a serious hindrance for collecting and transporting reindeer.

Reindeer husbandry and air pollution

Nordic reindeer husbandry areas are located far from great industrial centers. Nevertheless, reindeer husbandry is a livelihood severely threatened by atmospheric air pollution. Relatively high concentrations of radiocesium were found in the lichen–reindeer–man food chain already in the 1960s as a result of nuclear weapon tests in the atmosphere (Rissanen & Rahola 1990). However, a much more serious drawback for reindeer husbandry was the Chernobyl disaster in 1986 particularly in Sweden and southern Norway (see Rehbinder et al. 1990). In Sweden, tens of thousands of reindeer carcasses with high ^{137}cesium concentration were destroyed, and reindeer were shifted to uncontaminated ranges. Additionally, several methods of decreasing ^{137}cesium concentrations were developed (Åhman 1993). In Sweden, reindeer meat exceeding 1500 Bq ^{137}cesium/kg is not considered to be safe for consumption. The mean value of the meat offered for sale is 500 Bq/kg. If the weekly consumption of such meat is less than 3 kg, one does not risk exceeding the recommended maximum annual radiocesium dose which in Sweden is 1 mSv (Åhman 1993).

Atmospheric air pollution threatens the prerequisites of reindeer husbandry in other ways, too. Many lichen species are susceptible to sulphur dioxide. Helle et al. (1990c) reported a 50 % reduction in the biomass of arboreal *Alectoria* sp. and *Bryoria* sp. in two old spruce stands in Finnish Lapland between 1976 and 1988. The decreasing trend was noted also by Kautto et al. (1986) and Mattila (1988). Recently, Tarhanen et al. (1992) found changes in membrane permeability and cell ultra structure in *Bryoria fuscescens* in the Kola peninsula and eastern Finnish Lapland. These changes increased with proximity the nickel smelters in the towns of Nickel and Monchegorsk in the Kola area in Russia, producing annually a total of 700,000 tons of sulphur dioxide.

In the surroundings of these smelters, reindeer lichens are also affected. Tömmervik et al. (1991) reported that reindeer lichens had totally disappeared or drastically decreased in Sødre–Varanger, Norway, 30–50 km north of Nickel. In this area, the local population of semi–domesticated reindeer is decreasing although in other parts of Finnmark the number of reindeer has increased. In Monchegorsk, the growth rate of *Cladina rangiferina* (L.) Web. reaches the normal figures when 50–60 km distant from the smelters (Helle & Kojola 1992). Another potential threat is the concentration of heavy metals in reindeer.

Reindeer husbandry and outdoor recreation

Tourism and associated outdoor recreation activities have been, in northern Scandinavia during recent decades, the most rapidly increasing livelihood. Whereas they increase the demand for reindeer meat and souvenirs made of reindeer material, they also have negative effects on reindeer husbandry.

Quantitative data on these effects are available from the Saariselkä area, in northeastern Finnish Lapland. Finland's biggest holiday resort is situated on the western edge of the area. It comprises 8,000 beds, and offers visitors slalom, cross–country skiing, hiking, and other kinds of outdoor recreation. The first report on the deterioration of ground vegetation as a result of tourism appeared in the 1970s (Hoogesteger 1976). A few years later, Saastamoinen (1982) concluded that the most serious problem was the disturbance of reindeer by skiers and hikers, which resulted in the loss of a calving ground, which had earlier been important (Aikio 1977). Helle and Särkelä (1993) studied the spatial variation in the distribution of reindeer as a function of the distance from the holiday resort. In general, the relative reindeer densities increased when outdoor activities decreased. The disturbance effect reached about 10 km from the holiday resort. It was possible to determine the consequent range loss and its economic value. Within the "disturbance area", reindeer were not capable to graze in an optimal way in winter, i.e. to concentrate feeding on sites with abundant lichen vegetation or shallow snow. This possibly has a negative effect on the reindeer's energy balance, but quantification of the deficit appeared to be impossible.

In comparison to the gross income from tourism in Saariselkä, the economic value of the range loss appeared to be quite small. However, from the viewpoint of local reindeer owners the loss is not negligible. It is widely agreed that a distinction should be made between efficiency and equity, when the values of different social or cultural groups are relevant to the decision (e.g. Layard & Walters 1978). Consequently, better planning and restrictions of outdoor activities will be needed in order to minimize the harmful effects on reindeer management.

Table 7.1 The average annual kill of forest game in 1978–1991. Sources: Denmark: Denmarks Miljøundersøgelser; Finland: Metsästäjien keskusjärjestö; Norway: Norsk Institutt for Naturforskning; Sweden: Svenska Jägareförbundet.

	Denmark	Finland	Norway	Sweden
Mammals				
Red squirrel (*Sciurus vulgaris*)	900	2 000	3 600	1 400
Beaver (*Castor fiber/C.canadensis*)	-	1 100	-	5 200
Red fox (*Vulpes vulpes*)	48 800	30 600	17 400	23 200
Racoon dog(*Nyctereutes procyonoides*)	-	54 200	-	-
American mink (*Mustela vison*)	2 800	58 600	24 600	50 000
Pine marten (*Martes martes*)	4 300	10 700	9 800	15 700
Badger (*Meles meles*)	1 400	11 400	3 900	33 600
Hare (*Lepus timidus/L. europaeus*)	151 800	293 600	119 300	179 400
Fallow deer (*Dama dama*)	3 300	20	-	2 800
Red deer (*Cervus elaphus*)	1 840	-	10 100	600
Sika deer (*Cervus nippon*)	380	-	-	-
White-tailed deer (*Odocoileus virginianus*)	-	6 500	-	-
Moose (*Alces alces*)	-	48 700	27 400	130 400
Wild reindeer (*Rangifer tarandus*)	-	-	8 600	-
Roe deer (*Capreolus capreolus*)	66 200	3	29 000	197 000
Birds				
Willow grouse (*Lagopus lagopus*)	-	77 200	522 000	39 500
Capercaillie (*Tetrao urogallus*)	-	42 000	22 200	33 300
Black grouse (*Tetrao tetrix*)	-	170 000	55 500	44 400
Hazel grouze (*Bonasa bonasia*)	-	95 600	9 300	24 100
Pheasant (*Phasianus colchicus*)	802 400	9 100	-	14 100
Woodcock (*Scolopax rusticola*)	23 800	3 500	8 000	30 000
Woodpigeon (*Columba palumbus*)	310 200	88 200	78 600	90 400

7.3 Hunting

Hunting statistics

Hunting is one of the most popular hobbies in Scandinavia among men. The number of hunters in Norway is 160,000, Denmark 170,000, and in Finland and Sweden 300,000. Hunting is most popular in northern Scandinavia, where every third man hunts. In each country, the number of hunters has steadily increased during the last decades, but it seems to have stabilized during recent years. Public attitudes towards hunting are still quite liberal compared to many other European countries. One reason may be that in most parts of Scandinavia hunting has been traditionally a livelihood or hobby of the ordinary countryside people. On the other hand, one should be aware that hunting legislation has developed all the time into being more prohibitive or restricting to hunting methods, as criticism has been voiced among non–hunters. An English–speaking reader can receive an excellent overall picture of Scandinavian hunting from the book "Swedish Game – Biology and Management" (Svenska Jägareförbundet 1992).

Differences in the climate, proportion of forest land, tree species composition etc. between different parts of Fennoscandia are reflected in the abundance of different game species and the structure of the bag as well. Bag statistics for Denmark, Finland, Norway, and Sweden from recent years are given in *Table 7.1*.

In earlier times, Fennoscandia was famous for its rich sources of high quality furs. Excessive hunting caused extinction of the beaver (*Castor fiber* L.) (which has been later re–introduced), and reduced the number of pine marten (*Martes martes* L.), for instance. Economically the most important species was red squirrel (*Sciurus vulgaris* L.); the maximum bags in this century ranged, in Finland and Sweden, between 1–2 million individuals annually. The crash in fur prices in the 1960s almost wholly put an end to traditional professional or half–professional fur hunting. At present, trapping and hunting of red fox (*Vulpes vulpes* L.), badger (*Meles meles* L.), racoon dog (*Nyctereutes procyanoides* Gray) and pine marten are carried out for game management purposes. Large predators are protected, or their hunting is strictly limited.

The Danish game habitats with field, bush and forest mosaic are highly productive. The annual kill of hare (in Denmark only the brown hare (*Lepus europaeus* Pall.) is of the same magnitude as in Sweden, and bags of red fox, pheasant (*Phasianus colchicus* L.), wood pigeon (*Columba palumbus* L.), and woodcock (*Scolopax rusticola* L.) are even numerically highest of all the Nordic countries. Rather similar game habitats are to be found in southernmost Sweden. In the taiga zone game production is generally lower, but the species diversity higher. The importance of ungulates is strongly emphasized both numerically and as meat production (Mattsson & Kriström 1987, Mattsson 1989, 1994). Hunting of forest grouse seems to be most popular in Finland, where bags of black grouse and hazel grouse are four times larger than in Sweden.

The "case" of moose

Moose is the most important game species in Sweden both in terms of meat value and the recreational value of hunting (Mattsson 1994). The same applies to the most part of the Fennoscandian taiga forest area. The trends in the population dynamics reflected in the number of harvested animals have been quite similar in Finland, Norway and Sweden (*Figure 7.4*). Moose were close to extinction in wide areas in each country at the beginning of the 20th century. The main reason for that was excessive hunting. Due to strict hunting regulations, the number of moose began to increase, and its density peaked in Finland and Sweden in the 1980s. In Sweden, "the land of moose", the summer herd comprised during the peak years about 300,000 animals, from which about 150,000 are killed during the hunting season in autumn. These figures were about three–fold compared to Finland, and the status of moose in Sweden is also unique in a world wide comparison (Danell 1989).

Moose is in the focus of several and partly contradictory interests. Moose hunters prefer, understandably, relatively high moose density. However, it seems to be self–evident that the goal of moose management can not be ecologically determined maximum sustained yield. Such a policy neglects the negative consequences of moose, including damage to forestry and agriculture, as well as the costs of traffic accidents. A comprehensive cost–benefit analysis has never been carried out. The benefits can be determined quite reliably (Mattsson & Kriström 1987), whilst accurate data on damage are not available. The present moose densities in Sweden and Finland are a compromise between various interest groups and are lower than during the peak years in the 1980s (*Figure 7.4*).

The increase in moose density has been related, besides to hunting restrictions and manipulation of the herd structure, to improved food availability (Ahlén 1975, Myrberget 1979, Nygren & Pesonen 1989). In the absence of large predators, moose density is determined by winter food, comprising deciduous trees and young Scots pines. In forest renovation, original mixed stands have commonly been replaced by Scots pine, and the associated clearcutting has increased the amount of deciduous trees, too. Drainage has had, similarly, a positive effect on food availability (Lykke 1964). Thirdly, Ahlén (1975) mentioned that previously most of the domestic animals grazed in forests in the summer using largely the same food, except for pine seedlings, as moose in winter. Because the most intense period, both in forest renovation by clearcutting and in drainage, is over, one may predict that food resources of moose will decline in the future. This may be counterbalanced by new management orders giving space to deciduous trees.

Whereas moose likely have benefited from modern forestry, they are at the same time the greatest single damaging agent in forests. The first reports on moose damage to forestry appeared in the late 1800s, being associated with the increasing economic value of wood (see Lavsund 1987). Palo (1981) estimated that one moose uses twigs from 5,000 pine seedlings per year, and needs about 20 hectares of young stand in order to meet its energy requirement from year to year. If this area/year is smaller, the proportion of severely damaged seedlings begins to increase due to the resultant more intense browsing (Lavsund 1987).

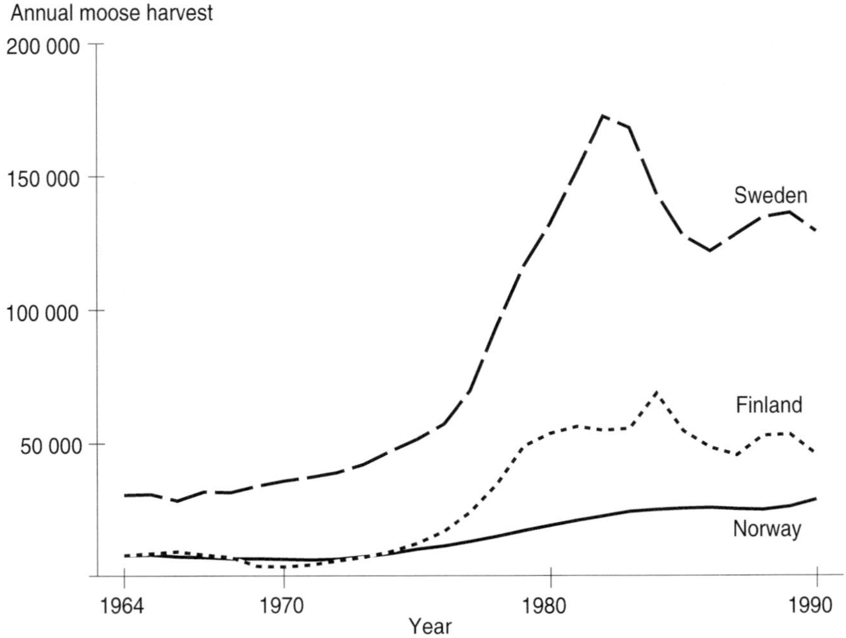

Figure 7.4 Annual moose harvest in Finland, Norway and Sweden in 1964–90. Sources: Finland: Nygrén and Pesonen 1989, 1993; Norway: Norsk Institutt For Naturforskning; Sweden: Swedish game... 1992.

Moose cause damage in young stands of Scots pine and silver birch (*Betula pendula* Roth), and to a lesser extent in young stands of Norway spruce (*Picea abies* L.). The most common type of damage is breakage of twigs and leader shoots. In another type of damage, moose break the main stem of seedlings 2–4 m in height. The frequency of stem breakage is commonly quite low, but it increases if the availability of twigs is reduced by intense browsing (Hjeljord & Fjellbakk 1982, Lavsund 1987, Heikkilä & Mikkonen 1992).

The effects of moose damage on growth and survival of pine seedlings vary with browsing intensity and soil fertility. A reduction of 75–95 % in the green parts will kill the seedlings (Hjeljord & Fjellbakk 1982, Bergström et al. 1983). Occasional heavy browsing has only a minor effect on the growth (Löyttyniemi 1983, Heikkilä & Löyttyniemi 1992), while repeated browsing may involve a complete cessation of growth (Lavsund 1987). It is a common finding that the detrimental effects of damage are most pronounced on poor soils. That is likely to be associated with nutritional deficiencies weakening compensatory growth, and such seedlings are thus subjected longer to repeated browsing (Danell 1989).

Much of the research on the habitat and food selection of moose has been stimulated by the idea of reducing the damage risk by silvicultural measures. The final goal might be the finding of poorly palatable progeny origins of forest trees to be planted in areas with a high damage risk. A lot of theoretical and empirical work on qualitative and quantitative aspects of food selection has been carried out (e.g. Löyttyniemi 1981, Haukioja et al. 1983, Danell & Ericsson 1986, Bergström & Danell 1987, Palo 1987, Danell et al. 1991a, 1991b). Clear differences in palatability between progeny origins of Scots pine have been reported (Niemelä et al. 1989), but practical applications are still lacking. One of the most debated questions deals with the effect of cleaning, i.e. do deciduous trees in a pine plantation increase or decrease the damage risk of pine seedlings? Under given circumstances, moose browse only on the most preferred aspen (*Populus tremula* L.) and rowan (*Sorbus aucuparia* L.) but leave pines untouched. The resultant retarded growth of the most preferred species (Heikkilä 1991) gives an advantage to pine in inter–species competition and is reflected also in the growth of pines (Lavsund 1987). Otherwise, the results on the effects of cleaning are variable. In some areas, high incidence of deciduous trees increases the damage risk (Westman 1958, Olsen & de Vibe 1981), while in some others no correlation has been found (Lykke 1964, Löyttyniemi & Piisilä 1983, Danell et al. 1991a, Heikkilä 1991). A relatively high damage risk was found in stands with high birch density and clear overgrowth of pine (Heikkilä 1993).

Similarly, the results on the effect of stand density are somewhat conflicting. Lavsund (1987) did not find any difference in the damage frequency between sparse and dense young stands. In a Finnish study, the number of twigs and twig biomass browsed by moose increased with stand density, but despite that the number of untouched seedlings was higher in dense stands (Heikkilä & Mikkonen 1992).

Improvement of habitat quality has been considered as one possible measure to reduce damage risk (Bergström & Hjeljord 1987). Lääperi (1990) established high quality feeding sites for moose in order to keep them away from seedling stands. In these feeding sites, deciduous trees were fertilized, and mineral licks as well as tops of aspen and pine were available for moose. Compared to adjacent control areas the feeding sites appeared highly preferred. However, the reduction in damage remained unclear, probably because the distance between established feeding sites and young stands was too short.

These various measures aimed at reducing damage risk have appeared relatively ineffective due to the close relationship between moose density and damage risk. Therefore, various kinds of repellents may be the only way to keep moose away from young stands. The main problem with chemical repellents is that the treatment should be repeated annually for 5–10 years, which makes the method expensive. That is why these can be recommended only for restricted areas with a high damage risk (Lavsund 1987). Fencing of the regeneration area is effective, but expensive, too.

Finally, the "moose question" can also be considered from a wider ecological point of view. In Sweden, Angelstam (1990a) related the conservation problems of woodpeckers to super–normal moose density. Where large predators are present, moose densities range between 0.1–0.4 individuals/km^2, whilst in Sweden their number is about 10 fold. In future, mature aspen will disappear due to long–lasting intense browsing of young trees. This results in a lack of nesting trees for woodpeckers, and is reflected also in the amount of birds using nesting holes made by woodpeckers.

Grouse habitats and forestry

The decreasing number of grouse has been documented in all Scandinavian countries. The decline of capercaillie has been the most drastic (Marcström 1979, Järvinen & Väisänen 1984, Lindén 1981, Rolstad & Wegge 1987). The causes underlying these decreasing trends may be various, but striking similarities in declines between different species and countries suggest that the causes may be the same. Is shooting one of these causes? In Finland that is quite a common public opinion. Indeed, evidence exists that overharvesting was apparent in northern Finland in the late 1970s; annual yields increased, despite the declining phase in the cyclically fluctuating grouse populations. This probably slowed the expected population increase after "poor" years (Lindén 1981, 1991a, Lindén & Raijas 1986). Another study showed that an extremely high harvesting rate affected both capercaillie and black grouse but not hazel grouse (Lindén & Sorvoja 1992). On the other hand, Lindén (1981) found that capercaillie populations censused in southern Finland declined despite protection from shooting for nine years. Thus, one may conclude that shooting is not the principal cause for decreasing population trends even in Finland, where shooting yields are essentially larger than in Sweden, for instance (see *Table 7.1*).

Grouse habitats have traditionally been described by means of stand characteristics (e.g. Pynnönen 1954, Seiskari 1962, Uusvaara 1963, Marcström et al. 1983, Winqvist 1983). From the 1970s onwards, increasing attention has been paid to the effects of forestry on grouse habitats. However, identification and quantification of these effects have appeared to be a difficult task, and explanations of the ultimate causes of the decline of grouse populations are still variable or even contradictory.

Abundance of winter food has been proposed as an important factor controlling the densities of many birds. The winter diet of grouse species commonly contains only one food species. Capercaillie is specialized in the needles of Scots pine, whilst black grouse and hazel grouse feed in winter mainly on catkins and buds from birch (*Betula* sp.) and alder (*Alnus* sp.), respectively.

Pulliainen (1979) and Lindén (1984) reported that capercaillie prefer needles with high nitrogen and energy, but low resin contents. Hunters from the old days knew that capercaillie likely fed on needles of pines damaged by a forest fire (Virkkula 1928). However, the assumption that present forests do not provide sufficiently high quality food is hardly plausible (Rolstad 1989, Angelstam 1990b).

The earlier abundance of black grouse in Finland and Sweden has been related to shifting cultivation increasing the area of birch forests (Seiskari 1962, Angelstam 1983). Since the 1960s, the area and volume of birch forests in Finland has decreased in some districts and increased in some others. However, these changes can not explain the recent decline in black grouse populations (Helle, P. & Helle, T. 1991).

Hazel grouse seems to be the only grouse species whose present abundance is clearly related to the amount of winter food. Swenson (1993) found in Sweden and Finland that alder had a positive effect on hazel grouse at various levels of the scale, including the feeding site, territory, landscape and region.

The result is understandable, because the winter food resources of hazel grouse are much smaller than those of capercaillie and black grouse. In Finland, for instance, the proportion (by area) of alder dominated forests is only 0.5 % compared to 62 % for Scots pine and 7 % for birch (Yearbook of forest statistics... 1992). By volume the difference is much greater, due to the small size of a single alder tree.

Deciduous trees are commonly most abundant in young forests, which likely explains why both hazel grouse and black grouse prefer relatively young forests. Winqvist and Ringaby (1989) suggested that the abundance of grouse in central Sweden at the beginning of this century was related to the high proportion of young forests, and the decline is a result of the maturation of these forests. Hazel grouse apparently benefited from shifting cultivation and in particular from the associated livestock grazing in the forests; domestic animals consumed other deciduous trees, but left alder untouched. However, it remains unclear whether alder has also decreased during recent decades.

Most of the recent capercaillie researchers have concentrated on lekking behaviour and its relation to forest characteristics. Capercaillies have a social lek, and the place is commonly used for several decades (Hjort 1970, Valkeajärvi & Ijäs 1986, Wegge & Larsen 1987). The lekking places are located most commonly in mature forests dominated by Scots pine or spruce. Winqvist (1983) described the forest characteristics of 100 capercaillie lekking places in Sweden, and considered that they did not deviate from average forests. The forest characteristics between the lek centres and random points were compared in Finnish Lapland in the 1950s and 1980s (Helle, P. et al. 1994). In both cases, most leks existed in forests more than 150 years old. In the 1950s, before the time of intense forest renovation, such forests were abundantly available, and no difference was found between lekking centres and random points. By the late 1980s, the proportion of old forests had drastically decreased, resulting in a clear difference in forest age between lekking centres and random points.

The number of capercaillie cocks participating in the lek is related to the patch size pattern of mature forests. The daytime territories of the cocks are located around the lekking centre like "pieces of cake", each of them with an area of 0.15 km^2 (Rolstad & Wegge 1987). The mean size of "forest island" having a capercaillie lek amounts to about 0.6 km^2 (Angelstam 1983, Rolstad & Wegge 1987, Lindén & Pasanen 1987). The greater the proportion of mature forests around the lek centre, the more cocks can participate in the lek (Wegge & Rolstad 1986, Rolstad & Wegge 1987, Hjort 1990). These studies on radio tracked cocks showed that the maximum distance of a daytime territory from the lek centre is about 1 km. That is in agreement with the finding that the proportion of mature forest as well as the forest age remained on quite a high level up to 1 km from the lek centre (Lindén & Pasanen 1987, Helle, P. et al. 1994). Consequently, the minimum distance between two leks is in an optimal situation about 2 km (Hjort 1982, Wegge & Larsen 1987). On the other hand, if the area of mature forest is less than 0.5 km^2, and isolated by more than 100–200 m from other forests, it does not support lekking males (Angelstam 1983, Rolstad & Wegge 1987). Advice for preservation and management of lekking sites is given in several papers (Rolstad 1989b, Rolstad & Wegge 1989a, 1989b, Valkeajärvi & Ijäs 1991).

Besides lekking sites, the home range of capercaillie consists of several other seasonal ranges (Wegge & Larsen 1987, Rolstad 1989a, Hjort 1990, Helle, P. et al. 1994). Finnish lekking site studies have revealed that some forest characteristics, such as the proportion of forest land or old forests, exceeded the average figures up to 4 km from the lekking centre (Lindén & Pasanen 1987, Helle, P. et al. 1994), i.e. clearly outside the actual lekking sites. This may indicate that lekking sites are located in remote and large, uniform forest areas, providing all seasonal ranges for the local capercaillie population. These special requirements may explain why capercaillie populations in southern Finland decline, although the total area of more than 90 years old forests has increased during recent decades (Helle, P. & Helle, T. 1991).

Even relatively small clearcutting can wholly destroy a hazel grouse territory or a lekking area of capercaillie. The birds are forced to move elsewhere, probably to habitats of lower quality. However, the essential process behind the changes in forest landscape is fragmentation (see Angelstam 1990a). At present, agricultural fields, settlements, clearcuts and other human activities commonly isolate large forest areas from each other. If a local population becomes extinct for some reason, the area does not get recruits from outside, and therefore it remains unpopulated (e.g. Fritz 1979).

Fragmentation also has other effects possibly contributing to the decline of grouse populations. It increases the amount of forest edges, and this is likely to increase predation pressure (Andrén et al. 1985, Angelstam 1986, Rolstad 1989a). Another hypothesis suggests that clearcutting also increases the number of predators. Hansson (1979) showed that clearcutting produces, except on dry sites, grass– and herb–rich areas, which in turn increase the amount of voles. Reforestation of old fields has a similar effect. Henttonen (1989) pointed out that on clearcut areas, field vole (*Microtus agrestis* L.) replaces the red–backed vole (*Cletrionomys glareolus* Schreb.), which is adapted to forests. Field voles are, in contrast to "forest voles", non–territorial and social, making high densities possible. Increased vole density is assumed to sustain high densities of small and medium size predators, whose predation is partly directed to ground nesting birds (Storaas & Wegge 1985, Henttonen 1989, Rolstad & Wegge 1989c). Another "school" considers that, for many predators, birds are alternative food to voles, and therefore predation of birds may be most intense after a crash in cyclically fluctuating vole populations (Angelstam et al. 1984b, Marcström el al. 1988). Lindén (1988, 1992) did not find such relationships from extensive Finnish survey data, but he did concede the possibility that predators affect prey populations.

Storaas and Wegge (1985) argue that nest losses of capercaillie and black grouse have increased during recent decades. Nest losses have not been related to stand structure at the actual site. Instead, the general predation pressure should be considered at the landscape level, because of the movements of predators (Storaas & Wegge 1987). Indeed, Rolstad and Wegge (1989c) found in Norway that in the present forests, no part of the landscape was situated more than 1.5 km away from an abundant food resource of the red fox, whose radius of movement is just of the same longitude (*Figure 7.5*). In addition to favorable vole habitats, sites where moose had been slaughtered and garbage dumps were considered as rich food resources for the fox.

Forest fires have been an integral part of the dynamics of boreal forests. Therefore, one may assume that grouse have adapted to forest fires and black grouse and hazel grouse, at least, have benefited from them, because the post–fire succession is characterized by deciduous trees. Why then have grouse not apparently adapted to clearcuttings, although their effects are, at first glance, rather similar to those of forest fires? One possible answer is associated with the predator regimen. Pine forests on dry sites experienced a forest fire in 80–100 years intervals (Zackrisson 1977), while the fire interval for spruce forests on fresher sites varied between 150–450 years (Haapanen & Siitonen 1978, Hyvärinen & Sepponen 1988). Therefore, grass–rich openings typical to fresh sites were, in natural conditions, quite rare, compared to the present situation. Additionally, there are also obvious differences in plant succession between a burned area and clearcutting. In his studies on prescribed burning, Viro (1970) found on fresh sites in southern Finland that grasses were maximally 2.5 times and herbs 5 times as abundant as on an unburned area. However, the most abundant species, *Calamagrostis* sp. and *Epilobium angustifolium* L. have only minor importance in the winter diet of voles. Other soil treatments associated with clearcutting, for example, ploughing, harrowing and scalping, leave more of the original vegetation, and commonly create a flush of common hair grass, which altogether may provide for voles more abundant winter food resources than burned areas.

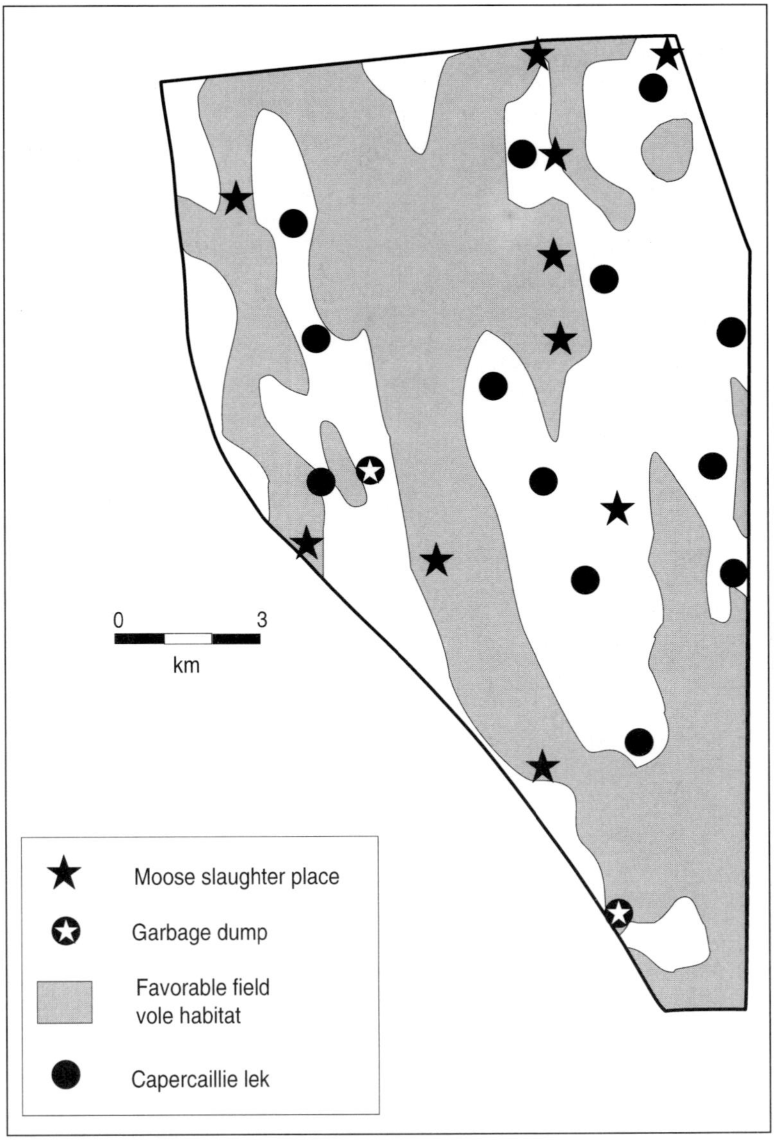

Figure 7.5 A sketch map on the distribution of capercaillie leks in relation to rich food sources of small and medium–size predators. Varaldskogen, Norway (Rolstad & Wegge 1989c).

7.4 Reindeer and game in forest planning

The forest planning systems applied to practical forestry have been based almost exclusively on timber production. In the 1980s, some trials were carried out to study the economic consequences when rotation time was increased in order to improve the quality of reindeer's winter ranges (Mattsson 1987, Varmola 1988). In both cases, the costs to timber production appeared so high that the "reindeer models" were considered at that time unrealistic from the viewpoint of timber production.

However, this does not mean that reindeer husbandry and game management have wholly been neglected in forest planning. For instance, the Finnish Forest and Park Service presents cutting plans in good time beforehand to local herding associations. Similar cooperation also exists in Sweden. Areas in cutting order can then be used intensively as reindeer ranges. It is also possible to alter the cutting order; reindeer owners require, for instance, that forests rich in arboreal lichens should be cut in late winter during the time of food shortage (Saastamoinen 1980). Negotiations are mostly informative, and it is the land owner who makes the final decision.

Many foresters are hunters themselves and thus interested in conserving and improving game habitats. On a small scale this has also been possible. Intensive forest renovation employing clearcutting has required more effective measures; first, recommendations on habitat management appeared in forest management directions in the 1970s. However, it is still a long way from those recommendations and orders to forest planning which routinely incorporates management of reindeer ranges and game habitats.

Planning of multiple–use forestry is faced at present with new opportunities. The shift from underproduction to overproduction of timber creates more space for other use forms, including reindeer ranges and game habitats. Recently, planning methods have been developed for taking the preferences of the decision maker more accurately into account (e.g. Kangas 1992). In the case of reindeer husbandry and game management, the basic problem concerns the lack of production functions and other evaluation models. However, the first steps have already been taken in Kangas et al. (1993) where wildlife expertise was used to incorporate the habitat preferences of black grouse into forest planning.

References

Aaltonen, V.T. 1919. Über die natürliche Verjüngung der Heidewälder im Finnischen Lappland I. Communicationes ex Instituto Quaestionum Forestalium Finlandiae Editae 1. 412 pp.

Ahlén, I. 1975. Winter habitats of moose and deer in relation to land use in Scandinavia. Viltrevy 9(3): 45–192.

Åhman, B. 1993. Vågar vi äta renskött efter Tjernobylolyckan? (Do we dare to eat reindeer meat after Tjernobyl accident?) Veterinärmedicin 5/1993. 4 pp. (In Swedish.)

Aikio, P. 1977. Saamelaisen ekosysteemin murtuminen Lapin paliskunnassa. Summary: Collapse of the Lapponian ecosystem in Lapin paliskunta. Suomen Luonto 36(2): 72–77.

Andrén, H., Angelstam, P., Lindström, E. & Widen, P. 1985. Differences in predation pressure in relation to habitat fragmentation: an experiment. Oikos 45: 273–277.

Angelstam, P. 1983. Population dynamics of tetraonids, especially the black grouse, in boreal forest. Abstracts of Uppsala Dissertations from the Faculty of Science 675. 33 pp.

Angelstam, P. 1986. Predation on ground nesting birds' nests in relation to predator densities and habitat edge. Oikos 47: 365–373.

Angelstam, P. 1990a. Factors determining the composition and persistency of local woodpecker assemblages in taiga forest in Sweden: a case for landscape ecological studies. In: Carlson, A. & Aulén, G. (eds.). Conservation and management of woodpecker populations. Swedish University of Agricultural Sciences, Department of Wildlife Ecology, Report 17: 147–164.

Angelstam, P. 1990b. Effects of modern forestry on the capercaillie: preserving the display arenas is not enough! In: Carlson, A. (ed.). The capercaillie and forestry. Swedish University of Agricultural Sciences, Department of Wildlife Ecology, Report 15: 51–62, 67–68. (In Swedish with English summary.)

Angelstam, P., Lindström, E. & Widén, P. 1984. Role of predation in short–term population fluctuations of some birds and mammals in Fennoscandia. Oecologia 62: 199–208.

Arnborg, T. 1955. Skador på tallplantor inom ett starkt renbetat område i Hede socken. (Damages in pine seedlings in an intensively used reindeer pasture in the community of Hede.) Norrlands Skogvårdsförbunds Tidskrift 25: 363–387. (In Swedish.)

Bergan, J. 1962. Reinskader på den naturlige gjenvekst av furu i Pasvik. (Damages caused by reindeer in the naturally born seedling stands of pine in Pasvik.) Tidsskrift for Skogbruk 70(3): 175–193. (In Norwegian.)

Bergström, R. & Danell, K. 1987. Moose winter feeding in relation to morphology and chemistry of six tree species. Alces 22: 91–112.

Bergström, R., Lavsund, S., Sandegren, F. & Stålfelt, F. 1983. Furndalsprojektet. (Furndals project.) Umeå. 80 pp. (In Swedish.)

Bjärvall, A., Franzén, R., Nordkvist, M. & Åhman, G. 1990. Renar och rovdjur. (Reindeer and predators.) Naturvårdsverket Förlag, Tryckindustri, Solna. 296 pp. (In Swedish.)

Brown, T. & Mikola, P. 1974. The influence of fruticose soil lichens upon the mycorrhizae and seedling growth of forest trees. Acta Forestalia Fennica 141. 23 pp.

Caughley, G. 1976. Wildlife management and the dynamics of ungulate populations. In: Coaker, T.H. (ed.). Applied biology, vol. 1. Academic Press, London. 183 pp.

Danell, K. 1989. Älgskador vårt största skogsskyddsproblem f.n. I. Älgens skadegörelse i tallbestånd: den biologiska bakgrunden. Summary: Moose browsing on pines. I. The biological background. Kungl. Skogs– och Lantbruksakademiens Tidskrift 128(2): 105–109.

Danell, K., Edenius, L. & Lundberg, P. 1991a. Herbivory and tree stand composition: Moose patch use in winter. Ecology 72: 1350–1357.

Danell, K. & Ericsson, L. 1986. Foraging by moose on two species of birch when these occur in different proportions. Holarctic Ecology 9: 79–84.

Danell, K., Niemelä, P., Varvikko, T. & Vuorisalo, T. 1991b. Moose browsing on Scots pine along a gradient of plant productivity. Ecology 72: 1624–1633.

Eriksson, O. 1975. Silvicultural practices and reindeer grazing in Sweden. In: Luick, J.R., Lent, P.C., Klein, D.R. & White, R.G. (eds.). Proceedings of the 1st International Reindeer/Caribou Symposium, Fairbanks 1972. Biological Papers of the University of Alaska, Special Report 1: 108–121.

Eriksson, O. 1980. Effects of forest fertilization on the cratering intensity of reindeer. In: Reimers, E., Gaare, E. & Skjenneberg, S. (eds.). Proceedings of the 2nd International Reindeer/Caribou Symposium, Røros, Norway, 1979. Part A: 26–40. Direktoratet for vilt og ferskvannfisk, Trondheim.

Eriksson, O. & Raunistola, T. 1990. Impact of soil scarification on reindeer pastures. Rangifer, Special Issue 3: 99–106.

Eriksson, O. & Raunistola, T. 1993. Impact of forest fertilizers on winter pastures of semi–domesticated reindeer. Rangifer (in press).

Eriksson, O., Sandewall, M. & Wilhelmsson, E. 1986. A model for analyzing influence of timber production on lichens for reindeer grazing. Rangifer 7(2): 15–32. (In Swedish with English summary.)

Fritz, R.S. 1979. Consequences of insular population structure: distribution and extinction of spruce grouse populations. Oecologia 42: 57–65.

Haapanen, A. & Siitonen, P. 1978. Kulojen esiintyminen Ulvinsalon luonnonpuistossa. Summary: Forest fires in Ulvinsalo strict nature reserve. Silva Fennica 12(3): 187–200.

Hansson, L.1979. Small mammal abundance in relation to environmental variables in the Swedish forest phases. Studia Forestalia Suecica 147. 40 pp.

Haukioja, E., Huopalahti, R., Kotiaho, R. & Nygren, K. 1983. Millaisia männyntaimia hirvi suosii? Summary: The kinds of pine are preferred by moose. Suomen Riista 30: 22–27.

Haukioja, E. & Salovaara, R. 1978. Summer weight of rendeer (Rangifer tarandus) calves and its importance for their future survival. Reports from Kevo Subarctic Research Station 14: 1–4.

Heikkilä, R. 1991. Moose browsing in a Scots pine plantation mixed with deciduous tree species. Acta Forestalia Fennica 224. 13 pp.

Heikkilä, R. 1993. Ravinnon määrän ja puulajikoostumuksen vaikutus hirven ravinnonkäyttöön ja taimituhoihin mäntytaimikoissa. Summary: The effect of food quantity and tree species composition on moose (Alces alces) browsing in Scots pine plantations. Folia Forestalia 815. 18 pp.

Heikkilä, R. & Löyttyniemi, K. 1992. Growth response of young Scots pines to artificial shoot breaking simulating moose damage. Silva Fennica 26(1): 19–26.

Heikkilä, R. & Mikkonen, T. 1992. Effects of density of young Scots pine (Pinus sylvestris) stand on moose (Alces alces) browsing. Acta Forestalia Fennica 231. 14 pp.

Heikura, K. Lindgren, E., Pulliainen, E., Sulkava, S. & Erkinaro, E. 1983. Grouping behavior of the forest reindeer in Kuhmo in 1978–81. In: Pulliainen, E. (ed.). Proceedings of the 3rd International Reindeer/Caribou Symposium, Saariselkä, Finland, 1982. Acta Zoologica Fennica 175: 25–28.

Helle, P. & Helle, T. 1991. Miten metsärakenteen muutokset selittävät metsäkanalintujen pitkän aikavälin kannanmuutoksia? Summary: How do changes in forest structure explain recent changes in Finnish grouse populations? Suomen Riista 37: 56–66.

Helle, P., Helle, T. & Lindén, H. 1994. Capercaillie (Tetrao urogallus) lekking sites in a fragmented Finnish forest landscape. Scandinavian Journal of Forest Research 9: 386–396.

Helle, P., Jokimäki, J. & Lindén, H. 1990. Metsokukkojen elinympäristövalinta Pohjois–Suomessa: radiometrinen tutkimus. Summary: Habitat selection of the male capercaillie in northern Finland: a study based on radiotelemetry. Suomen Riista 36: 72–81.

Helle, T. & Aspi, J. 1983. Effects of winter grazing by reindeer on vegetation. Oikos 40: 337–343.

Helle, T., Aspi, J. & Kilpelä, S.–S. 1990a. The effects of stand characteristics on reindeer lichens and range use by semi–domesticated reindeer. Rangifer, Special Issue 3: 107–114.

Helle, T., Kilpelä, S.–S. & Aikio, P. 1990b. Lichen ranges, animal densities and production in Finnish reindeer management. In: Rehbinder, C., Eriksson, O. & Skjenneberg, S. (eds.). Proceedings of the 5th International Reindeer/Caribou Symposium. Rangifer, Special Issue 3: 115–121.

Helle, T., & Kojola, I. 1992. Harmaaporonjäkälän kasvunvaihtelu Itä–Fennoskandiassa. Summary: The variation in the growth rate of Cladonia rangiferina in eastern Fennoscandia. In: Kauhanen, H. & Varmola, M. (eds.). Itä–Lapin metsävaurioprojektin väliraportti. (The Lapland Forest Damage Project Interim Report.) Metsäntutkimuslaitoksen tiedonantoja 413: 106–114.

Helle, T. & Kojola, I. 1993. Reproduction and mortality of Finnish semi–domesticated reindeer in relation to density and management strategies. Arctic 46(1): 72–77.

Helle, T. & Moilanen, H. 1993. The effects of reindeer grazing on natural regeneration of Scots pine. Scandinavian Journal of Forest Research 8: 395–407.

Helle, T., Norokorpi, Y. & Saastamoinen, O. 1990c. Reduction of arboreal lichens in two spruce stands in northern Finland between 1976 and 1988. In: Kinnunen, K. & Varmola, M. (eds.). Effects of air pollutants and acidification in combination with climatic factors on forests, soils, and waters in northern Fennoscandia. Report from a workshop held in Rovaniemi, Finland 17–19 October, 1988. Nordic Council of Ministers, Nord 2: 57–65.

Helle, T. & Nöjd, P. 1992. Poron laidunnuksen vaikutus männyn kasvuun ja kuntoon. (The impacts of reindeer grazing on the growth and condition of pine.) Metsäntutkimuslaitoksen tiedonantoja 437: 5–15. (In Finnish.)

Helle, T. & Särkelä, M. 1993. The effect of outdoor recreation on range use by semi–domesticated reindeer. Scandinavian Journal of Forest Research 8: 123–133.

Henttonen, H. 1989. Metsien rakenteen muutoksen vaikutuksesta myyräkantoihin ja sitä kautta pikkupetoihin ja kanalintuihin – hypoteesi. Summary: Does an increase in the rodent and predator density, resulting from modern forestry, contribute to the long–term decline in Finnish tetraonids? Suomen Riista 35: 83–90.

Hjeljord, O. & Fjellbakk, Å. 1982. Taksering av elgskader på skog: en metodevurdering. (The inventory of damages caused by moose in forest: method evaluation.) NF rapport 2/85. 55 pp. (In Norwegian.)

Hjort, I. 1970. Reproductive behavior in Tetraonids. Viltrevy 7(4).

Hjort, I. 1982. Attributes of capercaillie display grounds and the influence of forestry. Proceedings of the Second International Symposium on Grouse. p. 26–33.

Hjort, I. 1990. Relationship between the qualities of the capercaillie display arena and the recruitment area. In: Carlson, A. (ed.). The capercaillie and forestry. Swedish University of Agricultural Sciences, Department of Wildlife Ecology, Report 15: 25–38. (In Swedish with English summary.)

Hoogesteger, M. 1976. Changes in vegetation around the refuge huts in Koilliskaira forest area. Silva Fennica 10(1): 40–53.

Hytönen, M. 1995. History, evolution and significance of multiple–use concept. In: Hytönen, M. (ed.). Multiple–use forestry in the Nordic countries. The Finnish Forest Research Institute, Helsinki. p. 43–66.

Hyvärinen, V. & Sepponen, P. 1988. Kivalon alueen paksusammalkuusikoiden puulaji– ja metsäpalohistoriaa. Summary: Tree species history and local forest fires in the Kivalo area of northern Finland. Folia Forestalia 720. 26 pp.

Ingold, T. 1980. Hunters, pastoralists and ranchers. Cambridge University Press. 326 pp.

Itkonen, T. 1948. Suomen lappalaiset vuoteen 1945. Osa II. (The Lapps in Finland before 1945. Part II.) Porvoo–Helsinki. 629 pp. (In Finnish.)

Järvinen, O., Kuusela, K. & Väisänen, R.A. 1977. Metsien rakenteenmuutoksen vaikutus pesimälinnustoomme viimeisten 30 vuoden aikana. Summary: Effects of modern forestry on the numbers of breeding birds in Finland 1945–1975. Silva Fennica 11: 284–294.

Järvinen, O. & Väisänen, R.A. 1984. Metson, teeren ja pyyn pesimäaikainen kanta Suomessa 1973–77. Summary: Numbers of breeding capercaillie, black grouse and hazel grouse in Finland in 1973–77. Suomen Riista 31: 61–73.

Kangas, J. 1992. Multiple–use planning of forest resources by using the analytic hierarchy process. Scandinavian Journal of Forest Research 7: 259–268.

Kangas, J., Karsikko, J., Laasonen, L. & Pukkala, T. 1993. A method for estimating the suitability function of wildlife habitat for forest planning on the basis of expertise. Silva Fennica 27(4): 259–268.

Kärenlampi, L. 1973. Suomen poronhoitoalueen jäkälämaiden kunto, jäkälämäärät ja tuottoarvot vuonna 1972. (The condition of lichen ranges, the amount of lichen and production values in the reindeer herding area in Finland in 1972.) Poromies 40(2): 15–19. (In Finnish.)

Kautto, A., Kärenlampi, L. & Nieminen, M. 1986. Jäkäläisten talvilaidunten kunnon muutos Suomen poronhoitoalueella. (The change in condition of the lichen–rich winter ranges in the reindeer herding area in Finland.) Poromies 53(3): 28–34. (In Finnish.)

Kershaw, K. 1985. Physiological ecology of lichens. Cambridge University Press, Cambridge. 283 pp.

Kershaw, K.A. & Field, G.–F. 1975. Studies on lichen dominated systems. XV. The temperature and humidity profiles in a Cladonia alpestris mat. Canadian Journal of Botany 53: 2614–2620.

Kojola, I. 1993. Peura ja poroistutusten ekologiaa. Summary: Ecology of reindeer introductions. Suomen Riista 39:74–84.

Kojola, I. & Helle, T. 1991. Productivity of semi–domesticated reindeer in Finland. Rangifer 11: 53–63.

Kollström, R.E.S. 1988. Reindeer husbandry and forestry in Pasvik. Conflicts caused by common land use. Ökoforsk rapport 10: 1–34. (In Norwegian with English summary.)

Lääperi, A. 1990. Hoidettujen talvilaitumien vaikutus hirvituhoihin mäntytaimikoissa. Summary: Effects of winter feeding on moose damage to young pine stands. Acta Forestalia Fennica 212. 46 pp.

Lavsund, S. 1987. Moose relationships to forestry in Finland, Norway and Sweden. Swedish Wildlife Research, Suppl., part I: 229–244.

Layard, P.R. & Walters, A.A. 1978. Microeconomic theory. McGraw–Hill Book Company, New York. 498 pp.

Lenvik, D. 1988. Utvalgsstrategi i reinflokken. (Selection strategy in a reindeer herd.) Reindriftadministrasjonen, Alta. 26 pp. (In Norwegian.)

Lindén, H. 1981. Hunting and tetraonid populations in Finland. Finnish Game Research 39: 69–78.

Lindén, H. 1984. The role of energy and resin contents in the selective feeding of pine needles by the capercaillie. Annales Zoologici Fennici 21: 435–439.

Lindén, H. 1988. Latitudinal gradients in predator–prey interactions, cyclicity and synchronism in voles and small game populations in Finland. Oikos 52: 341–349.

Lindén, H. 1991a. Patterns of grouse shooting in Finland. Ornis Scandinavica 22: 241–244.

Lindén, H. 1991b. Mitä riistatiedustelut paljastavat pienriistasykleistämme? Summary: Analysis of game questionnaires: The validity of the alternative prey hypothesis in Finland. Suomen Riista 37: 67–78.

Lindén, H. & Pasanen, J. 1987. Metsien pirstoutuminen metsokantojen uhkana. Summary: Capercaillie leks are threatened by forest fragmentation. Suomen Riista 34: 66–76.

Lindén, H. & Raijas, M. 1986. Yliverotammeko metsäkanalintukantoja? Summary: Do we overharvest our grouse populations? An educated guess. Suomen Riista 33: 91–96.

Lindén, H. & Rajala, P. 1981. Fluctuations and long–term trends in the relative densities of tetraonid populations in Finland, 1964–77. Finnish Game Research 39: 13–34.

Lindén, H. & Sorvoja, V. 1992. Metsästyspaineen vaikutus kanalintukantoihin Suomessa: valtakunnallinen tarkastelu ja Oulaisten metsästystutkimus. Summary: Harvesting grouse in Finland: A detailed analysis of national statistics and an experimental harvesting study in Oulainen. Suomen Riista 38: 69–78.

Löyttyniemi, K. 1983. Männyn taimen kehitys latvan katkeamisen jälkeen. Summary: Recovery of young Scots pines from stem breakage. Folia Forestalia 560. 11 pp.

Löyttyniemi, K. & Piisilä, N. 1983. Hirvivahingot männyn viljelytaimikoissa Uudenmaan–Hämeen piirimetsälautakunnan alueella. Summary: Moose (Alces alces) damage in young pine plantations in the Forestry Board District Uusimaa–Häme. Folia Forestalia 553. 23 pp.

Lykke, J. 1964. Elg och skog. (Moose and forest.) Meddelelser fra Statens viltundersøkelser 2(17): 1–58. (In Norwegian.)

Marcström, V. 1979. A review of the tetraonid situation in Sweden. In: Lovel, T. (ed.). Woodland Grouse Symposium, Inverness 1978. p. 13–16.

Marcström, V., Brittas, R. & Engren, E. 1988. The impact of predation on boreal tetraonids during vole cycles: an experimental study. Journal of Animal Ecology 57: 859–872.

Marcström, V., Brittas, R., Engren, E. & Winqvist, T. 1983. Field form for description of woodland grouse habitat. Sveriges lantbruksuniversitetet, Uppsala, Sweden.

Mattila, E. 1979. Kangasmaiden luppometsien ominaisuuksia Suomen poronhoitoalueella 1976–1978. Summary: Characteristics of the mineral soil forests with arboreal lichens in the Finnish reindeer management area, 1976–1978. Folia Forestalia 417. 39 pp.

Mattila, E. 1981. Survey of reindeer winter ranges as a part of the Finnish National Forest Inventory in 1976–1978. Communicationes Instituti Forestalis Fenniae 99(6): 1–74.

Mattila, E. 1988. Suomen poronhoitoalueen talvilaitumet. Summary: The winter ranges of the Finnish reindeer management area. Folia Forestalia 713. 53 pp.

Mattsson, L. 1981. Relationen skogsbruk – renskötsel: om framväxten av en markanvändningskonflikt "Land use – north". (Relationship forestry – reindeer husbandry: the development of the land–use conflict "Land use – north".) University of Umeå, Geografiska Institutionen, Rapport 2. 269 pp. (In Swedish.)

Mattsson, L. 1984. Forestry and reindeer husbandry in Sweden: some aspects on the development of a douple land–use. In: Saastamoinen, O., Hultman, S.–G., Koch, N.E. & Mattsson, L. (eds.). Multiple–use forestry in the Scandinavian countries. Communicationes Instituti Forestalis Fenniae 120: 89–96.

Mattsson, L. 1989. The economic value of wildlife for hunting. In: Mattsson, L. & Sødal, D.P. (eds.). Multiple use of forests – economics & policy. Proceedings of a conference held in Oslo, May 1988. Scandinavian Forest Economics 30: 42–61.

Mattsson, L. & Kriström, B. 1987. The economic value of moose as a hunting object. In: Hänninen, R. & Selby, J.A. (eds.). Proceedings of the biennial meeting of the Scandinavian Society of Forest Economics, Porvoo, Finland, May 1987. Scandinavian Forest Economics 29: 27–37.

Moxness, E., Sara, A.N., Solbakken, J.I. & Stenseth, N.C. 1993. Reindrift og beitegrunnlag. En modellanalyse. (Reindeer husbandry and pastures. A model analysis.) Bajos, Utviklingsselskap A/S, Kautokeino. 59 pp. (In Norwegian.)

Müller–Wille, L. 1975. Changes in Lappish reindeer herding in northern Finland caused by mechanization and motorization. In: Luick, J.R., Lent, P.C., Klein, D.R. & White, R.G. (eds.). Proceedings of the 1st International Reindeer/Caribou Symposium, Fairbanks 1972. Biological Papers of the University of Alaska, Special Report 1: 122–126.

Myrberget, S. 1979. The Norwegian moose population 1945–1977. Meddelelser fra Norske viltforskning 3(8): 18–33.

Niemelä, P., Hagman, M. & Lehtilä, K. 1989. Relationship between Pinus sylvestris L. origin and browsing preference by moose in Finland. Scandinavian Journal of Forest Research 4: 239–246.

Nieminen, M. 1993. Kadot ja vasakuolemat. (Disappearances and deaths of young reindeer.) In: Huttu–Hiltunen, V., Nieminen, M., Valmari, A. & Westerling, B. Porotalous. (Reindeer husbandry.) Painatuskeskus Oy, Helsinki. p. 186–195. (In Finnish.)

Nieminen, M. & Heiskari, V. 1987. Diets for freely grazing and captive reindeer during summer and winter. Rangifer 9(1): 17–34.

Nieminen, M. & Leppäluoto, J. 1993. Petojen aiheuttamat porovahingot Suomessa vuosina 1976–92. (The losses of reindeer caused by predators in Finland, 1976–92.) Poromies 60(2): 38–45. (In Finnish.)

Nygrén, T. & Pesonen, M. 1989. Hirvisaaliit ja hirvenlihan tuotanto Suomessa vuosina 1964–87. Summary: Moose harvest and production of moose meat in Finland 1964–87. Suomen Riista 35: 128–153.

Nygrén, T. & Pesonen, M. 1993. The moose population (Alces alces L.) and methods of moose management in Finland, 1975–1989. Finnish Game Research 48: 46–53.

Oksanen, L. 1978. Lichen grounds of Finnmarksvidda, northern Norway, in relation to summer and winter grazing by reindeer. Reports from Kevo Subarctic Research Station 14: 64–72.

Oksanen, L. 1992. Renproblemet inom den svenska fjällvärlden i ekologiskt och historiskt perspektiv. (Reindeer problem in the Swedish mountains from the ecological and economic viewpoints.) WWFs Renbeteskonferens, Umeå. Världsnaturfonden. p. 3–13. (In Swedish.)

Olsen, P. & de Vibe, D. 1981. Beiting og bestandsutvikling hos elg i Norge. (Grazing and stock development of moose in Norway.) Zoologisk institutt, Oslo. 315 pp. (In Norwegian.)

Palo, T. 1981. Översikt över hjortdjurens näringsekologi, speciellt med avseende på älg. (Overview of the feeding ecology of deer, with special emphasis on moose.) Sveriges lantbruksuniversitet, Institutionen för viltekologi, Rapport 7. 71 pp. (In Swedish.)

Palo, T. 1987. Phenols as defensive compounds in birch (Betula spp.). Implications for digestion and metabolism in browsing mammals. Dissertation. Department of Animal Physiology, Swedish University of Agricultural Sciences, Uppsala.

Pelto, J. 1973. The snowmobile revolution: technology and social change in the Arctic. Cummings Publishing Company, California. 225 pp.

Pohtila, E. 1977. Reforestation of ploughed sites in Finnish Lapland. Communicationes Instituti Forestalis Fenniae 91(4): 1–98.

Pruitt, W.O., Jr. 1960. Behaviour of the barren–ground caribou. Biological Papers of the University of Alaska 3. 44 pp.

Pulliainen, E. 1979. Autumn and winter nutrition of the capercaillie (Tetrao urogallus) in the northern Finnish taiga. In: Lovel, T. (ed.). Woodland Grouse Symposium, Inverness 1978. p. 92–96.

Pynnönen, A. 1954. Beiträge zur Kenntnis der Lebenweise des Haselhuhns, Tetrastes bonasia (L.). Papers of Game Research 12. 90 pp.

Raitio, H. 1990. Decline of young Scots pines in a dry heat forest. Acta Universitatis Ouluensis, Series A, Scientiae Rerum Naturalium 216. 40 pp.

Rehbinder, C., Edquist, L., Lundström, K. & Villafone, F. 1982. A field study of management stress in reindeer (Rangifer tarandus L.). Rangifer 2(2): 2–21.

Rehbinder, C., Eriksson, O. & Skjenneberg, S. (eds.). 1990. Proceedings of the 5th International Reindeer/Caribou Symposium. Rangifer, Special Issue 3.

Reimers, E. 1986. Management of wild reindeer in Norway. In: Gunn, A., Miller, F. L. & Skjenneberg, S. (eds.). Proceedings of the 4th International Reindeer/ Caribou Symposium, Whitehorse, Canada, 1985. Rangifer, Special Issue 1: 241–246.

Rissanen, K. & Rahola, T. 1990. Radiocesium in lichens and reindeer after the Chernobil accident. Rangifer, Special Issue 3: 55–61.

Roll–Hansen, F. 1975. Phacidium infestans on dwarf plants of Pinus sylvestris. European Journal of Forest Pathology 5: 1–7.

Rolstad, J. 1989a. Habitat and range use of capercaillie, Tetrao urogallus L., in south-central Scandinavian boreal forests with special reference to the influence of modern forestry. Dr. Agric. thesis. Department of Nature Conservation, Agricultural University of Norway, Ås. 12 pp.

Rolstad, J. 1989b. Effects of logging on capercaillie (Tetrao urogallus) leks. I. Scandinavian Journal of Forest Research 4: 99–109.

Rolstad, J. & Wegge, P. 1989a. Effects of logging on capercaillie (Tetrao urogallus) leks. II. Scandinavian Journal of Forest Research 4: 111–127.

Rolstad, J. & Wegge, P. 1989b. Effects of logging on capercaillie (Tetrao urogallus) leks. III. Scandinavian Journal of Forest Research 4: 129–135.

Rolstad, J. & Wegge, P. 1987. Distribution and size of capercaillie leks in relation to old forest fragmentation. Oecologia 72: 389–394.

Rolstad, J. & Wegge, P. 1989c. Capercaillie (Tetrao urogallus) and modern forestry: case for landscape ecological studies. Finnish Game Research 46: 43–52.

Ruong, I. 1982. Samerna. (The Sámi people.) Bonniers Grafiska Industrier AB, Stockholm. 280 pp. (In Swedish.)

Saastamoinen, O. 1982. Economics of multiple–use forestry in the Saariselkä forest and fell area. Communicationes Instituti Forestalis Fenniae 104. 102 pp.

Saastamoinen, O. 1980. Cutting areas as reindeer pasturage. Communicationes Instituti Forestalis Fenniae 95(4): 1–28.

Seiskari, P. 1962. On the winter ecology of the capercaillie, Tetrao urogallus, and the black grouse, Lyrurus tetrix, in Finland. Papers of Game Research 22. 119 pp.

Semenov–Tian–Shanskii, O.I. 1975. The status of wild reindeer in the USSR, especially in the Kola peninsula. In: Luick, J.R., Lent, P.C., Klein, D.R. & White, R.G. (eds.). Proceedings of the 1st International Reindeer/Caribou Symposium, Fairbanks, Alaska, 1972. Biological Papers of the University of Alaska, Special Report 1: 155–161.

Siivonen, I. 1975. New results on the history and taxonomy of the mountain, forest and domestic reindeer in northern Europe. In: Luick, J.R., Lent, P.C., Klein, D.R. & White, R.G. (eds.). Proceedings of the 1st International Reindeer/Caribou Symposium, Fairbanks 1972. Biological Papers of the University of Alaska, Special Report 1: 33–40.

Skjenneberg, S. 1989. Reindeer husbandry in Fennoscandia. In: Hudson, R., Drew, K. & Baskin, L. (eds.). Wildlife production systems. Cambridge, Cambridge University Press. p. 207–222.

Skogland, T. 1986. Density dependent food limitation and maximal production in wild reindeer herds. Journal of Wildlife Management 50: 314– 319.

Sparrevik, E. 1983. Trädlevande tagellav som renbete: kvantitativa undersökningar av några arter tillhörande släktena Alectoria and Bryoria. (Arborial lichens as reindeer feed: quantitative studies on a few species belonging to genera Alectoria and Bryoria.) Medd. Växtbiol. Inst. Uppsala 3. (In Swedish.)

Staaland, H. & Nieminen, M. 1993. World reindeer herding: origin, history, distribution, economy. In: Proceedings of World Conference on Animal Production. Edmonton, Canada. p. 161–203.

Storaas, T. & Wegge, P. 1985. High nest losses in capercaillie and black grouse in Norway. Proceedings of the 3rd International Grouse Symposium, York 1984. p. 481–498.

Storaas, T. & Wegge, P. 1987. Nesting habitats and nest predation in sympatric populations of capercaillie and black grouse. Journal of Wildlife Management 51(1): 167–172.

Sulkava, S. & Helle, T. 1975. Range ecology of domesticated reindeer in the Finnish coniferous forest area. In: Luick, J., Lent, P., Klein, D. & White, R.G. (eds.). Proceedings of the 1st International Reindeer/Caribou Symposium, Fairbanks 1972. Biological Papers of the University of Alaska, Special Report 1: 308–315.

Swedish game – Biology and Management. 1992. Svenska Jägareförbundet, Uppsala. 159 pp.

Swenson, J.E. 1993. The importance of alder to hazel grouse in Fennoscandian boreal forest: evidence from four levels of scale. Ecography 16(1): 36–46.

Tanner, V. 1929. Antropogeografiska studier inom Petsamo–området. I. Skoltarna. (Anthropogeographic studies in the Petsamo region. I. Skolts.) Fennia 49. (In Swedish.)

Tarhanen, S., Oksanen, J. & Holopainen, T. 1992. Kaarnajäkälien kalvonvuototestit ja hienorakenne Kuolan päästöjen vaikutusten kuvaajina. Summary: The membrane permeability test and cell ultrastructure of epiphytic lichens as indicators of the effects of the Kola emissions. In: Kauhanen, H. & Varmola, M. (eds.). Itä–Lapin metsävaurioprojektin väliraportti. (The Lapland Forest Damage Project Interim Report.) Metsäntutkimuslaitoksen tiedonantoja 413: 115–127.

Thorisson, S. 1980. Status of Rangifer in Iceland. In: Reimers, E., Gaare, E. & Skjenneberg, S. (eds.). Proceedings of the 2nd International Reindeer/Caribou Symposium, Røros, Norway, 1979. Part B: 766–770. Direktoratet for vilt og ferskvannfisk, Trondheim.

Tömmervik, H.A., Johansen, B.E. & Pedersen, J.P. 1991. Mapping the air pollution impact on the natural environment in the border area of Norway/USSR using satellite remote sensing method. In: Putkonen, J. (ed.). 11th annual international geoscience & remote sensing symposium 1: 45–48.

Utsi, P.M. 1992. Renar och betning. (Reindeer and grazing.) WWFs Renbeteskonferens, Umeå. Världsnaturfonden. p. 14–20. (In Swedish.)

Uusvaara, O. 1963. Pyyn elinympäristön metsikkörakenteesta. Summary: The structure of the habitats of hazel grouse, Tetrastes bonasia. Suomen Riista 16: 31–45.

Valkeajärvi, P. & Ijäs, L. 1986. Metson soidinpaikkavaatimukset Keski–Suomessa. Summary: On the display ground requirements of capercaillie in central Finland. Suomen Riista 33: 5–18.

Valkeajärvi, P. & Ijäs, L. 1991. Soidinkeskusten hakkuun vaikutuksista metson soitimeen. Summary: The impact of clearcutting on a capercaillie lek centre. Suomen Riista 37: 44–55.

Varmola, M. (ed.). 1988. Lapin metsä 2000 –ohjelma. (Forest 2000 –program for Lapland.) Lapin lääninhallitus, Rovaniemi. 146 pp. (In Finnish.)

Virkkula, E. 1928. Valkoinen erämaa. (White wilderness.) WSOY, Porvoo/Helsinki. 176 pp. (In Finnish.)

Viro, P. 1970. Prescribed burning in forestry. Communicationes Instituti Forestalis Fenniae 67(7): 1–49.

Vørren, Ö. 1973. Some trends in transition from hunting to nomadic economy. In: Berg, G. (ed.). Circumpolar problems. Wenner–Gren Center International Symposium Series, vol. 21: 185–194. Stockholm.

Wegge, P. & Larsen, B. 1987. Spacing of adult and subadult male common capercaillie during the breeding season. The Auk 104: 481–490.

Wegge, P. & Rolstad, J. 1986. Size and spacing of capercaillie leks in relation to social behaviour and habitat. Behavioral Ecology and Sociobiology 14: 401–408.

Westman, H. 1958. Älgens skadegörelse på ungskogen. (The damages caused by moose in young forest stands.) Kungl. skogshögskolans skrifter 28: 1–148. (In Swedish.)

Winquist, T. 1983. 100 tjäderspelplatser. Summary: 100 capercaillie courtship display grounds. Sveriges Skogsvårdsförbunds Tidskrift 81(2): 5–25.

Winqvist, T. & Ringaby, E. 1989. Skogshönsen och skogsbestånd av olika ålder. Summary: How do grouse utilize forests of different age. Sveriges Skogsvårdsförbunds Tidskrift. 2: 39–48.

Wretlind, J.E. 1931. Bidrag till belysande av föryngringsbetingelserna på övre Norrlands tallhedsmarker. (Contribution to clarify the prerequisites of regeneration in the pine forests of Norrland.) Norrlands Skogsförbunds Tidskrift 18: 263–314. (In Swedish.)

Yearbook of forest statistics 1990–91 (Metsätilastollinen vuosikirja 1990–91). 1992. Folia Forestalia 790. 281 pp.

Zackrisson, O. 1977. Influence of forest fires in the North West Swedish boreal forest. Oikos 29: 5–10.

8 Fauna and flora management in forestry

Kjell Sjöberg[1]
Tommy Lennartsson[2]

Abstract

Many species are threatened by forestry activities in the Nordic countries. This has led to the development of silvicultural methods to meet the demands for maintaining biological diversity. There are differences in biodiversity of forests between the Nordic countries depending on the history of land use; the amount of plantations of exotic species is much bigger in Iceland and Denmark than in Norway, Sweden and Finland. The goals and methods of protecting biodiversity are different in areas where forests are still intact, in managed forests, in forests influenced by other activities than forestry and in plantations established on agricultural land. Forests are also affected by disturbances. Large–scale dicturbances caused, for example, by fire have created forests with different succession stages. Small–scale disturbances create forests with long continuity and gap dynamics. The main strategies for preserving biodiversity are the protection of forests in nature reserves, development of alternative silvicultural methods and environmental protection measures in daily forestry activities. Key and indicator species and key habitats are used in defining the protection value of a forest. The protection of biodiversity in forestry requires education, strategic planning and cooperation between organizations and people locally and internationally.

Keywords: biodiversity, silviculture, nature protection, fire, old–growth forests, key species, landscape ecology.

---

[1] Swedish University of Agricultural Sciences
Department of Animal Ecology
S–901 83 Umeå, Sweden

[2] Swedish University of Agricultural Sciences
Department of Ecology and Environmental Research
S–750 07 Uppsala, Sweden

8.1 Introduction

Fauna and flora management – biological diversity in focus

One of the main tasks of nature conservation is to maintain biological diversity (World Conservation Strategy 1980, Our Common Future... 1987, Wilcove 1989, Almered–Olsson et al. 1991). On a global scale there is an immense need of conservation work directed towards biodiversity, which includes finding methods and systems for evaluating entire ecosystems, designing reserves, etc. According to Ehrlich and Ehrlich (1992), conservation biologists have about a decade to develop and deploy conservation systems if they are to play a significant role in preventing the loss of more than half of present terrestrial biodiversity.

Hopefully, the situation is not that bad in the Nordic countries. One reason is that the ecosystems are fairly young. They have developed after the last glaciation period, and consequently species diversity is relatively low. Another reason for optimism is the fact that during the last few years there has been a breakthrough in understanding the biodiversity problems in forest ecosystems among the foresters responsible for the development of the economically and ecologically important forest resourses. That understanding was largely absent during the previous decades, with the rapid development of production oriented forestry. This new trend is backed up by political decisions, and there is an obvious development towards cooperation between biologists and foresters in solving the biodiversity problems in the Fennoscandian forests. For example, in Sweden the government's policy is that "biological diversity and genetic variation should be secured" and environmental aspects should be as important as production goals in forestry activities (En ny skogspolitik... 1992/93, SFS 1993:553). Finland, Norway and Denmark have also outlined policy statements which are directed towards multiple–use forestry, more or less stressing the importance of maintaining biodiversity (Rassi & Väisänen 1987, Flersidig skogbruk... 1989, Koch & Kristiansen 1991).

However, there are many threatened species even in the Fennoscandian forests. Forestry activities influence the forest habitats in many ways. Most of the land area in the region is covered by forests – in Finland 70 % (Wahlström et al. 1992), in Sweden 65 % (Sohlberg 1990), in Norway 21 % (Korsmo 1991) and in Denmark 10 % (Report on the monitoring... 1992). As most of the forests are used for timber production (about 95 % in Sweden), it is no wonder that the majority of all threatened species live in forest habitats. In Finland 43 % (Rassi et al. 1992), in Sweden 45–55 % (Ahlén et al. 1979, Ehnström et al. 1993), and in Norway 50 % (Hågvar 1993) of the endangered species are threatened by forestry activities and/or live in forest habitats (see also Rassi & Väisänen 1987, Wahlström et al. 1992, Bernes 1993).

During the last two decades, environmental aspects have become more to the fore. Forestry practices have been, and still are, subject to both confrontation between foresters and environmentalists, buyers' boycotts and demands, and to a dialog and joint research projects between biologists and foresters. The result has been increased interest in improving the prevailing methods (e.g. clearcutting) and development of new forestry methods to meet the present–day demands of maintaining biological diversity in the forests (*Figure 8.1*).

The reasons for maintaining biological diversity will not be discussed here, but the interested readers are referred to, for example, McNeely (1988), Reid and Miller (1989), Bunnell (1991), Randall (1991), Botkin and Talbot (1992), Ehrlich and Ehrlich (1992), Perrings et al. (1992) and Stebbins (1992) who discuss and evaluate aesthetic, ethical, economic and other aspects of the issue.

Biodiversity is one quality aspect of a forest, and in a way a measurement of authenticity. Of course, the quality of forest could also be treated in a wider sense. As discussed by Dudley (1992), criteria for a definition of forest quality could also cover measurement of how closely a forest stand mirrors the natural forests of the area. The health of the forest, i.e. its health with respect to disease or pollution damage and the health of forest flora and fauna are

Figure 8.1 Large–scale clearcuttings have contributed to the recent negative reaction by the public towards commercial forestry. Now foresters are increasingly aware of their responsibility for maintaining biological diversity. Consequently, nowadays the clearcutting method is developed to include also considerations for fauna and flora, and in forest stands with long continuity, alternative methods are used. Photo: K. Sjöberg.

also possible indicators. Environmental benefits, which include benefits that extend beyond the boundaries of the forest, for example, soil and watershed protection and local climatic benefits are another area of quality aspects. Values to humans, such as recreation and aesthetic and cultural dimensions, are also possible considerations (Koch & Kennedy 1991, Dudley 1992).

Multiple–use aspects of the Nordic forests

The forests in the Nordic countries have for centuries been used for different purposes (Saastamoinen et al. 1984, Esseen et al. 1992). For example, there have been regulations concerning game species since at least the 15th century in Sweden, Finland and Denmark (Rassi & Väisänen 1987). The Fennoscandian forests were also used, among other things, for tar, fuelwood and charcoal production, and, in the agricultural landscape, the forests were grazed by cattle at considerable distances around the settlements (Esseen et al. 1992). However, during the last century, the timber and pulp dependent forest industry has dominated the use of the forests, especially in Finland and Sweden. Consequently, in these countries, where still more than half of the land area is covered with forests, most forests are managed intensively. Nowadays there exist only a few percent of the original pristine forests in the Nordic countries.

Based on the Right of Public Access, people in the Nordic countries traditionally have been able to use forests for recreation, irrespective of ownership of the land. For example, in Sweden and Finland, everyone has access even to private forests to pick berries and mushrooms for private use. Also hunting in the forests has traditionally been an activity deeply rooted among the people living in these countries, although hunting rights are tied to land ownership. In Sweden, for example, there are more than 300,000 hunters, and the value of hunting (recreation value and the value of meat together) is estimated to be around 1.9 billion SEK per year (Mattsson 1992). The corresponding value in Norway is about 364 million NOK (Navrud 1993). In Finland, the economic value of forests for berry and mushroom picking, hunting and reindeer husbandry is about 10 % of its value for timber production (Kuusipalo 1992).

The right of access to the forests may be one reason for the public's increasing concern for the management of the forests during the last decades, although most people today live in urban areas. For example, partly as a result of public reaction, the large–scale use of herbicides (for reducing deciduous bushes) and DDT (for control of insect damage) in the forests was declared illegal in Sweden in 1977.

As mentioned earlier, at the end of the 1980s and continuing into the 90s, increasing concern over conservation biology is evident both in the governments and in the forest industry in all Nordic countries. Much of the policy is now focused on maintaining biological diversity, and includes an increase in the area of nature reserves, more frequent use of alternative forestry methods, and increased allowance for nature conservation considerations in daily forestry work (Eriksson & Hedlund 1993). Also such non–forestry products as recreation and aesthetic qualities are now beginning to be evaluated in economic terms. There exist estimates of how much the public is willing to pay for keeping certain quality aspects in forests (Mattsson & Li 1994). These values are an indication of how much society is ready to invest in methods or areas necessary to keep such values, for example by buying forests for nature reserves. Studies have also been made of how different forest management practices affect the non–timber values of forests (Mattsson & Li 1994, Johansson et al. 1994) and of the costs of environmental adjustments (Carlén 1994, Holgén & Lind 1994).

The Scandinavian forestry – sustainable use of a resource?

Scandinavian forestry is often regarded as one of the few examples of sustainable utilization of natural resources. However, this statement normally refers to the efficient use of forests for timber production. For example, as early as 1903 a reforestation law was passed in Sweden which led to planting of large areas of deforested areas and to restoration of low–productive over–utilized forests (Mattsson 1992).

For a better understanding of sustainable utilization in line with the current interpretation of the word, it is important to first of all define what a sustainable natural resource is. In the present context it can refer to both the yield and the productivity of forests, as well as to biodiversity in forest habitats. In terms of the yield of forest products such as timber or paper, the abovementioned statement is probably true, because, since 1973, the growth of the standing crop in Swedish forests has been larger than the harvest (Andersson et al. 1992, see also Linder & Östlund (1992) for a longer time perspective). The same is the case in Finland, where at least since the beginning of independence (1917), the standing crop has never been as large as at present, and cutting is currently about 70 % of annual growth (Wahlström et al. 1992, Kuusipalo 1992). This has been achieved by, among other things, planting fast–growing stands which are often developed as monocultures by management methods such as cleaning and thinning. However, when sustainable use is defined in accordance with the definition given in the convention on biological diversity presented in the UNCED in Rio de Janeiro 1992 (i.e. "... the use of components of biological diversity in a way and at a rate that does not lead to the long–term decline of biological diversity, thereby maintaining its potential to meet the needs and aspirations of present and future generations"), the present forestry is not in line with the definition.

In terms of forest ecosystem diversity, or plant and animal species diversity, the utilization of forest resources has not been sustainable so far. On the contrary, when a major criterion of ecological sustainability is that the original mixture of species remains viable (Haila 1994), the red data lists indicate no such sustainability.

The question of whether production really is sustainable is also at present often a subject of discussion, not least in connection with the overall problems caused by pollution, acidification and increased nitrogen concentration. In this situation, the importance of fauna and flora in keeping forest ecosystems in such a condition that they produce a sustainable yield of forest products is increasingly highlighted. Soil organisms are the main focus of interest in this context (Mladenoff & Pastor 1993).

Forest–living species form a dominant group of threatened species noted in the red data books documenting endangered species, and forestry activities are blamed for contributing the most to this situation (Rassi & Väisänen, Ehnström et al. 1993). Certainly the red data lists will still be modified when we know more about the actual reasons for species' disappearance. In a comparison between red data lists from 1977 and the end of the 80s, and from the beginning of the 90s, positive changes (i.e. species is removed from the list or considered less threatened) are more often motivated by improved knowledge, while negative changes (e.g. addition of species to the list), are more often motivated by a real, increased threat to the species (Eriksson & Hedlund 1993).

We certainly know that the majority of red–listed forest species have specific ecological demands and suffer from present forestry activities. It has not yet been evaluated to what extent present awareness of biodiversity problems in forestry and the consequent improvements in methods and strategies will have the desired positive effects on biological diversity. There is a crying need for more research, and at the moment alternative forestry methods are being employed before we have knowledge of how these methods will affect flora and fauna.

The differences in biodiversity aspects of forests and forestry in each Nordic country can be explained by their land–use history. Consequently, it is difficult to make a comparison between countries with regard to the biodiversity of managed forests and natural forests. In Iceland, for example, nowadays only 1 % of the land area is covered by woodland, while 25–30 % of the country is estimated to have been forested at the time of settlement. During the latter half of the 1980s, a considerable change in agricultural practices took place (e.g. a decrease from 950,000 to 530,000 winter–foddered sheep). This makes the present reforestation program possible. About 3 % of the land area in Iceland is considered climatically suitable for wood production (Blöndal 1991). New forests are being created and old ones restored. In 1990, the Iceland Forestry Association initiated a pilot project for reforesting eroded land. Birch (*Betula pubescens*) is the dominant species in the reforestation programme, but also several introduced exotics are involved, for example Russian larch (*Larix sukaczewii*), Lodgepole pine (*Pinus contorta*), mountain pine (*P. mugo*), Sitka spruce (*Picea sitchaensis*), black cottonwood (*Populus trichocarpus*) and several *Salix* species. Thus, the biodiversity of forest–living organisms will gradually increase, but in quite a new context.

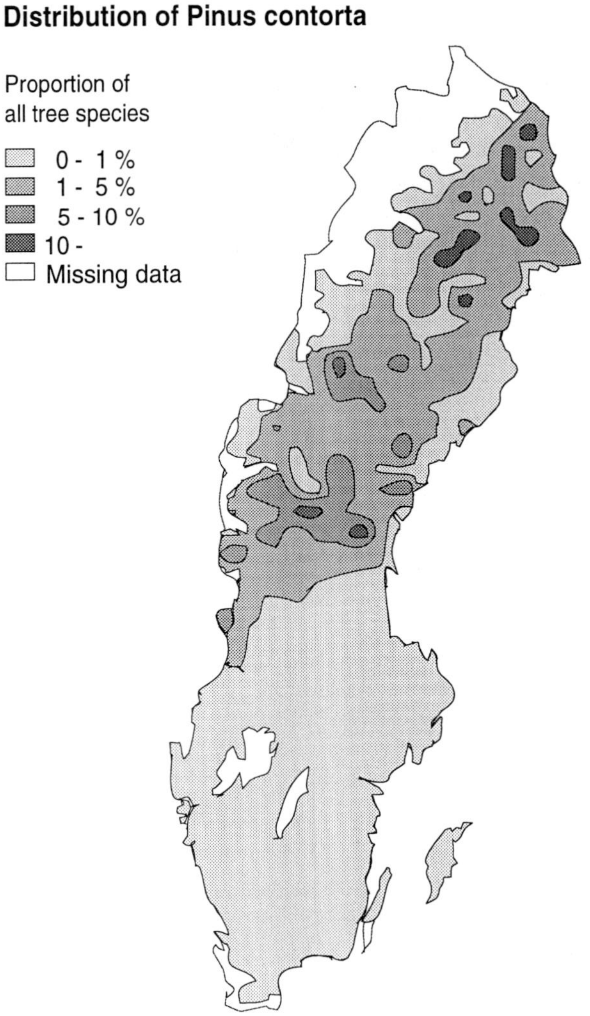

Figure 8.2 Map showing the present distribution of the introduced North American Lodgepole pine (*Pinus contorta*) in Sweden. Source: Department of Forest Survey, Swedish University of Agricultural Sciences.

The situation in Iceland is much different from the situation in, for example, Finland and Sweden, where the problem is rather how to maintain biodiversity in the remaining pristine forests, and how to maintain biodiversity in the dominating forest types, i.e. forests with different degrees of management impact, ranging from naturally renewed forests to planted forests at clearcut areas.

Even in Sweden, where most forests are already managed, an exotic coniferous tree species, the North American Lodgepole pine (*Pinus contorta*), is used in efforts to increase the forest production; about 540,000 hectares have been planted with this species (Contortatallen i Sverige... 1992). It should also be mentioned that a large proportion of the planted spruce seedlings originate from far away (sometimes even from central Europe; the central European spruce populations are consistently depauperate of genetic variability compared to Swedish populations of Norway spruce (Lagercrantz & Ryman 1990)). It has not yet been fully evaluated if, and how, the Lodgepole pine and the different provenances will influence the biodiversity. The answer might depend on the scale used: the stand, the forest or the landscape level (*Figure 8.2*)

In Norway, large areas of exotic conifers have been planted to improve forest production in the mountain areas in the northern part of the country, as well as to create new forests in old sheep–grazed areas in the western part. In Denmark, finally, the situation is partly different. There too, as in parts of Norway and Iceland, the intention is to increase the acreage of forests. In addition, plantations which bind shifting sand are important. A large proportion of the forests consists of plantations of spruce, which is not indigenous to Denmark. A small part of the forest area (less than 10 %) is planted with spruce and exotic conifers to be used as Christmas trees and decorations. Again, in such forests the biodiversity aspects are quite different from areas which have continuously been covered with indigenous tree species.

8.2 Basic concepts

What is biodiversity?

McNeely et al. (1990) defines the term biodiversity in the following way: "Biological diversity encompasses all species of plants, animals, and micro–organisms and the ecosystems and ecological processes of which they are parts. It is an umbrella term for the degree of nature's variety, including both the number and frequency of ecosystems, species or genes in a given assemblage. It is usually considered at three different levels: genetic diversity, species diversity, and ecosystem diversity". In the convention on biological diversity accepted at the UNCED conference at Rio de Janeiro in 1992, the following definition was used: "Biological diversity (or biodiversity) means the variability among living organisms from all sources including, inter alia, terrestrial, marine and other aquatic ecosystems and the ecological complexes of which they are part; this includes diversity within species, between species and of ecosystems". There are also other definitions of biodiversity (see e.g. McNeely 1988). Most of them include all three levels, i.e. genetic, species and ecosystem diversity.

Noss (1990), based on Franklin et al. (1981), recognizes three primary attributes of biodiversity, namely composition, structure and function. These can be arranged in a hierarchy that incorporates elements of each attribute at four levels of organization. These are regional landscape, community/ecosystem, population/species and genetic levels. Noss stresses the process as an important component of the diversity of interactions between species, natural disturbances and nutrient cycles (cf. McNeely 1994). For a more detailed survey of the biodiversity concept, see Haila and Kouki (1994).

The diversity concept can be used in different ways. For example, the quality aspect may be emphasized. Consequently, when discussing biodiversity, it ought to be defined precisely. One may argue that a clearcut area increases the biodiversity in a homogenous boreal coniferous landscape. However, although it may be true for a certain area, increased biodiversity is not, according to the conservationists, automatically an improvement of the biological quality. They rather compare the biodiversity in forests regenerated on clearcut areas with the original forest ecosystems. The number of more demanding or specialized, usually rare or threatened species, will decrease after clearcutting and thus lead to the trivialization of the area.

What are the goals in forestry policy and biodiversity?

There seems to be an increasing political consensus in the Nordic countries that biological diversity and genetic variation should be secured. To be able to do that, the general goal must be to maintain the fauna and flora in natural and viable populations (e.g. ensuring 95 % probability of survival for a population for at least 100 years into the future). This is also stated in the new Swedish Forestry Act (Svensk Författningssamling 1993:553), where the environmental goal is given the same weight as production goals. It is concluded that the forest is a national resource which should be treated in such a way that it gives a sustainable yield, and, at the same time, biodiversity should be maintained.

Since it is hardly realistic to maintain the biodiversity or the species composition at every single site or stand in areas with forestry activities, it is necessary to determine the scale on which these qualities actually should be preserved. The landscape or the regional perspective in forestry planning has clearly been a subject of attention recently (see for example Angelstam et al. 1990, Hansson & Angelstam 1991, Hansson 1992, Franklin 1993). Consequently, a forest stand may lose some of its diversity as a result of, for example, clearcutting, but, at the landscape or regional level, the biodiversity and the original fauna and flora composition should be kept intact in viable populations.

One way of looking at the problem of maintaining biodiversity in the landscape perspective is to divide the forest landscape into compartments according to current forestry and conservation practices, and to treat those parts in different ways. Ingelög and Lennartsson (1991) have discussed three area types: 1) managed forest stands, in which certain considerations to flora and fauna are shown, but within the frames of profitable forestry, 2) small, often low–productive or difficult to manage areas and corridors between the managed areas, which are preserved as intact as possible, and 3) large areas, usually in the form of nature reserves, completely without normal forestry. These three ways of maintaining biodiversity are partly replaceable, as long as the total potential for biodiversity over a landscape scale is large enough. When dealing with various types and levels of consideration to flora and fauna, it is important to take into account the history of each stand, area or region (e.g. McNeely 1994). Most forests have to some degree already been influenced by human impact.

If we want to maintain the forest flora and fauna, an effort must be made to refer to the original flora and fauna of the region. If the area is in an unexploited condition, the goal is to maintain the diversity. In most cases, as a result of earlier treatments, the goal is to restore degraded stands. Maximizing diversity seldom seems to be a desirable goal, as it could include, for example, introduction of exotic species to replace the lost old–growth species (Hunter 1990).

What is original fauna and flora? From a conservation point of view it seems practical to use the species composition and processes in natural forests within the region as a reference. Of course, species in natural successional stages which lead to mature climax forests must be included in the management process as well as ecological processes such as natural disturbances caused by fire and wind. However, in some cases it may be impossible to estimate the species composition, since the boreal forests may be inherently characterized by non–equilibrium disturbance regimes (Sprugel 1991). In other cases the species and processes typical of natural forests in a region simply must be estimated, since reference areas are lacking. This applies especially to regions in southern Fennoscandia, cultivated in prehistoric times, or to regions along the Bothnian land uplift coasts, which may have been cultivated as soon as they appeared. Some ecosystems exist only as a result of human activities (e.g. wooded pastures). There are no natural reference areas to be found for them.

A. Biodiversity goals in areas where forests are still intact

In natural forests not influenced by forestry (for example, in national parks and nature reserves), the goal must be to maintain the fauna and flora typical for the region intact in natural populations. Even here, however, the goal is not easy to attain. The original forest, even if not influenced by forestry, is influenced by air pollution, acid rain, reindeer management, hunting, etc. The diversity of ecological processes in the ecosystems may be lacking or reduced; for example, the forest fires have not been allowed to develop naturally in national parks and reserves. Isolation of reserves and efficient fire protection also decrease the possibilities of trees (and tree species composition) to develop naturally (Esseen et al. 1992).

Furthermore, the small size of reserves usually tightly limits natural processes and the reserve's value for maintaining biodiversity in general. Consequently, the goal must not only be to maintain a diversity of species, but also to manage for specific ecosystem and process diversity. The goal must also be to maintain biodiversity by securing area configuration and spatial distribu-

Figures 8.3 and *8.4* During several decades, more or less natural forests with multi–layer structures and natural forest qualities have been, via radical methods such as soil preparation, converted into plantations (sometimes with exotic species like *Pinus contorta* with a single–layer structure and without the qualities of natural forest, such as dead wood). Photos: K. Sjöberg.

tion both on a regional and a national scale. Even temporal distribution must be considered, especially concerning successional biotopes.

B. Biodiversity goals in managed forests

Managed forests in the Nordic countries vary in degree of management influence from almost natural forests (naturally regenerated and slightly influenced by forestry activities) to man–made plantations of exotics after clear-felling – or even plantations of long since deforested land, for example, abandoned fields (*Figures 8.3, 8.4*). Since the managed forests cover a far larger area than the pristine ones, species and biodiversity cannot be maintained without involving managed areas. The quality of the managed forests (which normally create the matrix in the present forest landscape) is of great importance, because it influences the effectiveness of reserves and controls landscape connections (Franklin 1993). The goal could be to create a diverse landscape of stands which are managed with ecologically adapted methods and surrounded by corridors designed for various purposes (Ingelög & Lennartsson 1991).

An important part of the maintaining or restoration of biodiversity is to allow dispersal of organisms from remaining small fragments still retaining specialized species and pristine forest qualities (key biotopes), for example, by giving such areas key positions in the network of unmanaged habitats. Rare species may require the restoration of ecosystems to be able to survive in viable populations, or they need special species–adapted activities, such as addition of logs and other types of wood from outside the managed area.

C. Biodiversity goals in forests influenced by other activities than forestry

Substantial efforts are also being made to preserve many man–influenced, or even man–made habitats (such as pastures, grazed areas, etc.) in the conservation programs in the Nordic countries. Many of these habitats are the result of agricultural, cattle–based activities. Now, when agricultural influence is decreasing, coniferous plantations are spreading over older meadows and pastures, thus replacing deciduous trees. Even without plantations, the light open habitats disappear due to bush encroachment, once management stops.

In the Forestry Act of Sweden, such areas are considered as forest land, and in the red data lists, the species living there are often regarded as influenced by forestry activities. At least in Sweden, the habitats involved (e.g. meadows with scattered trees) are nowadays treated according to special management programs. The landowner can get financial support for maintaining such threatened habitats in their original status. The program covers about 50,000 hectares of mainly meadows and pastures. Because of the program, the rate of decrease of such originally grazed or mowed habitats may be slowing down. A large group of red–listed species are confined to such habitats (Lennartsson 1991, Ehnström et al. 1993). The maintaining of these habitats is a difficult problem, but some species groups living in them could possibly find suitable habitats in, for example, fire–induced successional forests (Ahnlund & Lindhe 1992, Gärdenfors & Baranowski 1992, Nilsson et al. 1994).

D. Biodiversity goals in forest plantations on agricultural land

Forest plantations on agricultural land, abandoned fields, heaths, etc., can in a way be regarded as restoration of former forests, as forests probably normally preceded the agricultural areas. Although a realistic goal can hardly be to restore species diversity to a situation normal for forested areas of the region, it is indeed possible to increase diversity in the habitats by including some natural forest qualities, for example by avoiding single–layer monocultures, and by creating a mix of different tree species, as well as diverse forest edges.

Conclusion

Apparently, to fulfil the biodiversity goals presented in the sections A–D, both preservation and management activities as well as restoration activities will be necessary. Hunter (1990) expresses the differences between managing for diversity, maintaining diversity and maximizing diversity: "Managing for diversity simply assumes that having a variety of wildlife is a management goal; it is a somewhat vague phrase. To maintain species diversity is to ensure that viable populations of all the native species of flora and fauna characteristic of the management area will be present. Maximizing diversity should probably not be a goal, because it could lead to the aforementioned prospect of trying to raise eight species of game bird on the same area, or even trying to import exotic species. In cases where the natural diversity has been reduced by human interference, it would be preferable to speak of restoring diversity, rather than maximizing it".

The problem of scale

In the biodiversity discussion, scale is an important aspect. The spatial and temporal scales with respect to habitat area requirements are different for species belonging to different taxa (e.g. soil organisms, plants, beetles, birds, large mammalian predators). For example, as shown by Helle (1986), not even a large protected virgin forest of 70 km^2 is a "closed" unit for bird populations, as the bird populations even here are influenced by large–scale changes in the surrounding areas (cf. Glenn & Nudds 1989). To conserve the vast array of insects, it is also important to have a clear picture of spatial scales and processes on which conservation research and management are carried out (Samways 1993).

Species of different size will certainly react differently to the fragmented landscape. The landscape elements such as matrix, patches and corridors, are visual, anthropocentric interpretations, where as, in reality, the fragmented landscape perspective is not appropriate for describing the local limits of distribution of many animals (Bruce & Howard 1990, Samways 1994). Their distribution or dispersal patterns may not coincide with the borders of landscape elements. Further, each species has a characteristic response to landscape patterns. The landscape, in effect, acts like a differential filter for different species, and even for different sexes and life stages (Samways 1994).

Many problems of spatial and temporal scale in conservation relate to disturbance. The area affected by a disturbance (patch size) is of primary consideration in studies of disturbance ecology, and determines the grain of a landscape (Noss 1992). Small disturbances such as individual tree falls create a fine–grained pattern, whereas large disturbances create a coarse–grained pattern. Different types of organisms certainly perceive a landscape grain quite differently (Noss 1992, Rolstad & Wegge 1987, 1989). Most landscapes are characterized by overlaying fine and coarse–grained patterns, which reflect the diverse history of disturbances (Noss 1992). Therefore, no single scale of observation will provide full understanding. Thus, use of a broad spectrum of spatial and temporal scales is recommended (Wiens et al. 1986, Morris 1987, Wiens 1989). Consequently, in conservation practice, the need to manage at multiple scales and levels of organization is becoming obvious (Noss 1992, Noss & Harris 1986). This is also the case when selecting or evaluating conservation strategies, for example, size and distribution of networks of protected areas of high conservation value (Prendergast et al. 1993).

Biodiversity and conservation biology – theoretical framework

The biodiversity concept is an accepted part of the fast–growing discipline called conservation biology. Actually, Soulé (1986) defines conservation biology as the science of scarcity and diversity, and regards it partly as a response to the present biological crisis. It is supposed to help us decide what facets of nature to preserve, how to avoid extinction, and how to restore ecological damage (Western 1989).

Conservation biology deals with many aspects of ecology, such as habitat selection, and breeding conditions in the habitats. These influence the population levels of plants and animals, leading to analyses of the viability of the populations (Soulé 1986, Western & Pearl 1989, Fiedler & Jain 1992, Haila & Kouki 1994). For forest–living species, all these aspects are involved in the present lively discussion about the habitat fragmentation of the forests (for a recent review of this theme based on Fennoscandian conditions, see Hansson 1992, and articles therein, for example, by Angelstam and Esseen et al.). Habitat fragmentation, in turn, could be discussed in relation to three main subjects, namely island biogeography, metapopulation dynamics and landscape ecology.

Considerable attention has been paid to island biogeography in nature conservation work (Ås et al. 1992.). Although the ideas about species/area relationship were developed for a real island (MacArthur & Wilson 1963, 1967), they have repeatedly been applied to the conditions of fragmented forest landscape, with forest islands in a matrix of mainly clearcut areas. It has been suggested that the species/area relationships are an equilibrium between extinctions and colonizations, with extinction rate determined primarily by the area of the island, and colonization rate by the isolation of the island. The size of and distance between reserves in the landscape has been one of the core problems in the debate on how to design nature reserves (Diamond 1976, Simberloff & Abele 1976).

Metapopulation theory, based on a model for spatial population dynamics, presented by Levins (1969), and developed by Hanski (1982), Hastings (1991) and others, has also been much used in the conservation biology and habitat fragmentation discussion (Ås et al. 1992). Metapopulations are spatially structured subpopulations separated by space or barriers and connected by dispersal movements (Opdam 1991). Such local populations can become extinct and then be re–established, resulting in a distribution pattern that shifts over time (Opdam 1991). The possibility of a subpopulation surviving in a patch or a fragmented area is influenced by stochastic and deterministic (e.g. forest fragmentation by man) factors and by the size of the patch. The smaller the area, the lesser the chance of survival. The possibility that a patch which has lost a subpopulation will be recolonized depends on the degree of isolation of the patch.

Population viability analysis (PVA), i.e. the process of estimating the probability of the persistence of a population for some arbitrary time into the future (Soulé 1987, Boyce 1992), is a part of the metapopulation theory, and has become one of the key ideas of conservation biology (Wagner 1989, Boyce 1993). It has been used to estimate the probability of a certain population becoming extinct (Soulé 1986, 1987, Gilpin & Soulé 1986). Examples of such analyses are given in North America for the grizzly bear (Schaffer 1983) and for the spotted owl (Boyce & Irvin 1990), and in Fennoscandia for the pool frog (*Rana lessonae*) (Sjögren 1991).

The connectivity problem, i.e. the interconnection of functionally related ecological elements of a landscape (Merriam 1984), is an important part of landscape ecology. It can be illustrated, for example, by a species' possibility of moving between forest islands. The connectivity of the landscape depends on such landscape elements as corridors and stepping stones, but also on the landscape matrix. According to the reviews made by Simberloff and Cox (1987), Simberloff et al. 1992, and Hobbs (1992), there are few good data on the importance of these landscape elements. Furthermore, Bennett et al. (1994) have shown that there is no empirical evidence that corridors reduce extinction risk or enhance recolonization rates. However, there are studies carried out in agricultural landscapes with forest patches which show that corridors are functioning for some species (Fritz & Merriam 1993). Even if the importance is not yet proved accurately in forest landscape (cf. Simberloff et al. 1992), without doubt such corridors may constitute an important habitat in their own right (Simberloff & Cox 1987, Naiman et al. 1993). They will certainly act as habitat for many species with limited dispersal habits, and thus be useful as a part of the network of forest patches of the original ecosystem (Ingelög & Lennartsson 1991, Lindenmayer & Nix 1993, Mladenoff et al. 1993).

8.3 How forestry influences biodiversity

Forestry activities influence species composition, ecosystems and ecological processes in the forest in many different ways. It is reasonable to assume that, on the species level, the species which will be most severely affected are those with specific demands on the substrata or microhabitat, and those depending on habitats that are decreasing due to forestry or general reduction of natural processes, such as forest fires (*Table 8.1*).

Consequently, from a biodiversity point of view, it ought to be beneficial to discuss the detailed demands of the decreasing species, and what were the main influences in the forests before man began to be the main "impact force" in boreal forests; what influenced regeneration, tree species composition, age structure of trees, canopy layer structure, patchiness, mosaic structure and other habitat qualities of a forest which, in turn, influence the availability of habitats and conditions needed by the forest species (*Figures 8.5, 8.6*). The primary question, i.e. which habitats and conditions are needed by the more sensitive, demanding, and decreasing forest species is, as long as we do not insist on details, easy to answer. In Swedish conditions, the most threatened species (at least 80 % of them) demand substrata typical of natural forests but occuring very rarely in exploited stands; 30 % demand dead wood, 25 % old trees, 35 % unexposed soil, 70 % are sensitive to light exposure, etc. (Lennartsson 1991).

Concerning the main influences listed above, human activity has greatly altered the size and frequency of disturbances. This has resulted in major differences between natural forests and managed forests. For example, wildfires are quickly extinguished today.

Depending on the scale of the disturbance, two main types can be separated:

- Large–scale disturbance patterns caused by forest fires, which create forest with marked succession stages.
- Small–scale disturbance patterns caused by, for example, wind, which create forests with long continuity and internal stand dynamics (gap dynamics).

Table 8.1 Examples of species and habitat qualities which are heavily influenced by today's forestry activities.

I Species connected to natural forest qualities

1. Species connected to stable habitats (e.g. fire refugia). These species groups demand habitats with:

   a. Specific substrata typical in fire refugia:

   * Old trees – *Calicium parvum, Usnea longissima, Lecanactis abietina*
   * Moist shaded ground and stone block substrata – *Anastrepta orcadensis, Hylocomium umbratum, Epipogium aphyllum*
   * Dead wood at different stages of decay (often shaded sites) – *Anastrophyllum hellerianum, Calypogeia suecica, Herzogiella turfacea, Amylocystis lapponica*

   b. Continuity:

   * Old trees/ snags/logs continuosly occuring within a limited area – species with limited dispersal ability
   * Shaded, moist substrata (e.g. trees, logs, stone, ground) under stable conditions (<u>long stand continuity</u>) – species sensitive to rapid microclimate changes

2. Species occurring in disturbed habitats. These species groups demand habitats with:

   a. Specific substrata formed by disturbance:

   I. Disturbance caused by fire:

   * Burned ground and wood – *Geranium bohemicum, Fayodia carbonaria, Hypocenomyce anthracophila*
   * Deciduous trees – *Neckera pennata, Ficedula parva, Dendrocopos leucotos, Collema subnigrescens, Lobaria pulmonaria, Lentinellus vulpinus*
   * Wood of deciduous trees – *Clavicorona pyxidata, Radulodon eriksonii, Gloiodon strigosus, Inonotus rheades, Flammulaster limulata*
   * Brown earth types – *Astragalus glycyphyllus, Platanthera bifolia*

   II. Disturbance caused by grazing:

   * Semi–open or patchy deciduous forests, often with old trees and herb–grass dominated field layer – *Aurantioporus croceus, Clavaria rosea, Polyporus umbellatus, Perenniporia medulla–panis*

   b. Continuity:

   * Trees/snags/logs continuously occuring within a large area (<u>long forest continuity</u>) – species with the ability of surviving unfavourable conditions or finding suitable substrata or conditions far away

II Forest habitats which have decreased or are still decreasing:

* Alder swamps and other deciduous wet forests
* Some types of wet or moist spruce dominated forests
* Deciduous and mixed coniferous/deciduous forests
* Broadleaved deciduous forests and mixed broadleaved deciduous/ coniferous forests
* Grazed forests, especially forests dominated by deciduous trees

Figures 8.5 and *8.6* Fire and storm felling has created natural disturbance processes in the original forest landscape. Up till recently, also cattle grazing has affected forests. Both fire and cattle grazing have practically disappeared as ecological factors in forests, and after storm felling the trees are normally removed immediately. Photos: K. Sjöberg.

Forest fire as a large–scale disturbance pattern

In dry and mesic forests, fire seems to have played an important role in structuring the forest stands within the boreal region (Streijlen & Zackrisson 1986, Bonan & Shugart 1989, Zackrisson & Östlund 1991). Actually, natural disturbances may have been so common that they prevented forests from reaching a stable state (Sprugel 1991).

Forest fires were up till the end of the last century an important disturbance process in the Fennoscandian forests. In a study area in northern Sweden, fires swept over the forest stands at a mean interval of 80 years (Zackrisson 1977). However, there is great variation in fire frequency from region to region (Bradshaw & Zackrisson 1990). Apparently, about 70 % of the forests in large parts of northern Sweden were influenced by repeated forest fires, while about 30 % became more or less protected from fire, forming fire refugia (Zackrisson & Östlund 1991, Liljelund et al. 1992).

Forest fires and other natural disturbances create succession stages of higher complexity compared to a clearcut area. Monitoring dead wood at the end of the 19th century in Sweden showed that the standing, but dead, trees could constitute 15–20 % of the tree volume, or 10–30 large trees per hectare. The forests were unevenly–aged and had a multi–layer structure. The fires to some extent also determined the tree composition; Scots pine and deciduous trees could survive better, compared to Norwegian spruce (Zackrisson & Östlund 1991).

Natural disturbance patterns in forest ecosystems are in many respects very different from the disturbances caused by forestry activities (Pettersson 1991, Haila et al. 1994). Different forest types and regions show different types of disturbance, as well as different frequency and intensity of the disturbance due to differences in terrain morphology, etc. These variations create natural landscape mosaics of different succession stages. Thus, parts of the landscape are seldom disturbed in the natural stage and create habitats with long tree layer continuity, for example forested patches in mire areas. Sensitivity to disturbances caused by forestry activities is therefore most pronounced in such areas.

Fires occur more seldom in wet forests and forests close to mountain ranges. In the natural state, forests with long continuity will be the result. Even if a fire appears, its character is different; the fire gets a more patchy distribution and less trees are killed (Haapanen & Siitonen 1978, Hörnberg et al. 1992).

The animals and plants living in burned areas show adaptation to the natural disturbance patterns in which they have evolved (Pettersson 1991). For example, grazing by megaherbivores in the deciduous forests of southern Sweden and Denmark in prehistoric times may have shaped forest structure before present–day large–scale forestry (Andersson & Appelqvist 1990). The species adapted to semi–open deciduous forests may, in historical times, have benefited from grazing by the wild herbivores.

Small–scale disturbance pattern – old–growth forests

The dynamic within the tree layer in forests without fire is caused by putrefaction and other mortality factors caused by natural aging. Regeneration in such forests occurs continuously, and appears mainly in the patches generated by windthrows, and on the decaying logs. Such gap disturbances are also caused by snow, ice, etc. The resulting microsites are favourable for seedling establishment, which creates heterogenous forest structures and contributes to the coexistence of coniferous and broadleaved tree species (Hytteborn et al. 1987, Hörnberg et al. 1992, Kuuluvainen 1994). This also maintains high diversity for other organisms in the boreal forests, such as bryophytes (Jonsson & Esseen 1990).

In this type of old–growth forests, i.e. mature forests with long forest continuity, certain specific qualities can be defined as important for maintaining the natural biodiversity. There are standing dead trees or snags, and fallen trees or logs at different stages of decay are abundant. As regards the living trees, there are also old trees interspersed with young ones, creating an uneven multi–layer forest structure. High humidity is another feature of many old–growth forests, as well as wood, trunks and soil substrata seldom exposed to the sun.

In Sweden, only about 5 % of the forests are regarded as natural old-growth forests. For example, in the province of Hälsingland, where forests cover 80 % of the land area, only 0.01 % thereof is virgin old–growth forests. Just 10 % of the forests consist of stands older than 120 years, and only 0.4 % are older than 160 years (Delin 1992). This can be compared to the possible biological age of mature trees (*Figures 8.7, 8.8*). The Norway spruce, for example, can reach the age of about 400 years, and Scots pines can be much older (Engelmark & Hofgaard 1985). A managed forest will normally reach the age of just 80–130 years before cutting (fast growing spruce on abandoned agricultural land can be harvested at an even lower age).

There are few vascular plant species in Fennoscandia exclusively dependent on old forest stands. In total, about 15 % of the red–listed vascular plants in Sweden occur in forest habitats. Species such as *Calypso bulbosa* and *Listera cordata* normally require a period of about 60–80 years before they become re–established after clearfelling (Ingelög et al. 1987), and bilberry (*Vaccinium myrtillus*) needs about 80 years to reach full development (Kardell 1980). Factors that affect vascular plants besides clearing, thinning and clearfelling are drainage of forest wetlands, soil scarification, transformation of deciduous forests – especially broadleaved types – to coniferous forests, and abandonment of grazed (or mowed) forest types.

Bryophytes are more dependent than vascular plants on forest habitats. About 41 % of the threatened bryophytes in Sweden are dependent on dead trees (Söderström & Jonsson 1992), and many bryophytes grow on such large–sized logs that they only occur in very late successional stands (Söderström 1987). For these species even the degree of exposure to wind is a criti-

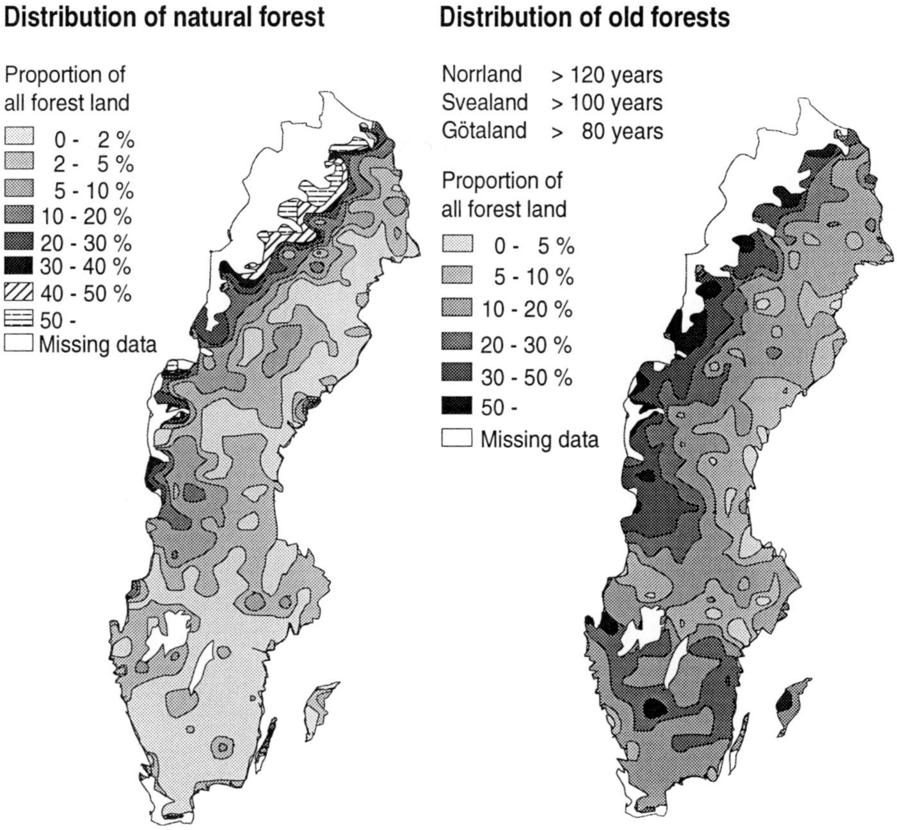

Figures 8.7 and *8.8* Maps showing the location of remaining natural forests and the age distribution of the forests in Sweden. Source: Department of Forest Survey, Swedish University of Agricultural Sciences.

cal point. However, studies are at present being carried out to test if it is possible to keep, or even bring in from elsewhere, old wood at logged areas to compensate for the loss of dead wood (Ås 1993). Bader et al. (1994) believe that it is possible to restore or create substrata for polypores and other wood–inhabiting organisms in intensively managed forests.

Many of the cryptogam and invertebrate species associated with old–growth utilize specific old–growth components such as decaying logs and snags (e.g. Esseen et al. 1992, Söderström & Jonsson, 1992, Solbraa & Grönvold 1992; for a review, see Samuelsson et al. 1994). Bader et al. (1994) have shown that polypore fungi species richness is negatively affected by logging activities because they reduce substrata availability and quality.

Many lichens are dependent on old trees (Topham 1977, Esseen 1981, Hallingbäck 1986, Ingelög et al. 1987, Søchting & Christensen 1989). For example, the mean stand age of 13 sites with *Usnea longissima* was 194 years (range 120–260 years) (Esseen & Ericson 1982). Of the 900 lichen species in

Denmark, about 550 are particularly worth protecting (184 of these species appear in broadleaved forests and 100 in old natural forests) (Søchting 1992). As these species have disappeared from northern Germany and are decreasing in southern Sweden, Denmark seems to have a particular responsibility to protect them (Søchting 1987).

Also a large number of insects occur in old–growth stands, for example, the beetle *Pytho kolwensis*, which utilizes logs of old age and size (Ehnström & Waldén 1986). About 880 species of beetles in Sweden are dependent on dead trees (Andersson et al. 1987). In Denmark, the figure is about 500 (Naturpleje i skov 1989, Hubertz 1991). One third to half of Denmark's 18,000–20,000 insect species are related to trees and bushes (Martin 1992). Økland (1992) stresses the important relation between natural forests and many of Norway's 13,000 insect species. Most of the forest insects which are regarded as threatened and vulnerable are dependent on dead wood. In Finland, invertebrate species associated with primeval forests, especially with the wood decomposition system in them, seem to have declined drastically (Heliövaara & Väisänen 1984).

The relation between old forests and bird species is well documented in Fennoscandia, particularly in Finland (e.g. Helle 1984, Helle & Järvinen 1986, Virkkala 1987, Virkkala 1990). The black woodpecker normally uses trees of a certain size for its nest cavities (Sandström 1992, Johnsson 1993) and many other birds utilize secondarily the same holes (e.g. jackdaw and stock dove). Some birds of prey (e.g. golden eagle) construct nests of such a weight that only old–growth trees are suitable. The mean age of 121 nesting trees in northern Sweden is 330 years (range: 120–540 years) (Tjernberg 1983). Again, other birds secondarily use their nests for breeding. For example, some birds of prey are dependent on these primary nest constructors, as they do not build nests of their own (Esseen et al. 1992). Some hole–nesting birds (e.g. owls and goosander), can, because of their size, only use holes or cavities formed in old trees. In a study from central Sweden, Sandström (1992) found on average a 2.8 times higher density of hole–nesting birds in natural forests compared to managed forests.

Several studies have confirmed that the changes in the bird species composition are consequences of changes in forest tree and stand ages (e.g. Ahlén 1975, Brøgger–Jensen 1992). For example, the structure of the forests in northern Finland has changed remarkably during the past decades. The most dramatic changes are the increase of young tree stands and forest fragmentation, which have favoured bird species related to open land and bush layer (e.g. *Emberiza citrinella*, *Lanius collurio*), but has been detrimental for species of old coniferous forests, i.e. *Parus montanus*, *P. cristatus*, *P. cinctus*, *Certhia familiaris*, *Perisoreus infaustus* (Järvinen et al. 1977, Helle 1985, Helle & Järvinen 1986, Virkkala 1987). In a study of forest birds in Denmark, the mean number of breeding pairs in natural forests was about twice the number found in managed forests (Brøgger–Jensen 1987). Bird species breeding in tree cavities made up one third to half the number of all birds in natural forests, but their share was only about 10 % in managed forest (Brøgger–Jensen 1987, Hubertz 1979, Hansen 1986). For frogs, forests are important habitats for parts of the year (Sjöberg & Ericson 1982).

As an example of a possible indirect consequence of changes in forest structure and age, studies on microtine rodents can be mentioned. The many clearcuts and fields with forest plantations increase the number of graminids in forested areas, thus creating better conditions for grassland species among the rodents, which can reach higher densities. As a consequence, the predatory species, for example the red fox, could increase in number, which, in turn, may have contributed to the declining of tetraonid populations by increasing the predation pressure (e.g. Christiansen 1979, Henttonen, 1989), which also may influence the predation pressure in remaining old–growth forests (*Figures 8.9, 8.10*).

When speaking about the importance of age of forests as the reason for changes in fauna and flora, the following aspects should be taken into account:

- The age of single trees influences species which are dependent on tree dimension and bark structure, and species living on dead wood of long–living tree species, such as oak and pine. Examples are found among epiphytic bryophytes and lichens, and birds (e.g. black woodpecker, *Dryocopus martius*) (Johnsson 1993), and insects such as click beetles (*Elatheridae*) (Martin 1989).

- The age of the forest stand influences species with weak dispersal capacity or species which are dependent on late successional stages. These include species living on dying large aspens in dense spruce forest, or logs of Norway spruce in old spruce stands.

- The age of the forest area, i.e. how long a particular area has been forested, influences species sensitive to exposure and drying. Such species are found among species living on, for example, stone blocks.

- The age of the forest structure, i.e. how long the region has had the natural dynamics of fire, storm felling etc., which create structures such as fire refugia, burned areas and early succession broadleaved trees, influence many species adapted to successional habitat gradient. For example, some *Coleoptera* species require burned wood for reproduction.

It is important to distinguish these factors when making decisions in relation to daily forestry activities.

Figures 8.9 and 8.10 Habitats created by silviculture during past decades favour some species (e.g. moose) which can utilize the dense bush layer of mainly birch on clearcut areas. Many species, particularly those adapted to old–growth forests (e.g. capercaillie), are negatively influenced. Photos: K. Sjöberg.

8.4. Strategies for maintaining biodiversity

To achieve the goal of maintaining all naturally occurring species in viable populations, both general and specific activities must be included in the basic strategy. As the biodiversity concept also covers genetic and ecosystem diversity, as well as ecological processes, the most efficient strategy must be to preserve as much as possible of original forests, and to imitate the dynamics, structure and function of the natural ecosystems in a given region in managed forests.

The main strategies for preserving biodiversity are the following: 1) protection of forests in nature reserves, 2) alternatives to the current dominant forestry methods (e.g. clearcutting), and 3) environmental protection measures, or considerations for the fauna and flora in the daily activities within all forestry practices.

According to the first strategy, the number of nature reserves should be increased up to the acreage expected to be enough to maintain biodiversity, while the rest of forest land could be used for intensive forestry. For example, 10–30 % of the forested land could be set aside as reserves, or restored to a more original state (Liljelund et al. 1992, Bernes 1993). In the Swedish Environmental Protection Agency's opinion, about 5 % of the forest acreage below the border of montane forests ought to be totally excluded from forestry activities. The second strategy requires limited increase in the acreage of nature reserves, and a differentiated intensity of forestry in the rest of the forest land. The third strategy is also based on a limited increase in the acreage of nature reserves, but all the rest of forest land is utilized, giving production forestry and environmental considerations equal value.

Regardless of the strategy, the goal must be to achieve the maintenance of biodiversity at the desired scale. Setting aside only a small area of nature reserves thus increases the need for biodiversity considerations in the remaining forest areas (Ingelög & Lennartsson 1991).

The capacity to create new nature reserves is at least at present limited by economic factors, as well as by the fact that there are only few natural forests left in those regions where the reserves are most needed. Therefore, we can not rely on the first strategy alone to maintain biodiversity.

The second strategy includes a limited expansion of reserves, and different levels of intensity of management on the rest of the forest land. With this strategy, the forests should be managed differently, for example, depending on production potential and transport distance to the factories. The most productive forests close to factories should be intensively managed, while low productive forests at substantial distance from factories should be managed less intensively. However, this strategy seems difficult to implement because of land–owner structure and distribution. In Sweden, for example, private non–company owners own half of the forest land (Grayson 1993). They are not as flexible as regards geographic allocation of management intensity as the big forest companies can be. There will also be an unacceptable geographic gradient from south to north and from west to east in productivity and subsequent production costs for land owners with this strategy.

The third strategy thus remains, i.e. a limited increase in nature reserve acreage, and strongly modified and ecologically adapted forestry activities on all other forest land. This is in accordance with the new forest policy in Sweden, and independent of land–owner category and geographic distribution. It requires an integrated combination of unmanaged nature reserves and key habitats, improvement of present forestry methods, introduction of new forestry methods, and consideration given to fauna and flora in all forestry activities.

Utilization of a landscape planning perspective, including preservation of key habitats, corridor linkages and stepping stone habitats between reserves, is a part of this strategy. Actually, the corridors and stepping stone habitats will act as small–scale reserves and additions or substitutes for more formal nature reserves.

The present, dominant forestry methods such as clearcutting can be a part of this future strategy, but in a strongly modified form. As mentioned earlier, at least in parts of northern Fennoscandia, about 70 % of forest land has been regularly disturbed by forest fires. If the forestry methods were as similar as possible to the natural dynamics, function and structure of the forests within the region, modified clearcutting can be closer to the natural disturbance caused by fire than forestry based on stand regeneration.

Key habitats, in the form of corridors or small "microreserves", must be preserved to a substantial extent, if the area of nature reserves and the amount of modifications in forestry methods are to be kept on an economically realistic level. Key habitats must not only serve as linkages between reserves, but must make up a considerable part of the total amount of the "improvement".

We hope that the combination of modified forestry methods and an undisturbed network of key habitats and nature reserves in a landscape perspective will help the preservation of even demanding species. This could also be a strategy on which both foresters and biologists can agree, and which they can develop together.

8.5 Tools for maintaining biodiversity

Protected forests

Worldwide, about 3 % of the land area is protected as nature reserves, parks and refuges (Reid & Miller 1989, Wilcove 1989). In Sweden, 2.6 million hectares of land, representing 6 % of the land surface receive some degree of legal protection under the auspices of the Nature Conservancy Act, 1991 (Skyddad natur 1992). Of this area, 694,000 hectares, which corresponds to less than 3 % of the total productive forest land area, are protected within national parks and nature reserves, land purchased for the nature conservation foundations, as well as crown forest reserves. About 85 % of these areas are above the border line of montane forests. Only 0.5 % of the productive forest land area below the border line is protected. In the important forest region in Västernorrland county, only 0.11 % of the land surface is protected within nature reserves, and in the province of Östergötland only 0.09 % (Kardell & Ekstrand 1990). According to the Swedish Environmental Protection Agency, about 5 % of the forested area below the border of montane forests ought to be withdrawn from forestry activities.

In the other Nordic countries, the situation is much the same. In Finland, with 23 million hectares of forests, in all, the acreage of natural forests protected from any commercial use is 4 % of the total forested area. In another 4 % of wilderness forests, utilization is restricted. Most of the protected forests are in the northern part of the country (5 % protected in northern Finland, but only some 0.1 % in the south) (Kuusipalo 1992). Most attention relating to forest protection is focused on conserving old climax–stage forests, which are the habitat for some 40 % of Finland's endangered species (Kuusipalo 1992).

In Norway, less than a quarter of the land area is covered by productive coniferous forests. Most of the forest area is on private land, and the web of nature reserves does not cover the diversity of available forest habitats (Korsmo 1991). The share of the protected productive forests is 0.6 % (of which 400 km^2 are coniferous forests) (Bernes 1993).

In Denmark, where originally about 90 % of the land area was covered with forests, the coverage is now only about 12 % – and more than 50 % of the forested area consists of plantations with exotic species. Only 35,000 hectares are regarded as natural forests. Since 1805, all forests are protected as forests, but only about 500 hectares are protected specifically as natural or untouched forests (Koester 1984). Most of the forests are small, fragmented islands in a matrix of agricultural land. Only very few of these are set aside as reserves – and even in forests within nature reserves ordinary silvicultural measures are often allowed. However, the National Forest and Nature Agency has recently proposed a protection strategy which aims at protecting 10 % of the forest area (Skov– og Naturstyrelsen 1992).

In summary, only small areas of productive forests are protected in the Nordic countries, and they are not evenly distributed over the forested regions. Furthermore, many forest reserves established in forests with the aim of preserving the natural forest flora and fauna, are not in a natural state (Lindholm 1987). They have experienced some human impact, but are left for free development and to recover. In some places, measures are taken to speed up the recovery of the ecosystems, for example, by filling in ditches.

If the preservation of species diversity is to rely on nature reserves, protected areas should represent the forest types and diversity shown all over the country. Esseen et al. (1992) have listed a number of forest types and habitats which are underrepresented or even missing in the present nature reserve sys-

Figure 8.11 The natural ecological succession after forest fire includes such ecosystem types as dense stands of broad-leaved trees, mainly birch. As the natural successions nowadays normally are broken, such ecosystems seldom occur in the landscape. Photo: K. Sjöberg.

tem in Sweden. Among those are: 1) early to intermediate post–fire successions, particularly those dominated by deciduous trees, 2) fire refugia forests, 3) herb–rich conifer forests with intermediate to high productivity, 4) swamp forests, 5) forested ravines, 6) forest margins along water courses, and 7) forested islands in archipelagoes (cf. Nilsson & Götmark 1992) (*Figure 8.11*).

Biological diversity will probably, as discussed above, not be conserved effectively in nature reserves alone because of the present insufficient acreage of reserves (Wilcove 1989, Hansen et al. 1991, Franklin 1993). Therefore, the reserves must be complemented by a matrix of "seminatural" land where ecological principles are used to manage both for commodity production and conservation of species diversity, as Harris (1984) expressed the problem (see also Franklin 1993, Mladenoff et al. 1993, Mladenoff & Pastor 1993). Also ecologically functioning patches in the forest landscape with seemingly intact biodiversity must be set aside as a complement to nature reserves. As an example, productive swamp forest patches can be mentioned. How to design and effectively manage such multipurpose landscapes are the concerns of many ecologists and foresters today (e.g. Harris 1984, Franklin et al. 1986, Angelstam et al. 1990, Hunter 1990, Andersson et al. 1992).

Modified and alternative forestry methods

Clearcutting has been the dominating cutting method in Fennoscandia since the middle of the present century. Even in Denmark, it has been the dominating silvicultural practice in spruce forests, while seed tree cutting and natural regeneration have been used in pine forests (Haveraaen 1992). When discussing how forestry activities in managed forests could be combined with maintaining biodiversity, concensus is emerging among foresters and biologists that one way is to try to reflect the natural disturbance patterns as much as possible when using the clearcut method (Haila et al. 1994).

In most of the Nordic forests, fire is regarded as being the main natural disturbance. Nowadays, however, forest fires are normally stopped quite rapidly. In Finland, for example, during the ten–year period 1979–1988, a mean number per year of 466 forest fires produced a burnt area of only 348 ha. (Metsätilastollinen vuosikirja... 1989). A strongly modified clearcutting method could be a reasonable substitute for natural disturbance by forest fire on forest grounds which were frequently exposed to large–scale disturbances in older times, even if there are distinct differences between the dynamics created by clearcutting and fire disturbance (Hunter 1990, 1993, Hansen et al. 1991, Esseen et al. 1992). Management strategies, based on estimates of the

Figure 8.12 Burning as a conservation tool is introduced and tested in small scale to maintain biological diversity. Photo: K. Sjöberg.

frequency and type of natural forest fires, are now being tested in many forest organizations and companies (e.g. Angelstam et al. 1993) (*Figure 8.12*).

The clearcutting method ought to be modified in such a way that it better reflects the burned area after a fire. A difference more in line with the traditional way could be natural regeneration of a proportion of the clearcut area to preserve the local genetic variation in the landscape (Liljelund et al. 1992). Burning of some clearcut areas would improve the conditions for those species specialized in such habitats. Enough wood must be left in the logged areas to secure the natural processes and to supply hole–trees, snags and logs in the future succession stages, as a large proportion of the threatened species are confined to such substrata (Liljelund et al. 1992). Liljelund et al. (1992) have estimated that up to 10 % of the wood has to be left after clearcutting to maintain sustainable biological values. Deciduous trees must be saved to a much larger extent than at present in logging operations in coniferous forests and in the resulting successional stands. Also a proportion of other tree species must be saved and allowed to age and die according to the natural life span of each species. The clearcut areas must be considerably smaller to allow migration and recolonization from surrounding stands. Patches and corridors of fire refugial character should not be clearfelled.

When a well managed stand is ready for logging most of these qualities have normally already been removed during, for example, thinning at earlier stages. This causes problems when we want to increase the amount of deciduous trees and dead wood. Improvement of such quality aspects must be based on saving trees, snags, etc. right from the first stages after logging.

Biodiversity in a forest is influenced not only by the age of the trees, tree species composition and tree layer composition, but also by the flow of material, disturbance patterns caused by wind and fire, fragmentation, etc. Special attention should be paid to the most vulnerable species groups. One such well defined group is the species dependent on old trees or dead wood. Another important group of species is dependent on deciduous trees. In a forest clearcut with a rotation period of about 100 years, few, if any, of those species will survive within a certain forest stand without being given particular attention. It will be even more evident when considering the time needed for recolonization of a forest stand. Specialized species will disappear after every clearcut rotation period. Consequently, special steps must be taken to save such species.

The interest in alternative methods and practices in Fennoscandia is increasing rapidly (see, for example, Hagner 1992, and the following articles therein: Haveraaen, Larsen, Lähde, Bradshaw & Gemmel). One such method is dimension felling or single tree felling (Lundqvist 1990, 1992, Lindhe & Drakenberg 1992, Vadla 1992, Andreasson 1992). Also different variants of shelterwoods, natural regeneration of all–sized spruce dominated stands treated by single tree selection, and selective felling combined with enrichment planting are discussed and tested nowadays (Hagner 1992ab, Persson

1992, Solbraa 1991, 1992, Solbraa & Grönvold 1992, Hånell 1992, Mielikäinen et al. 1992). There are also experiments going on with intermittent two–storied silviculture with Scots pine and Norway spruce (Falck & Rydberg 1992).

In Sweden, selection or single tree cutting has been traditionally used in the coniferous forest belt close to the mountain range area in the northwest part of the country in earlier times, but is nowadays almost abandoned. At present, however, these methods are being discussed again as alternatives to clearcutting (Lundqvist 1990). Studies on selection cutting are also going on in Norway (Solbraa 1993). However, it seems that this system can be used only on a small proportion of the forested area. The method requires a multi–layer tree structure. Only a limited acreage of forests comply with that structure nowadays, which limits the immediate use of the method. It is not regarded as realistic that the type of forest now in the majority (a single–layer forest prepared for the use of the clearcutting method), can be transformed to a structure suitable for dimension cutting.

It is not clear to what extent dimension cutting will improve the possibility of keeping the original species diversity intact. For example, Lindhe and Drakenberg (1992) state that such qualities in the natural forest as dead wood, snags and logs become limited with the selective cutting method as well as with clearcutting. Also windthrows, which are essential in small–scale disturbances, will probably be rare. Another important aspect is that the trees are cut at a young age. However, when selective felling is used in combination with biological conservation measures (e.g. logs and snags are left behind and some trees are allowed to reach maturity and die), this method is regarded as a more natural one in areas normally protected from forest fires (Lindhe & Drakenberg 1992).

Another aspect of importance is that the new plants are from the local genetic material. A recently started research program will be partly directed to these questions (Atlegrim et al. 1992, Sjöberg & Atlegrim 1994).

Other adjustments to present methods are also discussed nowadays, such as changed tree species composition, modified stand configuration, and type and distribution of ditches. Planning of felling in a landscape perspective and better adaptation of forestry activities to local forest site and ecological conditions are other such adjustments (Lundmark 1986).

In conclusion: the rather large–scale monoculture clearcutting forestry that still dominates in virtually all forest types (but is particularly pronounced on land owned by forest companies and by the state), offers few possibilities for preserving biodiversity or threatened species. However, keeping the pre–human forest dynamics in mind, it ought to be possible to develop forestry methods that imitate the characteristics of the ecological conditions of each forest type in an acceptable way, at least when combined with nature reserves and other nature conservation measures. For example, in large areas of Fennoscandia, a modified clearcutting method ought to be an acceptable method in those areas where natural forest fires once caused large–scale disturbances, while, for example, in the mountain range and in parts of southern Fennoscandia, other forestry methods ought to predominate, or the clearcutting method should be used only in small–scale operations.

Considerations at each forestry operation

For reasons discussed earlier, nature reserves will not solely solve the problem of maintaining biodiversity. The alternative and modified forestry methods now available are not enough to fulfil the goal, either. Therefore, small–scale measures to protect fauna and flora at each forestry operation will be important in future work to maintain biodiversity, independent of the forestry method used.

In Sweden, rules for the protection of fauna and flora in connection with normal forestry activities were included in the Forestry Act of 1979. It stated, among other things, that consideration of the fauna and flora must be a part of normal forestry activities. Examples of such activities are saving trees for woodpeckers and birds of prey, as well as trees with woodpecker holes, and deciduous trees, leaving forested zones along edges of bogs and mires and along lake and stream shores, avoiding draining of small wet forest patches, and saving such patches intact in clearcut areas. By such rather simple changes in forestry practices, the diversity and population density of many common species, as well as generalist species, could be improved (*Figures 8.13, 8.14, 8.15, 8.16, 8.17*).

Some years after the introduction of the rules in the Forestry Act, Eckerberg (1988, 1990) made a study where she checked the number of protection measures taken (compared to the potential within the area before the felling). She concluded that only about half of the potential consideration had actually been given to flora and fauna at each forestry operation. Later tests of the same type have resulted in the same conclusions. The National Board of Forestry has made a similar study later, which showed that practices were changed in only about 1/2–2/3 of the potential possibilities (Sandström 1991). Since then, a large number of private forest owners, and most personnel in forest companies, have attended courses arranged by the National Board of Forestry in, among other things, forest ecology, including conservation biology.

Up till now, only a few studies actually present the biological consequences, or impacts on biological diversity, of the activities mentioned in the Forestry Act. It has been claimed that the conservational impact has been insufficient (Natur '90 1990). However, data confirming positive effects even with small–scale measures are now being published in increasing numbers (Angelstam et al. 1990, Gärdenfors & Baranowski 1992, Ahnlund & Lindhe 1992).

(a)

(b)

(c)

(d)

(e)

Figures 8.13, 8.14, 8.15, 8.16 and 8.17 Examples of small–scale measures to protect fauna and flora in forestry: saving a) old *Salix* trees, b) snags, c) old aspen and other broadleaved trees, d) standing dead trees and e) logs in different stages of decay. Photos: K. Sjöberg.

Figures 8.18 and *8.19* Pristine wet forests sometimes remain as fire refugia, thus creating forests with long continuity. Many of these forest stands have been drained. Photos: K. Sjöberg.

Also the knowledge of the impacts of forestry on single species is steadily increasing. For example, saving dead wood in a clearcut area in central Sweden has resulted in the appearance of a number of red–listed invertebrates (Ahnlund & Lindhe 1992). Apparently, they were attracted by the light, open and warm habitat. They were able to utilize it because the substrata, the dead wood, was present (for a review of the importance of dying and dead wood, see Samuelsson et al. 1994).

In Sweden, the basic knowledge of the consequences of forestry for different species has been summarized by Ahlén (1977) and Ahlén et al. (1984) for vertebrates, by Ehnström and Waldén (1986) for invertebrates, and by Ingelög et al. (1987) for plant species. These studies form a joint data bank for endangered species including not only species affected by forestry, but also species red–listed for other reasons, for example, by agricultural activities (Ingelög & Lennartsson 1991). Despite these works, there is much need for further research on the impact of different types of measures, as well as tests of new and modified forestry methods.

The use of key and indicator species and key habitats

A <u>key habitat</u> has great nature conservation value because rare, endangered and care–demanding species are likely to occur there (Nitare & Norén 1992). A woodland <u>key area</u> is a concept used for a forest stand holding one or more key habitats. Wet forests are an example of such a key habitat in boreal zone. They often function as fire refugia, i.e. they are situated in areas or in positions in the landscape where forest fires seldom appear. The result has been forests with long continuity, and with natural forest qualities such as dead wood, logs and snags in good quantities. They are rich in species because many micro–habitats are available within limited areas, and they have a multi–layer structure. Although separated in several types, they cover a substantial part of the forested land, particularly in Finland and Sweden (Wahlström et al. 1992, Bernes 1993) (*Figures 8.18, 8.19*).

Classified endangered species are supposed to be found in key areas, but not necessarily at the same time with the identification of the place, which is normally based on indirect criteria (Nitare & Norén 1992). Old–growth forest stands of all types, but particularly wet spruce forests rich in herbaceous species and wet alder forests, as well as a succession of dense stands of deciduous bushes and trees on a burned area, primary forest bordering lake shores, and riverine forests, are all examples of woodland key habitats. Certain ecosystem elements can be found in such habitats. They include old deciduous trees (e.g. aspen, oak, hazel, linden), decaying logs of aspen and spruce, a spring or a sunlit sand surface.

Certain selected species can be used to indicate high nature conservation values (see Landres et al. 1988 for critical aspects and Saetersdal & Birks 1993, Saetersdal et al. 1993 and Williams & Gaston 1993 for limitations). In the USA, the Forest Service has to compile a list of five groups of management indicator species (MIS) for each national forest (Wilcove 1989). They are threatened or endangered species, species commonly hunted, fished or trapped, non–game species of special interest, species sensitive to intended management practices and species selected because their population changes are believed to indicate the effect of management activities on other species.

Noss (1990) lists five categories of species that may warrant special conservation effort, including intensive monitoring. They are indicator species (signal the effects of perturbations on a number of other species with similar habitat requirements), key species (upon which the diversity of a large part of a community depends), umbrella species (with large area requirements which, if they are given protection, will also protect many other species), flagship species (species with symbol values), and vulnerable species (e.g. rare species).

Indicator species indicating high nature values are called signal species (e.g. Hallingbäck 1991). *Moneses uniflora*, *Lobaria pulmonaria*, *Phellinius pini*, *Fomitopsis rosea* and *Clavicorona pyxidata* can be mentioned as examples in Fennoscandia (Søchting & Christensen 1989, Nitare & Norén 1992). Indicator or signal species are used systematically to survey valuable woodland habitats (Larsson et al. 1991, Karström 1992a,b, Solbraa 1993, Hågvar 1993, Bredesen & Gaarder 1993). Rolstad (1992) suggests that woodpeckers ought to be used as indicators of biodiversity in Norway. The black woodpecker has been used as a key species and another woodpecker, the white spotted one, has been used as a flagship species in Sweden (Johnsson 1993).

Education

In Sweden many education programs and books dealing with conservation biology and biodiversity aspects have been carried out and produced recently (Ehnström & Walden 1986, Lundmark 1986, 1988, Ingelög et al. 1987, Aldentun et al. 1991). They have been planned for the personnel of the County Forestry Boards (the subdivisions of the National Board of Forestry), and for private foresters. The aim of the campaigns has been to reach forest owners and professionals, such as forest workers, contractors and employees. Also the main private forest companies, as well as the State Forest Company run education programs at all levels in the organizations. This has increased the knowledge of indicator species, key species, key habitats, and other biodiversity key aspects. One of the main objectives of the education is to ensure the implementation of the new forestry policy in daily work practices.

Strategic planning and landscape ecology

Not least as a result of the abovementioned education programs, interest in landscape level planning has also increased among foresters. It is easiest to use the landscape approach in big forest companies with large acreages of forest land. The application of landscape ecology thinking in daily forestry work is still in early development, although some forest companies have established study areas, where they are practicing landscape–ecology planning which involves such aspects as arranging stepping stone patches, corridor linkages, key habitats and nature reserves in an optimal way (Bratt et al. 1993).

In general, there is an increasing awareness in the biodiversity discussion of the importance of involving the whole landscape in a conservation program (Angelstam et al. 1990, Hansson & Angelstam 1991, Angelstam & Holmer 1993). This change in approach has led to many suggestions for measures at landscape level, but it also denotes some problems. For example, a lot of work is being done to arrange corridor linkages without knowing to what extent sensitive and threatened species (which are the ones we want this measure to preserve) can use these corridors or even survive at all in them. This applies especially for cryptogams, which comprise about 40 % of the red–listed forest species in Sweden. However, the landscape–planning concept does not include only the designing of a combination of habitat reserves, key habitats and corridor systems. Franklin (1993) regards understanding and appropriately manipulating the landscape matrix as of at least equal importance to reserves, since the matrix itself maintains diversity and controls landscape connectivity.

Not only forest companies, but also private landowners with smaller acreages of forest land, are supposed to plan a cutting with respect to available natural forest qualities such as old broadleaved trees, snags, logs, trees with woodpecker holes, wet forest patches, etc.. Thus, there is much need for landscape–planning functions in the organizations both of companies and of the National Board of Forestry: in forest companies in order to get a regional perspective on conservation measures on their own land and in the National Board of Forestry subdivisions in order to coordinate many private land owners' measures into a regionally sound combination. In the future, as a result of the new Forestry Act in Sweden, the officers at the National Board of Forestry and its subdivisions will be equally responsible for nature conservation and for forest production when supplying advice to the land owners.

8.6 Who should act?

Forests and mires create specific habitat mosaics and species combinations in the boreal zone (Sjöberg & Ericson 1992). The Nordic countries are divided into 66 nature regions according to these qualities (including Iceland) (Naturgeografisk regionindelning... 1977, Bernes 1993). Consequently, the Nordic countries have a special responsibility for maintaining biodiversity in all these regions.

Among species groups, bryophytes are important components of the boreal ecosystems. The boreal forests have a rich bryophyte flora even when compared to tropical regions and rain forest (Hallingbäck 1992). In 1991, the

Bern Convention recommended strict protection of about 10 bryophytes occurring in Sweden (Hallingbäck 1992). Habitats of special interest in this respect in the boreal zone are wet forests, old spruce and pine forests, and dead wood.

Many species living on dead wood which are decreasing or have disappeared from south and central Europe are still present in the boreal forests of Fennoscandia (Hallingbäck 1992, Söderström & Jonsson 1992). Consequently, the Scandinavian countries are internationally responsible for protecting and preserving the pristine boreal and boreo–nemoral flora. Also the nemoral forests of Denmark, southern Sweden and southern Norway are of international importance because of the lower pressure from air pollutants than in the nemoral zones in central Europe. The same applies for broadleaved deciduous forests in the boreo–nemoral regions.

Not only foresters and forest owners will be involved in the work of maintaining biodiversity. Cooperation with organizations and people who can support foresters with information on the actual situation within a region or area is also needed. Such organisations include ornithological or other nature societies, and state authorities of different kinds. For example, community authorities and local forestry agencies which collect information on rare and engangered species should work together at provincial level.

8.7 The future

From the biodiversity and nature conservation point of view, there are at present many positive trends in forestry in the Nordic countries. Drainage of wet forests and mires has decreased. The mean size of clearcut areas has decreased from about 20 hectares to about half that size in northern Sweden. The trend is the same in Finland, where clearcut areas in the state forests are not allowed to exceed 30 hectares. In the southern parts of Finland and Sweden, the mean size of clearcut areas is less than 2 hectares (Bernes 1993). Alternatives to the dominant clearcutting method are discussed and introduced.

At the same time, nitrogen fertilization in Finnish and Swedish forests has decreased, while in Norway it has never been of importance. Also the use of other chemical substances (e.g. herbicides and insecticides) in forestry has decreased. On the other hand, the amount of forest roads to increase the accessibility of forests is still increasing by 3–4000 km/year in Finland and Sweden (Bernes 1993).

This on–going change as regards biodiversity aspects of forestry also includes restoration programs. New methods are used to solve the future need for nature conservation areas when it is not possible to buy land. For example, stands with valuable deciduous trees can be protected for a certain time period and economic support to the land owners for maintaining valuable habitats is under development.

As a consequence of the new forestry policy in Sweden, advanced education programs will be available to both private land owners and personnel in forest companies, and new ways of cooperation between foresters and biologists are being developed. There is reason to believe that the degradation trend in biodiversity values shown during the last decades will end in the near future. The officers of forestry boards have been educated in conservation biology, and are now equally responsible for giving advice in both productivity and environmental matters to meet the new demands. However, there are also weaknesses in the new policy, for example, the compliance of the land–owner with the nature conservation considerations is voluntary.

There were once virgin forests all over Europe. In some countries they have all disappeared a long time ago through conversion to agricultural land, while in others there still exist large forest areas. However, most of the existing forests are managed. Now, at the end of the 20th century, nature conservation discussion has advanced so far, that the remaining virgin forests are regarded as being so valuable that they should be totally preserved for future generations. Problems with the preservation include lack of money to preserve all forests which ought to be protected, or alternatively, lack of possibilities to compensate the land owner with substitutional land. A program for compensating and encouraging the forest owners who really promote nature conservation and biodiversity on their land is needed. Now they are treated the same way as land owners who have already depleted their forest resources or have managed their forests in such a way that few, if any, natural qualities are left. Finally, there is also a risk that the conservation efforts described in this paper will be spoiled by other factors, for example air pollutants. Consequently, increased international cooperation is necessary for saving the last pristine forests of western Europe.

References

Ahlén, I. 1975. Forestry and the bird fauna in Sweden. Ornis Fennica 52: 39–44.

Ahlén, I. 1977. Faunavård. (Fauna management.) Liber, Stockholm. (In Swedish.)

Ahlén, I., Boström, U., Ehnström, B. & Pettersson, B. 1984. Faunavård i skogsbruket. (Fauna management in forestry.) Skogsstyrelsen, Jönköping. 60 pp. (In Swedish.)

Ahnlund, H. & Lindhe, A. 1992. Hotade vedinsekter i barrskogslandskapet: några synpunkter utifrån studier av sörmländska brandfält, hällmarker och hyggen. Summary: Endangered wood–living insects in coniferous forests: some thoughts from studies of forest–fire sites, outcrops and clearcuttings in the province of Sörmland, Sweden. Entomologisk Tidskrift 113(4): 13–23.

Aldentun, Y., Drakenberg, B., & Lindhe, A. 1991. Naturhänsyn i skogen. (Nature aspects in forest.) Forskningsstiftelsen Skogsarbeten, Kista. 123 pp. (In Swedish.)

Almered–Olsson, G., Liljelund, L.–E. & Hedlund, L. 1991. A strategy for the conservation of biodiversity. Ambio 20(6): 269–270.

Andersson, F., Andersson, S., Bäckström, P.–O. & Sjöberg, K. 1992. Skogsbruk med miljömål. (Forestry with environmental objectives.) In: Elmberg, J., Bäckström, P.–O. & Lestander, T. (eds.). Vår skog: vägvalet. (Our forest: choosing the way.) LTs förlag, Stockholm. p. 184–202. (In Swedish.)

Andersson, H., Coulianos, C.C., Ehnström, B., Hammarstedt, O., Imby, L., Janzon. L.–Å., Lindelöw, Å. & Waldén, H.W. 1987. Hotade evertebrater i Sverige. Summary: Threatened invertebrates in Sweden. Entomologisk Tidskrift 108(3): 65–75.

Andersson, L. & Appelqvist, T. 1990. Istidens stora växtätare utformade de nemorala och boreonemorala ekosystemen: en hypotes med konsekvenser för naturvården. Summary: The influence of the pleistocene megafauna on the nemoral and boreo–nemoral ecosystems: a hypothesis with implications for nature conservation strategy. Svensk Botanisk Tidskrift 84(6): 355–368.

Andersson, L.I. & Hytteborn, H. 1991. Bryophytes and decaying wood: a comparison between managed and natural forest. Holarctic Ecology 14: 121–130.

Andreasson, K. 1992. Volume production and stability problems in uneven–aged and multilayered spruce (Picea abies Karst.) forest in Norway. In: Hagner, M. (ed.). Silvicultural alternatives. Proceedings from an internordic workshop, June 22–25, 1992. Swedish University of Agricultural Sciences, Department of Silviculture, Reports No. 35: 23–28.

Angelstam, P. 1992. Conservation of communities: the importance of edges, surroundings and landscape mosaic structure. In: Hansson, L. (ed.). Ecological principles of nature conservation: applications in temperate and boreal environments. Elsevier Applied Science, London and New York. p. 9–70.

Angelstam, P. & Holmer, M. 1993. Vårda hela landskapet. (Managing the whole landscape.) Skog & Forskning 1: 4–12. (In Swedish.)

Angelstam, P., Rülcker, C., Rosenberg, P. 1993. Aldrig, Sällan, Ibland, Ofta. (Never, seldom, sometimes, often.) Skog & Forskning 1: 34–41. (In Swedish.)

Angelstam, P., Welander, J., Andrén, H. & Rosenberg, P. 1990. Ekologisk planering av skogsbruk. (Ecological planning of forestry.) Miljöprojekt Sundsvall–Timrå. Delrapport 8. (In Swedish.)

Ås, S. 1993. Are habitat islands islands? Woodliving beetles (Coleoptera) in deciduous forest fragments in boreal forest. Ecography 16: 219–228.

Ås, S., Bengtsson, J. & Ebenhard, T. 1992. Archipelagoes and theories of insularity. In: Hansson, L. (ed.). Ecological principles of nature conservation: applications in temperate and boreal environments. Elsevier Applied Science, London and New York. p. 201–251.

Atlegrim. O., Granström, A., Lundkvist, L., Mattsson, L., Ohlsson, M. & Sjöberg, K. 1992. Land use, timber production and biodiversity in forest ecosystems. Interdisciplinary research for approaching optimal forest resource management. In: Hagner, M. (ed.). Silvicultural alternatives. Proceedings from an internordic workshop, June 22–25, 1992. Swedish University of Agricultural Sciences, Department of Silviculture, Reports No. 35: 20–21.

Bader, P., Jansson, S. & Jonsson, B.G. 1994. Changes in the flora of wood–inhabiting fungi and substrate decline caused by selective logging in Boreal spruce forest. Manuscript.

Bennett, A.F., Henein, K. & Merriam, G. 1994. Corridor use and the elements of corridor quality: chipmunks and fencerows in a farmland mosaic. Biological Conservation 68: 155–165.

Bernes, C. 1993. Nordens miljö: tillstånd, utveckling och hot. Naturvårdsverket, Monitor 13. 211 pp. Also available in English: The Nordic environment: present state, trends and threats. The Nordic Council of Ministers, Nord 1993:12. 211 pp.

Blöndal, S. 1991. Socioeconomic importance of forests in Iceland. In: Alden et al. (eds.). Forest development in cold climates. Plenum Press, New York. 13 pp.

Bonan, G.B. & Shugart, H.H. 1989. Environmental factors and ecological processes in boreal forests. Annual Review of Ecology Systematics 20: 1–28.

Botkin, D.B. & Talbot, L.M. 1992. Biological diversity and forests. In: Sharma, N.P. (ed.). Managing the World's forests. Looking for balance between conservation and development. Kendall/Hunt Publishing Company, Dubuque, Iowa. p. 47–74.

Boyce, M.S. 1992. Population viability analysis. Annual Review of Ecology Systematics 23: 481–506.

Boyce, M.S. 1993. Population viability analysis: adaptive management for threatened and endangered species. Trans. 58th N. A. Wildl. Natur. Resour. Conf.

Boyce, M.S. & Irvin, L.L. 1990. Viable populations of spotted owls for management of old growth forests in the Pacific Northwest. In: Mitchell, R.S., Sheviak, C.J. & Leopold, D.J. (eds.). Ecosystem management: rare species and significant habitats. New York State Museum, Bulletin 471: 133–135.

Bradshaw, R. & Zackrisson, O. 1990. A two thousand year history of a northern Swedish boreal forest stand. Journal of Vegetation Science 1: 519–528.

Bradshaw, R. & Gemmel, P. 1992. Diversity in systems of forest management: a southern Swedish perspective. In: Hagner, M. (ed.). Silvicultural alternatives. Proceedings from an internordic workshop, June 22–25, 1992. Swedish University of Agricultural Sciences, Department of Silviculture, Reports No. 35: 29–38.

Bratt, L., Cederberg, B., Hermansson, J., Lundqvist, R., Nordin, A. & Oldhammar, B. 1993. Särnaprojektet. Inventeringsrapport från en landskapsekologisk planering. (Särna –project. Inventory report from a landscape ecological planning.) DALA–NATUR, Mora. (In Swedish.)

Bredesen, B. & Gaarder, G. 1993. Skogbrukets sjanse til å sikre miljøer for truede arter. (The possibilities of forestry to provide habitats for endangered species.) Norsk Skogbruk 1: 28–29 (In Norwegian.)

Bruce, R.G.H. & Howard, D.C. (eds.). 1990. Species dispersal in agricultural habitats. Belhaven, London. 288 pp.

Brøgger–Jensen, S. 1992. Fuglesamfund i naturskove. (Bird communities in a natural forest.) In: Danmarks naturskove. (Natural forests of Denmark.) Rapport fra symposium på Aarhus Universitet, 28 marts 1992. 2. udgave. Regnskovsgruppen Nepenthes med støtte fra Skov– och Naturstyrelsen. (In Danish.)

Bunnell, F.L. 1991. Biodiversity: what, where, why, and how. In: Chambers, A. (ed.). Wildlife forestry symposium: a workshop on resource integration for wildlife and forest managers. Prince George, British Columbia, March 7 and 8, 1990.

Caring for the Earth: a strategy for sustainable living. 1991. IUCN, Gland. 228 pp.

Carlén, O. 1994. Kostnader för naturvårdshänsyn inom privatskogsbruket. (Costs of nature protection in private forestry.) Sveriges lantbruksuniversitet, Institutionen för skogsekonomi, Arbetsrapport 182. (In Swedish.)

Christiansen, E. 1979. Skogs– og jordbruk, smågnagare og rev. (Forestry and agriculture, small rodents and foxes.) Tidskrift for Skogbruk 87: 115–119. (In Norwegian.)

Contortatallen i Sverige: en lägesrapport. (Lodgepole pine in Sweden: a status report.) 1992. Skogsstyrelsens contortautredning. SLU Reprocentralen, Umeå. 226 pp. (In Swedish.)

Delin, A. 1992. Kärlväxter i taigan i Hälsingland: deras anpassningar till kontinuitet eller störning. Summary: Vascular plants of the taiga: adaptations to continuity or to disturbance). Svensk Botanisk Tidskrift 86(3): 147–176.

Diamond, J.M. 1976. Island biogeography and conservation: strategy and limitations. Science193: 1027–1029.

Dudley, N. 1992. Forests in trouble: a review of the status of temperate forests worldwide. WWF, Gland. 260 pp.

Eckerberg, K. 1988. Clearfelling and environmental protection: results from an investigation in Swedish forests. Journal of Environmental Management 27: 237–56.

Eckerberg, K. 1990. Environmental protection in Swedish forestry. Avebury Studies in Green Research, Aldershot. 179 pp.

Ehnström, B., Gärdenfors, U. & Lindelöw, Å. 1993. Rödlistade evertebrater i Sverige 1993. (Red–listed invertebrates in Sweden 1993.) Sveriges lantbruksuniversitet, Databanken för hotade arter, Uppsala. (In Swedish.)

Ehnström, B. & Waldén, H. 1986. Faunavård i skogsbruket. Del 2. Den lägre faunan. (Fauna management in forestry. Part 2. The invertebrate fauna.) Skogsstyrelsen, Jönköping. 351 pp. (In Swedish.)

Ehrlich, P.R. & Ehrlich, A.H. 1992. The value of biodiversity. Ambio 21(3): 219–226.

Elmberg, J., Bäckström, P.–O. & Lestander, T. 1992. Vår skog: vägvalet. (Our forest: choosing the way.) LTs förlag, Stockholm. 207 pp. (In Swedish.)

En ny skogspolitik. (New forest policy.) Regeringens proposition 1992/93: 226. (In Swedish.)

Engelmark, O. & Hofgaard, A. 1985. Sveriges äldsta tall. Summary: The oldest Scots pine, Pinus sylvestris in Sweden. Svensk Botanisk Tidskrift 79(6): 415–416.

Eriksson, M.O.G. & Hedlund, L. (eds.) 1993. Biologisk mångfald. (Biological diversity.) Naturvårdsverket, Rapport 4138. (In Swedish.)

Esseen, P.–A. 1981. Host specificity and ecology of epiphytic macrolichens in some central Swedish spruce forests. Wahlenbergia 7: 73–80.

Esseen, P.–A., Ehnström, B., Ericson, L. & Sjöberg, K. 1992. Boreal forests: the focal habitats of Fennoscandia. In: Hansson, L. (ed.). Ecological principles of nature conservation: applications in temperate and boreal environments. Elsevier Applied Science, London and New York. p. 252–325.

Esseen, P.–A. & Ericson, L. 1982. Granskogar med långskägglav i Sverige. (Spruce forests in Sweden with the lichen Usnea longissima.) Statens naturvårdsverk, Rapport pm 1513. (In Swedish.)

Falck, J. & Rydberg, D. 1992. Intermittent two–storied silviculture with Scots pine and Norway spruce. In: Hagner, M. (ed.). Silvicultural alternatives. Proceedings from an internordic workshop, June 22–25, 1992. Swedish University of Agricultural Sciences, Department of Silviculture, Reports No. 35: 96.

Fiedler, P.L. & Jain, S.K. (eds.) 1992. Conservation biology: the theory and practice of nature conservation, preservation and management. Chapman & Hall, New York and London.

Flerbrukshensyn i skogbruksplanen. (Multiple–use approach in forestry plans.) 1987. Direktoratet for naturforvaltning, Rapport nr. 8a. (In Norwegian.)

Flersidig skogbruk: skogbrukets forhold til naturmiljø og friluftsliv. (Multiple–use forestry: the relationship of forestry with natural environment and outdoor recreation.) 1989. Norges offentlige utredninger, NOU 1989:10. 139 pp. (In Norwegian.)

Flersidig skogbruk: veiledende retningslijner for det praktiske skogbruk. (Multiple–use forestry: guidelines for practical forestry.) 1986. Landbruksdepartementet & Det Norske Skogselskap, Oslo. (In Norwegian.)

Forest nature conservation guidelines. 1990. Forestry Commission, London. 36 pp.

Franklin, J.F. 1988. Structural and functional diversity in temperate forests. In: Wilson, E.O. (ed.). Biodiversity. National Academy Press, Washington, D.C. p. 166–175.

Franklin, J.F. 1993. Preserving biodiversity: species, ecosystems or landscape? Ecological Applications 3(2): 202–205.

Franklin, J.F., Cromack, K., Denison, W., McKee, A., Maser, C., Sedell, J., Swanson, F. & Juday, G. 1981. Ecological characteristics of old–growth Douglas–fir forests. USDA Forest Service General Technical Report, PNW–GTR–118. 48 pp.

Franklin, J.F. & Forman, R.T. 1987. Creating landscape patters by forest cutting: ecological consequences and principles. Landscape Ecology 1(1): 5–18.

Franklin, J.F., Spies, T., Perry, D., Harmon, M. & McKee, A. 1986. Modifying Douglas–fir management regimes for nontimber objectives. In: Oliver, C.D., Hanley, D.P. & Johnson, J.A. (eds.). Modifying Douglas–fir management regimes for nontimber objectives. USDA Forest Service, Seattle, WA. p. 373–379

Fritz, R. & Merriam, G. 1993. Fencerow habitats for plants moving between farmland forests. Biological Conservation 64: 141–148.

Gilpin, M.E. & Soulé, M. 1986. Minimum viable populations: problems of species extinction. In: Soulé, M.E. (ed.). Conservation Biology: the science of scarcity and diversity. Sinauer Associates, Inc., Sunderland, MA. p. 19–34.

Gärdenfors, U. & Baranowski, R. 1992. Skalbaggar anpassade till öppna respektive slutna ädellövskogar föredrar olika trädslag. Summary: Beetles living in open deciduous forests prefer different tree species than those living in dense forests. Entomologisk Tidskrift 113(1–2): 1–11.

Glenn, S.M. & Nudds, T.D. 1989. Insular biogeography of mammals in Canadian parks. Journal of Biogeography 16: 261–268.

Grayson, A.J. 1993. Private forestry policy in western Europe. CAB International, Wallingford. 329 pp.

Haapanen, A. & Siitonen, P. 1978. Kulojen esiintyminen Ulvinsalon luonnonpuistossa. Summary: Forest fires in Ulvinsalo strict nature reserve. Silva Fennica 12(3): 187–200.

Hagner, M. (ed.). 1992a. Silvicultural alternatives. Proceedings from an internordic workshop June 22–25, 1992. Swedish University of Agricultural Sciences, Department of Silviculture, Reports No. 35. 214 pp.

Hagner, M. 1992b. Naturkultur. Befriande gallring kombinerad med berikande plantering. (Nature culture. Thinning combined with enrichment planting.) Skog & Forskning 4: 10–16. (In Swedish.)

Hågvar, S. 1993. Nøkkelbiotoper, kontinuitet og ikke–hogst. (Key biotopes, continuity and non–cutting.) Norsk Skogbruk 12: 20–21. (In Norwegian.)

Haila, Y. 1994. Preserving ecological diversity in boreal forests: ecological background, research, and management. Annales Zoologici Fennici 31(1): 203–217.

Haila, Y. & Hanski, I.K., Niemelä, J., Punttila, P., Raivio, S. & Tukia. H. 1994. Forestry and the boreal fauna: matching management with natural forest dynamics. Annales Zoologici Fennici 31(1): 187–202.

Haila, Y. & Kouki, J. 1994. The phenomenon of biodiversity in conservation biology. Annales Zoologici Fennici 31(1): 5–18.

Hallingbäck, T. 1986. Lunglavarna, Lobaria, på reträtt i Sverige. Summary: The decline of three species of Lobaria in Sweden. Svensk Botanisk Tidskrift 80(6): 373–381.

Hallingbäck, T. 1991. Mossor som indikerar skyddsvärd skog. Summary: Bryophytes indicating high nature conservation values in Swedish woodland sites. Svensk Botanisk Tidskrift 85(5): 321–332.

Hallingbäck, T. 1992a. Sveriges boreala mossflora i ett internationellt perpektiv. Summary: The boreal bryophyte flora of Sweden in an international perspective. Svensk Botanisk Tidskrift 86(3): 177–184.

Hallingbäck, T. 1992b. Hotade mossor i Sverige: rapport till konferensen "Flora– och faunavård 92". (Threatened mosses in Sweden: report for the conference "Flora– och faunavård 92".) Databanken för hotade arter 1992–04–29. (In Swedish.)

Hånell, B. 1992. Skärmföryngring av gransumpskog. (Shelterwood regeneration in wet spruce forest.) Skog & Forskning 4: 40–43. (In Swedish.)

Hansen, A.J., Spies, T.A., Swanson, F.J. & Ohmann, J.L. 1991. Conserving biodiversity in managed forests. BioScience 41: 382–392.

Hansen, K. (ed.). 1986. De danske skove: Dansk vildforskning 1985–86. (The Danish forests: Danish wilflife research 1985–86.) Meddelelse nr. 207 fra Landbrugsministeriets Vildbiologiske Station, Kalö. (In Danish.)

Hanski, I. 1982. Dynamics of regional distribution: the core and satellite species hypothesis. Oikos 38: 210–221.

Hansson, L. (ed.). 1992. Ecological principles of nature conservation: applications in temperate and boreal environments. Elsevier Applied Science, London. 436 pp.

Hansson, L. & Angelstam, P. 1991. Landscape ecology as a theoretical basis for nature conservation. Landscape Ecology 5: 191–201.

Harris, L.D. 1984. The fragmented forest: island biogeography theory and the preservation of biotic diversity. The University of Chicago Press, Chicago. 211 pp.

Hastings, A. 1991. Structural models of metapopulation dynamics. Biological Journal of the Linnean Society 42: 57–71.

Haveraaen, O. 1992. The background of the new trend in silviculture in Norway. In: Hagner, M. (ed.). Silvicultural alternatives. Proceedings from an internordic workshop, June 22–25, 1992. Swedish University of Agricultural Sciences, Department of Silviculture, Reports No. 35: 13–14.

Heliövaara, K. & Väisänen, R. 1984. Effects of modern forestry on northwestern European forest invertebrates: a synthesis. Acta Forestalia Fennica 189. 32 pp.

Helle, P. 1984. Observations on some taiga forest birds with respect to forest fragmentation. Ornis Fennica 61: 121–122.

Helle, P. 1985. Effects of forest fragmentation on bird densities in north boreal forests. Ornis Fennica 62(2): 35–41.

Helle, P. 1986. Bird community dynamics in a boreal forest reserve: the importance of large scale regional trends. Annales Zoologici Fennici 23: 157–166.

Helle, P. & Järvinen, O. 1986. Population trends of North Finnish land birds in relation to their habitat selection and changes in forest structure. Oikos 46: 107–115.

Henttonen, H. 1989. Metsien rakenteen muutoksen vaikutuksesta myyräkantoihin ja sitä kautta pikkupetoihin – hypoteesi. Summary: Does an increase in the rodent and predator densities, resulting from modern forestry, contribute to the long–term decline in Finnish tetraonids? Suomen Riista 35: 83–90.

Hobbs, R.J. 1992. The role of corridors in conservation: solution or bandwagon? TREE 7: 389–392.

Holgén, P. & Lind, T. 1994. How do adjustments in the forest landscape resulting from environmental demands affect the costs and revenues to forestry? Sveriges lantbruksuniversitet, Institutionen för skogsekonomi, Arbetsrapport 186.

Hörnberg, G., Ohlson, M. & Zackrisson, O. 1992. Struktur och dynamik i naturliga sumpskogsekosystem. (Structure and dynamics in natural wet forest ecosystems.) Sveriges lantbruksuniversitet, Avdelningen för skoglig vegetationsekologi. Rapporter och uppsatser Nr. 2. 28 pp. (In Swedish.)

Hübertz, H. 1979. Skovens struktur og fuglelivet med særlig henblik på ynglefuglenes forhold i kulturskoven. (Forest structure and birds; special emphasis on young birds in culture forests.) Dansk Skovforenings Tidskrift 64: 31–62. (In Danish.)

Hübertz, H. 1991. Naturhensyn i skovdriften. (Nature considerations in forestry.) SKOV–info Nr. 6. (In Danish.)

Hunter, M.L. 1990. Wildlife, forests and forestry: principles of managing forest for biological diversity. Prentice Hall, Englewood Cliffs, NJ. 370 pp.

Hunter, M.L. 1993. Natural fire regimes as spatial models for managing boreal forests. Biological Conservation 65: 115–120.

Hytteborn, H., Packham, J.R. & Verwijst, T. 1987. Tree population dynamics, stand structure and species composition in the montane virgin forest of Vallibäcken, northern Sweden. Vegetatio 72: 3–19.

Ingelög, T. & Lennartsson, T. 1991. Åtgärder för att stärka biologisk mångfald i odlings– och skogslandskap: idékatalog för beslutsfattare. (Measures to strengthen biological diversity in cultivated and forest landscape: handbook for decision–makers.) Sveriges lantbruksuniversitet. (In Swedish.)

Ingelög, T., Thor, G. & Gustafsson, L. (eds.). 1984. Floravård i skogsbruket. Del 2. Artdel. (Flora management in forestry. Part 2. Species.) Skogsstyrelsen, Jönköping. 407 pp. (In Swedish.)

Järvinen, O., Kuusela, K. & Väisänen, R. 1977. Metsien rakenteen muutoksen vaikutus pesimälinnustoomme viimeisen 30 vuoden aikana. Summary: Effects of modern forestry on the numbers of breeding birds in Finland 1945–1975. Silva Fennica 11(4): 284–294.

Johansson, P.O., Kriström, B. & Mattsson, L. 1995. Economic value of non–timber forest goods and services. In: Hytönen, M. (ed.). Multiple–use forestry in the Nordic countries. The Finnish Forest Research Institue, Helsinki. p. 343–355.

Johnsson, K. 1993. The black woodpecker Dryocopus martius as a keystone species in forest. Ph.D. Dissertation. Swedish University of Agricultural Sciences, Department of Wildlife Ecology.

Jonsson, B.G. & Esseen, P.–A. 1990. Tree fall disturbance maintains high bryophyte diversity in a boreal spruce forest. Journal of Ecology 78: 924–36.

Kardell, L. 1980. Occurrence and production of bilberry, lingonberry and raspberry in Sweden's forests. Forest Ecology and Management 2: 285–298.

Kardell, L. & Ekstrand, A. 1990. Skyddad skog i Sverige. 1. Areal och virkesförråd inom nationalparker, naturreservat och domänreservat. (Protected forests in Sweden. 1. Area and timber resources in National Parks, nature reserves and state reserves.) Sveriges lantbruksuniversitet, Institutionen för skoglig landskapsvård, Rapport 48. 74 pp. (In Swedish.)

Karström, M. 1992a. Steget före: en presentation. Summary: The project One step ahead: a presentation. Svensk Botanisk Tidskrift 86(3): 103–113.

Karström, M. 1992b. Steget före i det glömda landet. Summary: Habitats and rare species in virgin forests of northernmost Sweden. Svensk Botanisk Tidskrift 86(3): 115–146.

Koch, N.E. & Kennedy, J.J. 1991. Multiple–use forestry for social values. Ambio 20(7): 330–333.

Koch, N.E. & Kristiansen, L. 1991. Flersidigt skovbruk: et idékatalog. (Multiple–use forestry: a handbook of ideas.) Skov– og Naturstyrelsen, Hørsholm. 39 pp. (In Danish.)

Koester, V. 1984. Conservation legislation and general protection of biotopes in an international perspective. Environmental Policy and Law 12: 106–116.

Korsmo, H. 1991. Conserving coniferous forest in Norway: a critical time of international environmental obligations. Ambio 20(6): 238–243.

Kuuluvainen, T. 1994. Gap disturbance, ground microtopography, and the regeneration dynamics of boreal coniferous forests in Finland: a review. Annales Zoologici Fennici 31(1): 35–51.

Kuusipalo, J. 1992. Finnish multiple–use foresty research programme: towards an integrated approach. Nordic Forest Research Cooperation Committee (SNS), Nordic multiple–use research newsletter. p. 9–15.

Lagercrantz, U. & Ryman, N. 1990. Genetic structure of Norway spruce (Picea abies): concordance of morphological and allozymatic variation. Evolution 44: 38–53.

Lähde, E. 1992. The background ideas to the new trends in silviculture. In: Hagner, M. (ed.). Silvicultural alternatives. Proceedings from an internordic workshop, June 22–25, 1992. Swedish University of Agricultural Sciences, Department of Silviculture, Reports No. 35: 17–19.

Landres, P.B., Verner, J. & Thomas, J.W. 1988. Ecological uses of vertebrate indicator species: a critique. Conservation Biology 2: 316–328.

Larsen, B. 1992. New trends and on–going research in alternative silvicultural methods in Denmark. In: Hagner, M. (ed.). Silvicultural alternatives. Proceedings from an internordic workshop, June 22–25, 1992. Swedish University of Agricultural Sciences, Department of Silviculture, Reports No. 35: 15–16.

Larsson, T.-B., Ebenhard, T., Sjögren, P., Andrén, H., Angelstam, P. & Widén, P. 1991. Rekommendationer om hur mätbara naturvårdsmål skall formuleras. (Recommendations for the formulation of measurable nature conservation goals.) In: Larsson, T.–B. (ed.). Mål för naturvården: en strategi för bevarande av den biologiska mångfalden. (Goals for nature conservation: a strategy for the protection of biological diversity.) Naturvårdsverket, Rapport 3986. (In Swedish.)

Lennartsson, T. 1991. Alla skogsarter kan bevaras. (All forest species can be maintained.) Skog & Forskning 1: 9–14. (In Swedish.)

Levins, R. 1969. Some demographic and genetic consequences of environmental heterogeneity for biological control. Bulletin of the Entomological Society of America 15: 237–240.

Liljelund, L.–E., Pettersson, B. & Zackrisson, O. 1992. Skogsbruk och biologisk mångfald. (Forestry and biological diversity.) Svensk Botanisk Tidskrift 86(3): 227–232. (In Swedish.)

Lindermayer, D.B. & Nix, H.A. 1993. Ecological principles for the design of wildlife corridors. Conservation Biology 7: 627–630.

Linder, P. & Östlund, L. 1992. Förändringar i norra Sveriges skogar 1870–1991. Summary: Changes in the boreal forests of Sweden 1870–1991. Svensk Botanisk Tidskrift 86(3): 199–215.

Lindhe, A. & Drakenberg, B. 1992. Modifierade skogsbruksmetoder och biologisk mångfald. (Modified forestry methods and biological diversity.) Skog & Forskning 4: 26–31. (In Swedish.)

Lindholm, T. 1987. Luonnonsuojelualueittemme metsäluonto: mitä se on ja millaiseksi se kehittyy? Summary: How natural are the forests in Finnish nature reserves and what is their future? Luonnon Tutkija 91(1): 13–19.

Lundmark, J.–E. 1986. Skogsmarkens ekologi: ståndsortsanpassat skogsbruk. Del 1. Grunder. (Ecology of forest soil: site adapted forestry. Vol 1. Basics.) Skogsstyrelsen, Jönköping. 158 pp. (In Swedish.)

Lundmark, J.–E. 1988. Skogsmarkens ekologi: ståndsortsanpassat skogsbruk. Del 2. Tillämpning. (Ecology of forest soil: site adapted forestry. Vol 2. Application.) Skogsstyrelsen, Jönköping. 320 pp. (In Swedish.)

Lundqvist, L. 1990. Blädning i granskog: strukturförändringar, volymtillväxt, inväxning och föryngring på försöksytor skötta med stamvis blädning. (Use of the selection system in Norway spruce forests: changes in the stand structure, volume increment, ingrowth and regeneration on experimental plots managed with single–tree selection). Dissertation, Swedish Sveriges lantbruksuniversitet, Institutionen för skogskötsel. (In Swedish.)

Lundqvist, L. 1992. Blädning. (Selection cutting.) Skog & Forskning 4: 4–9. (In Swedish.)

MacArthur, R.H. & Wilson, E.O. 1963. An equilibrium theory of insular zoogeography. Evolution 17: 373–387.

MacArthur, R.H. & Wilson, E.O. 1967. The theory of island biogeography. Princeton Univ. Press, Princeton, New Jersey.

Martin, O. 1989. Smaeldere (Coleoptera, Elateridae) fra gammel løvskov i Danmark. (Click beetles (Coleoptera, Elateridae) from old broadleaved forests in Denmark.) Entomologiske Meddelelser 57: 1–110. (In Danish.)

Martin, O. 1992. Økologiske krav hos smaeldere fra skov. (Ecological demands of click beetles in forests.) In: Danmarks naturskove. (Natural forests of Denmark.) Rapport fra symposium på Aarhus Universitet, 28 marts 1992. 2. udgave. Regnskovsgruppen Nepenthes med støtte fra Skov– och Naturstyrelsen. p. 32–37. (In Danish.)

Mattsson, L. 1992. Skogens nyttigheter förutom virke. (Other forest benefits in addition to timber.) In: Elmberg, J., Bäckström, P.–O. & Lestander, T. (eds.). Vår skog: vägvalet. (Our forest: choosing the way.) LTs förlag, Stockholm. p. 178–183. (In Swedish.)

Mattsson, L. & Li, C.–Z. 1993. The non–timber value of northern Swedish forests: an economic analysis. Scandinavian Journal of Forest Research 8(3): 426–434.

Mattsson, L. & Li, C.–Z. 1994. How do different forest management practices affect the non–timber value of forests? – An economic analysis. Journal of Environmental Management. 41: 79–88.

McNeely, J.A. 1988. Economics and biological diversity: developing and using economic incentives to conserve biological resources. IUCN, Gland, Switzerland. 232 pp.

McNeely, J.A. 1994. Lessons from the past: forests and biodiversity. Biodiversity and Conservation 3: 3–20.

McNeely, J.A., Miller, K.R., Reid, W.V., Mittermeier, R.A. & Werner, T.B. 1990. Conserving the world's biological diversity. IUCN, Gland, Switzerland/WRI, CI, WWF–US/World Bank, Washington, D.C. 139 pp.

Merriam, G. 1984. Connectivity: a fundamental ecological characteristic of landscape pattern. In: Brand, J. & Agger, P. (eds.). Proceedings of the First International Seminar on Methodology in Landscape Ecological Research and Planning. International Association for Landscape Ecology, Roskilde University, Roskilde. Denmark. p. 5–15.

Metsätilastollinen vuosikirja 1988. (Yearbook of forest statistics 1988.) 1989. Folia Forestalia 730. 243 pp. (In Finnish and English.)

Mielikäinen, K., Kolström, T., Ojansuu, R., Saksa, T., Valkonen, S. & Valsta, L. 1992. Management of all–aged Norway spruce stands. Research program. In: Hagner, M. (ed.). Silvicultural alternatives. Proceedings from an internordic workshop, June 22–25, 1992. Swedish University of Agricultural Sciences, Department of Silviculture, Reports No. 35: 74–75.

Mladenoff, D.J. & Pastor, J. 1993. Sustainable forest ecosystems in the northern hardwood and conifer region: concepts and management. In: Aplet, G.H., Johnson, N., Olson, J.T. & Sample, V.A (eds.). Defining sustainable forestry. Island Press, Washington, D.C./Covelo, California. p. 145–180.

Mladenoff, D.J., White, M.A., Pastor, J. & Crow, T.R. 1993. Comparing spatial pattern in unaltered old–growth and disturbed forest landscapes for biodiversity design and management. Ecological Applications 3: 293–305.

Morris, D.W. 1987. Ecological scale and habitat use. Ecology 68: 362–369.

Naiman, R.J., DeCamps, H. & Pollock, M. 1993. The role of riparian corridors in maintaining regional biodiversity. Ecological applications 3.

Natur '90: aktionsprogram för naturvård. (Nature 90: action program for nature management.) 1990. Naturvårdsverket. (In Swedish.)

Naturgeografisk regionindelning av Norden. (Nature geographic regional classification in the Nordic countries.) 1977. NU B 1977: 34. (In Swedish.)

Naturpleje i skov. (Nature management in forest.) 1989. Skov– og Naturstyrelsen, Hørsholm. 212 pp. (In Danish.)

Navrud, S. 1993. Economic value of biological diversity in Norway. In: Proceedings of the Workshop "Valuing biodiversity: on the social costs of and benefits from preserving endangered species and biodiversity of the boreal forests", Espoo, Finland, October 1992. Scandinavian Forest Economics No. 34: 74–97.

Nilsson, C. & Götmark, F. 1992. Protected areas in Sweden: is natural variety adequately represented? Conservation Biology 6: 232–242.

Nilsson, S.G., Arup, U., Baranowski, R. & Ekman, S. 1994. Trädbundna lavar och skalbaggar i ålderdomliga kulturlandskap. Summary: Tree–dependent lichens and beetles in old–fashioned agricultural landscapes. Svensk Botanisk Tidskrift 88(1): 1–12.

Nitare, J. & Norén, M. 1992. Nyckelbiotoper kartläggs i nytt projekt vid Skogsstyrelsen. Summary: Woodland key–habitats of rare and endangered species will be mapped in a new project of the Swedish National Board of Forestry. Svensk Botanisk Tidskrift 86(3): 219–226.

Noss, R.F. 1990. Indicators for monitoring biodiversity. Conservation Biology 4: 355–364.

Noss, R.F. 1992. Issues of scale in conservation biology. In: Fiedler, P.L. & Jain, S.K. (eds.). Conservation biology: the theory and practice of nature conservation, preservation and management. Chapman & Hall, New York and London.

Noss R.F. & Harris, L.D. 1986. Nodes, networks and MUMSs: preserving diversity at all scales. Environmental Management 10: 299–309.

Økland, B. 1992. Østlandets barskoger. (The coniferous forests in Østland.) In: Solbraa, K. & Grønvold, S. (eds.). Skogøkologi og flersidig skogbruk III. Del A. Truete og sårbare arter. (Forest ecology and multiple use III. Part A. Threatened and vulnerable species.) Research paper of Skogforsk 13: 23–32. (In Norwegian.)

Opdam, P. 1991. Metapopulation theory and habitat fragmentation. Landscape Ecology 5: 93–106.

Our common future. World Commission on Environment and Development. 1987. Oxford University Press, Oxford. 383 pp.

Perrings, C., Folke, C. & Mäler, K.–G. 1992. The ecology and economics of biodiversity loss: the research agenda. Ambio 21(3): 201–211.

Persson, P. 1992. Skogsskötsel i förändring. (Silviculture is changing.) Skog & Forskning 4: 18–25. (In Swedish.)

Pettersson, B. 1991. Bevarande av faunans och florans mångfald vid skogsbruk. (Preservation of the diversity of fauna and flora in forestry.) Rapport från skogsstyrelsens arbetsgrupp för översyn av föreskrifter och allmänna råd till 21 paragraf skogsvårdslagen, Skogsvårdsstyrelsen. 27 pp. (In Swedish.)

Prendergast, J.R., Quinn, R.M., Lawton, J.H. Evershamn, B.C. & Gibbons, D.W. 1993. Rare species, the coincidence of diversity hotspots and conservation strategies. Nature 365: 335–337.

Randall, A. 1991. The value of biodiversity. Ambio 20: 64–68.

Rassi, P., Kaipainen, H., Mannerkoski, I. & Ståhls, G. 1992. Report on the monitoring of threatened animals and plants in Finland. Committee for the monitoring of threatened animals and plants in Finland. Committee report 1991:30. Ministry of the Environment. 328 pp. ISSN 0356–9470.

Rassi, P. & Väisänen, R. (eds.). 1987. Threatened animals and plants in Finland. English summary of the report of the Committee for the Conservation of Threatened Animals and Plants in Finland. Maa– ja metsätalousministeriö, komiteamietintö 1985:43. 82 pp.

Reid, W.V. & Miller, K.R. 1989. Keeping options alive: the scientific basis for conserving biodiversity. World Resources Institute, Washington D.C. 128 pp.

Report on the monitoring of threatened animals and plants in Finland. 1992. Ministry of Environment, Committee report 1991:30. 328 pp.

Ripple, W.J., Bradshaw, G.A. & Spies, T.A. 1991. Measuring forest landscape patterns in the Cascade Range of Oregon, USA. Biological Conservation 57: 73–88.

Rolstad, J. 1992. Prinsipper for bevaring av truete arter: hakkespetter som eksempel. (Principles for the preservation of endangered species: woodpecker as an example.) Aktuelt Skogforsk 16/92: 62–65. (In Norwegian.)

Rolstad, J. & Wegge, P. 1987. Distribution and size of capercaillie leks in relation to old forest fragmentation.Oecologia 72: 389–394.

Rolstad, J. & Wegge, P. 1989. Capercaillie (Tetrao urogallus) populations and modern forestry: a case for landscape ecological studies. Finnish Game Research 46: 43–52.

Saastamoinen, O., Hultman, S.–G., Koch, N.E. & Mattsson, L. (eds.). 1984. Multiple–use forestry in the Scandinavian countries. Communicationes Instituti Forestalis Fenniae 120. 142 pp.

Saetersdal, M. & Birks, H.J.B. 1993. Assessing the representativeness of nature reserves using multivariate analysis: vascular plants and breeding birds in deciduous forests, western Norway. Biological Conservation 65: 121–132.

Saetersdal, N., Line, J.M. & Birks, H.J.B. 1993. How to maximize biological diversity in nature reserve selection: vascular plants and breeding birds in deciduous woodlands, western Norway. Biological Conservation 66: 131–138.

Samuelsson, J., Gustafsson, L. & Ingelög, T. 1994. Dying and dead trees: a review of their importance for biodiversity. Swedish Threatened Species Unit, Swedish University of Agricultural Sciences, Uppsala, Sweden.

Samways, M.J. 1993. A spatial and process sub–regional framework for insect and biodiversity conservation research and management. In: Gaston, K.J., Newn, I.R. & Samways, M.J. (eds.). Perspective on insect conservation. Inercept Ltd, Andover.

Samways, M.J. 1994. Insect conservation biology. Chapman and Hall, London. 358 pp.

Sandlund, O.T., Hindar, K. & Brown, A.H.D. (eds.). 1992. Conservation of biodiversity for sustainable development. Scandinavian University Press, Oslo. 324 pp.

Sandström, E. 1991. Tagen hänsyn vid slutavverkningar 1989–91. (Nature aspects in the final cuttings 1989–91.) Skogsstyrelsen, Meddelande nummer 4. (In Swedish.)

Sandström, U. 1992. Cavities in trees: their occurrence, formation and importance for hole–nesting birds in relation to silvicultural practise. (Licentiate dissertation.) Swedish University of Agricultural Sciences, Department of Wildlife Ecology, Uppsala.

Schaffer, M.L. 1983. Determining minimum viable population sizes for the grizzly bear. Int. Conf. Bear Res. and Manage. 5: 133–139.

SFS 1993:553. Lag om ändring i skogsvårdslagen (1979:429); utfärdad den 3 juni 1993. (Law about the change in forestry law (1979:429); given 3 June 1993.) Svensk Författningssamling. (In Swedish.)

Simberloff, D.S. & Abele, L.G. 1976. Island biogeographic theory and conservation practice. Science 191: 285–286.

Simberloff, D. & Cox, J. 1987. Consequences and costs of conservation corridors. Conservation Biology 1: 63–71.

Simberloff, D., Farr, J.A., Cox, J. & Mehlman, D.W. 1992. Movement corridors: conservation bargains or poor investments? Conservation Biology 6: 493–504.

Sjöberg, K. & Atlegrim, O. 1994. Selective felling as a potential tool for maintaining biodiversity in managed forests. Manuscript.

Sjöberg, K. & Ericson, L. 1992. Forested and open wetland complexes. In: Hansson, L. (ed.). Ecological principles of nature conservation: applications in temperate and boreal environments. Elsevier Applied Science, London and New York. p. 326–351.

Sjögren, P. 1991. Extinction and isolation gradients in metapopulations: the case of the pool frog (Rana lessonae). Biological Journal of the Linnean Society 42: 135–147.

Skyddad natur. (Protected nature.) 30 juni 1991. 1992. Naturvårdsverket. (In Swedish.)

Søchting, U. 1992. Naturskovens laver: indikatorer for økologisk kontinuitet. (Lichens in natural forest: indicators for ecological continuity.) In: Danmarks naturskove. (Natural forests of Denmark.) Rapport fra symposium på Aarhus Universitet, 28 marts 1992. 2. udgave. Regnskovsgruppen Nepenthes med støtte fra Skov– och Naturstyrelsen. p. 46–51. (In Danish.)

Søchting, U. & Christensen, S.N. 1989. Overvågning af laver i danske naturskove 1988. (Lichen survey in natural forests in Denmark 1988.) Skov– och Naturstyrelsen, Hørsholm. (In Danish.)

Söderström, L. 1987. The regulation of abundance and distribution patterns of bryophyte species on decaying logs in spruce forests. (Ph.D. thesis.) University of Umeå, Department of Ecological Botany, Umeå.

Söderström, L. & Jonsson, B.G. 1992. Naturskogarnas fragmentering och mossor på temporära substrat. Summary: Fragmentation of old–growth forests and bryophytes on temporary substrates. Svensk Botanisk Tidskrift 86(3): 185–198.

Sohlberg, S. 1990. Skogskartläggning. (Forest mapping.) In: Nilsson, N.–E. (ed.). Skogen/Sveriges Nationalatlas. (The forests/National Atlas of Sweden.) SNA Förlag, Stockholm. p. 134–137.

Solbraa, K. 1991. Skogøkologi og flersidig skogbruk I. Summary: Forest ecology and multiple use I. Research paper of Skogforsk 10. 29 pp.

Solbraa, K. 1993. Skogøkologi og flersidig skogbruk. (Forest ecology and multiple–use forestry.) Aktuelt fra Skogforsk, Nr 3. (In Norwegian.)

Solbraa, K. & Grønvold, S. 1992. Skogøkologi og flersidig skogbruk III. Del B. Skoglige tilpasninger, friluftsliv og konsekvensvurderinger. Summary: Forest ecology and multiple use III. Part B. Silvicultural adaptations, open air recreation, and consequence evaluation. Skogforsk 14. 54 pp.

Soulé, M.E. 1986. Conservation biology and the "Real World". In: Soulé, M.E. (ed.). Conservation Biology: the science of scarcity and diversity. Sinauer Associates, Inc., Sunderland, MA. p. 1–12.

Soulé, M.E. 1987. Viable populations for conservation. Cambridge University Press, Cambridge. 189 pp.

Spies, T.A., Franklin, J.F. & Thomas, T.B. 1988. Coarse woody debris in douglas–fir forests of Western Oregon and Washington. Ecology 69: 1689–1702.

Sprugel, D.G. 1991. Disturbance, equilibrium, and environmental variability: what is "natural" vegetation in a changing environment. Biological Conservation 58: 1–18.

Stebbins, G.L. 1992. Why should we conserve species and wildlands? In: Fiedler P.L. & Jain S.K. (eds.). Conservation biology: the theory and practice of nature conservation, preservation and management. Chapman and Hall, New York and London. p. 453–470.

Streijlen, I. & Zackrisson, O. 1986. Long–term regeneration dynamics and successional trends in a northern Swedish coniferous forest stand. Canadian Journal of Botany 65: 839–848.

Stridsberg, E. 1984. Multiple–use forestry in former days. In: Saastamoinen, O., Hultman, S.–G., Koch, N.E. and Mattsson, L. (eds.). Multiple–use forestry in the Scandinavian countries. Communicationes Instituti Forestalis Fenniae 120: 14–18.

Tjernberg, M. 1983. Breeding ecology of the golden eagle, Aquila chrysaetos (L.), in Sweden. Swedish University of Agricultural Sciences, Department of Wildlife Ecology, Report 10.

Topham, P.B. 1977. Colonization, growth, succession and competition. In: Seaward, M.R.S. (ed.). Lichen Ecology. Academic Press, London. p. 31–68.

Urban, D.L., O'Neil, R.V. & Shugart, H.H. 1987. Landscape ecology. BioScience 37: 119–127.

Vadla, K. 1992. Virkeskvalitet i bledningsskog og skog etablert etter skjermstillingshogst: en litteraturstudie. Summary: Quality of wood with selection cutting and shelterwood cutting: a literature survey. Norsk Institutt for Skogforskning, Rapport fra Skogforsk 10. 21 pp.

Vesterholt, J. 1992. Om bevaring af truede svampe. (Protection of endangered mushrooms.) In: Danmarks naturskove. (Natural forests of Denmark.) Rapport fra symposium på Aarhus Universitet, 28 marts 1992. 2. udgave. Regnskovsgruppen Nepenthes med støtte fra Skov– och Naturstyrelsen. p. 42–45. (In Danish.)

Virkkala, R. 1987. Effects of forest management on birds breeding in northern Finland. Annales Zoologici Fennici 24(4): 281–294.

Virkkala, R. 1990. Ecology of the Siberian Tit, Parus cinctus, in relation to habitat quality: effects of forest management. Ornis Scandinavica 21(2): 139–146.

Wagner, F.H. 1989. American wildlife management at the crossroads. Wildlife Society Bulletin 17: 354–360.

Wahlström, E., Reinikainen, T. & Hallanaro, E.–L. 1992. Miljöns tillstånd i Finland. Vatten och miljöstyrelsen, Miljödatacentralen, Helsinki. 364 pp. Also available in English: Wahlström, E., Reinikainen, T. & Hallanaro, E.–L. 1993. The state of the Finnish environment. National Board of Waters and the Environment, Environment Data Centre, Helsinki. 163 pp.

Western, D. 1989. Conservation biology. In: Western, D. & Pearl, M.C. (eds.). Conservation for the twenty–first century. Oxford University Press, New York. p. 31–36.

Western, D. & Pearl, M.C. 1989. Conservation for the twenty–first century. Oxford University Press, New York. 365 pp.

Wiens, J.A. 1989. Spatial scaling in ecology. Functional Ecology 3: 385–397.

Wiens, J.A. 1990. Habitat fragmentation and wildlife populations: the importance of autecology, time, and landscape structure. In: Transaction of the 19th IUGB Congress, 1989, Trondheim. Norwegian Institute of Nature Research, Trondheim.

Wiens, J.A., Addicott, T.J. & Diamond, J. 1986. Overview: the importance of spatial and temporal scale in ecological investigations. In: Diamond, J. & Case T.J. (eds.). Community ecology. Harper & Row, Cambridge. p. 145–53.

Wilcove, D.S. 1989. Protecting biodiversity in multiple–use lands: lessons for the US Forest Service. Trends in Ecology & Evolution 4: 385–388.

Williams, P.H. & Gaston, K.J. 1994. Measuring more of biodiversity: can higher–taxon richness predict wholesale species richness? Biological Conservation 76: 211–217.

World Conservation Strategy: living resource conservation for sustainable development. 1980. IUCN, Gland. 73 pp.

Zackrisson, O. 1977. Influence of forest fire on the north Swedish boreal forest. Oikos 29: 22–32.

Zackrisson, O. & Östlund, L. 1991. Branden formade skogslandskapets mosaik. (Fire formed the landscape mosaic.) Skog & Forskning 4: 13–21. (In Swedish.)

9 Forest recreation

Frank Søndergaard Jensen[1]

Abstract

Comparisons between the Nordic countries show similarities as well as differences between the countries. In general, Norway, Sweden and Finland have the most in common in forest recreation. Denmark forms a link between central Europe and rest of Scandinavia, while Iceland takes up a special position. Comparable information is available on the following aspects of recreation: the relationship of forest recreation to other leisure activities, the number of visits to the forests, length of stay and transport time, time patterns, variations through the week, geographical use distribution within the forest, main activities, group size and means of transportation to the forest. Information on the importance of forests for recreation has been provided by studies on people's preferences concerning the quality of the environment, the facilities provided by management and the attitudes of recreationists towards other people in recreation areas. Nature schools give children education on nature and the environment in all the Nordic countries. Nature interpretation is considered to be an essential part of both recreation and tourism. The importance of urban forests and green spaces is increasing, because the amount of people living in towns is growing. Future research and management challenges include the problems connected with area competition and user fees.

Keywords: recreation, preferences, nature interpretation, nature schools, urban forests, research.

---

[1] Danish Forest and Landscape Research Institute
Skovbrynet 16, DK–2800 Lyngby, Denmark

9.1 Introduction

A detailed examination of the historical aspects of recreation in the Nordic countries is outside the scope of this text, so only an overview of the past is given, to get a feeling of the development of recreational behaviour. Recreation in northern Europe follows patterns found in many other Western societies, where the nature based recreational life of today can be traced back to the late 19th century (*Table 9.1*, *Figure 9.1*).

When talking about forest recreation of today, it is valuable to know the existing legal rights of access. In Norway, Sweden, Finland and Iceland the "Right of Public Access" ("Everyman's Right") is found. This right is a very old traditional privilege in these sparsely populated countries. It gives the right of access to private and public land for everybody, with some obligations implied. Within certain limits, visitors are allowed to move about freely. As long as they do not cause damage, they can walk or hike; ski; bicycle or go horse-riding; pick berries, flowers and mushrooms; go bathing and boating; and pitch a tent for a short period of time. The Right of Public Access was laid down as a written law in 1957 in Norway, and in 1960 in Iceland. In Sweden and Finland, it is called consuetudinary law, or a time–honored right.

In more than one way, Denmark forms a link between sparsely populated northern Europe and the more densely populated central Europe. This is evidenced also by the rules for public access.

Table 9.1 Rough outline of some of the trends in Nordic outdoor recreation during the last two centuries (partly based on Kardell 1979).

Year	Development in transportation	Movement in society	Recreation area	Groups of participants
1800				
	Railway system	Conservation org.	Beach/ archipelago	
		Tourism org.	Mountains	Expeditions
	Bicycles	Skiing clubs	Forests	Organized interest groups
		Scout groups		
	Public buslines	Rambling ass.		Families
	Private cars			
		Jogging movement		Special interest
		Equipment industry		
2000				

Figure 9.1 Outdoor recreation in the 1880s in Sweden (Nordiska museet, Stockholm).

It is believed that the Right of Public Access has been in force also in Denmark. Presumably, this was the case until an act in 1781 forbid the access of all unauthorized persons to the national (royal) forests. Legally, but not in practice, this situation prevailed, until the first Nature Conservation Act in 1917 re–established public access to national forests. A 1969 amendment to that act established a limited right of access to private forests. In the new Nature Protection Act, which came into force in 1992, these rights of access have just been enlarged and now include, among other things, the use of bicycles in private forests and access to the countryside.

9.2 Recreational use of the forest

The recreational use of the forest can be measured in various units: number of visitors, number of visits, or number of visitor/visit hours – and all these units can then be related to the size of the visited area (e.g. number of visits per hectare). However, it is important to keep in mind that the observable recreational use of forests is a result of people's choice between, for example, the known and available opportunities for forest recreation. In this way, the forestry sector itself also influences the amount of recreation. The potential demand for forest recreation is very difficult to estimate, and will not be covered in this context; only the results of people's recreation choice will be described.

The following sections will try to cover a wide range of aspects of the observable use of the forest for recreation. Comparisons between the Nordic countries will be presented when possible. When making such comparisons, it is important to remember that the results derive from different survey methods and cover a relatively long period. (Koch (1984) gives the main characteristics of about 30 different surveys). Therefore, some of the mutual comparisons are relatively difficult and uncertain.

Forest recreation compared to other leisure activities

It is of interest to get an idea of how important forest recreation is compared to other leisure activities. The forest plays a very important role in the leisure pattern of the Nordic people (*Table 9.2*). In general, more than 80 % of the adult population visit the forest at least once every year. This is a rather high percentage compared to other leisure activities, such as movies, concerts, museums, etc.

Number of visits to the forests

Only Denmark has tried to estimate and describe the recreational use of the individual forest and the variation in use between *all* forest areas in Denmark

Table 9.2 Participation rates for different recreation/leisure activities in four Nordic countries.

	Norway[a] %	**Sweden**[b] %	**Finland**[c] %	**Denmark**[d] %
Forests/nature areas	91	82	85	91
Libraries	-	51	66	64
Art exhibitions	17	31	41	37
Museums (excl. art)	23	45	-	34
Theatres	22	33	45	13
Sport grounds	33	44	15	32
Concert halls	15	33	20	12

Sources: a) Friluftsliv -undersøkelse 1975, Tveit 1979.
b) Levnadsförhållanden 1987.
c) Ulkoilututkimus 1980, Vuolle et al.1986, Niemi et al. 1991, Sievänen 1993.
d) Koch 1978, Fridberg 1989.

The data covers a very wide time period, and the definitions of the different activities may vary between the countries.

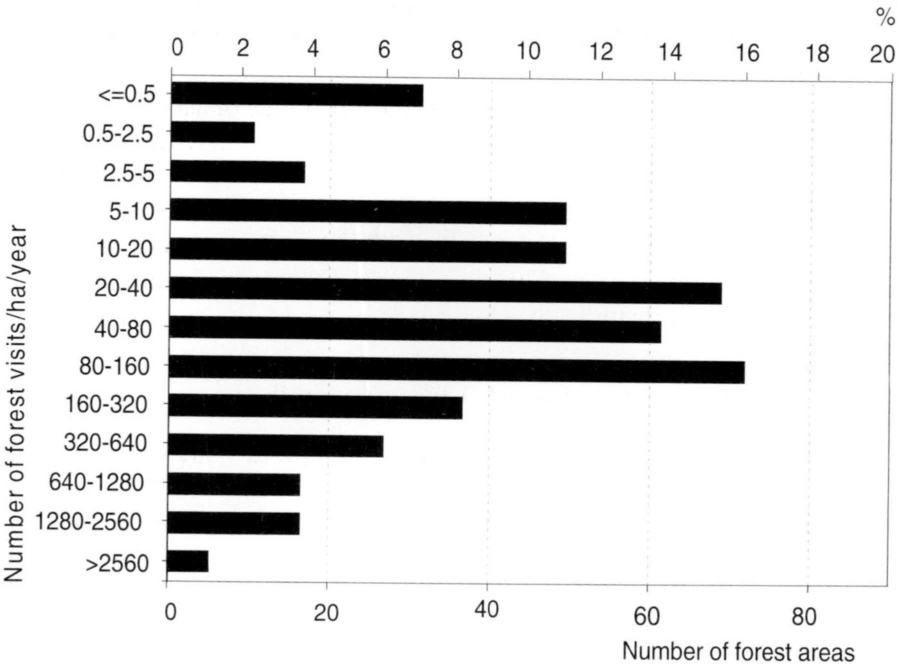

Figure 9.2 Classification of the Danish forest areas according to the yearly number of forest visits per hectare (Koch 1980).

(Koch 1980). The main result of these investigations is a classification of altogether 601 forests according to the intensity of recreational use (visits/ha/year). A description, based on questionnaires, is also given of how the different forests are used for recreation. *Figure 9.2* illustrates one of the important results of the investigations. It appears that there is a very large variation between the different forest areas in the yearly number of forest visits per hectare. In each of the 14 counties into which Denmark is divided, it can be found that some forests are used up to about one thousand times more intensively than other forests. The most intensively used forest in the investigation is used about twenty thousand times more than the least intensively used forests.

In contrast to the other countries, Denmark may be characterized by having a rather uniform geographical distribution of a relatively small forest area per inhabitant. Therefore, it must be expected that an even greater variation between forests in the intensity of recreational use would be found in the other Nordic countries. However, investigations carried out in the other countries have only dealt with rather few and relatively intensively used forests (*Table 9.3*). The most important reason for this is probably that in many areas it is difficult and costly to measure the amount of recreational use.

Table 9.3 Some measured use figures in selected investigations from Denmark, Norway, Sweden and Finland.

	Reference	Forest area	Number of visits per ha per year
Denmark	Koch 1984	Jægersborg Dyrehave	2021
		Store Dyrehave	96
		Åtte Bjerge	416
		Gl. Kjøgegaard	247
	Jensen 1992	Vestamager	235
Norway	Haakenstad 1975	Frognerseter/Sognvanns	223
		Ringkollen	23
Sweden	Kardell 1972	Bogesundslandet	34
	Heglebäck 1978	Nackareservatet	1500
		Järvafältet	300
		Lovön	100
Finland	Jaatinen 1973	Keskuspuisto	2200
		Luukkaa	315
	Pouta 1990	Luukkaa	260
		Salmi	60
		Tarusjärvi	20
	Sievänen 1992a	Aulanko	2000
		Ahvenisto	4000

Length of stay and transport time

Figure 9.3 shows comparisons of results from a Danish survey and results from a quite similar survey in Sweden. Based on this information, it may be concluded that the Swedes spend more time on each visit compared to the Danes, and spend less time on transport to the forest in general. These differences are in accordance with the distribution of the forest and living areas in Sweden compared to Denmark.

The length of stay is of course very dependent on, for example, what kind of forest we are talking about. This can be well illustrated by a Finnish survey, where Sievänen (1993) finds the average length of a visit to urban forests to be 1.5 hours, while the average for other forest types is 2.6 hours.

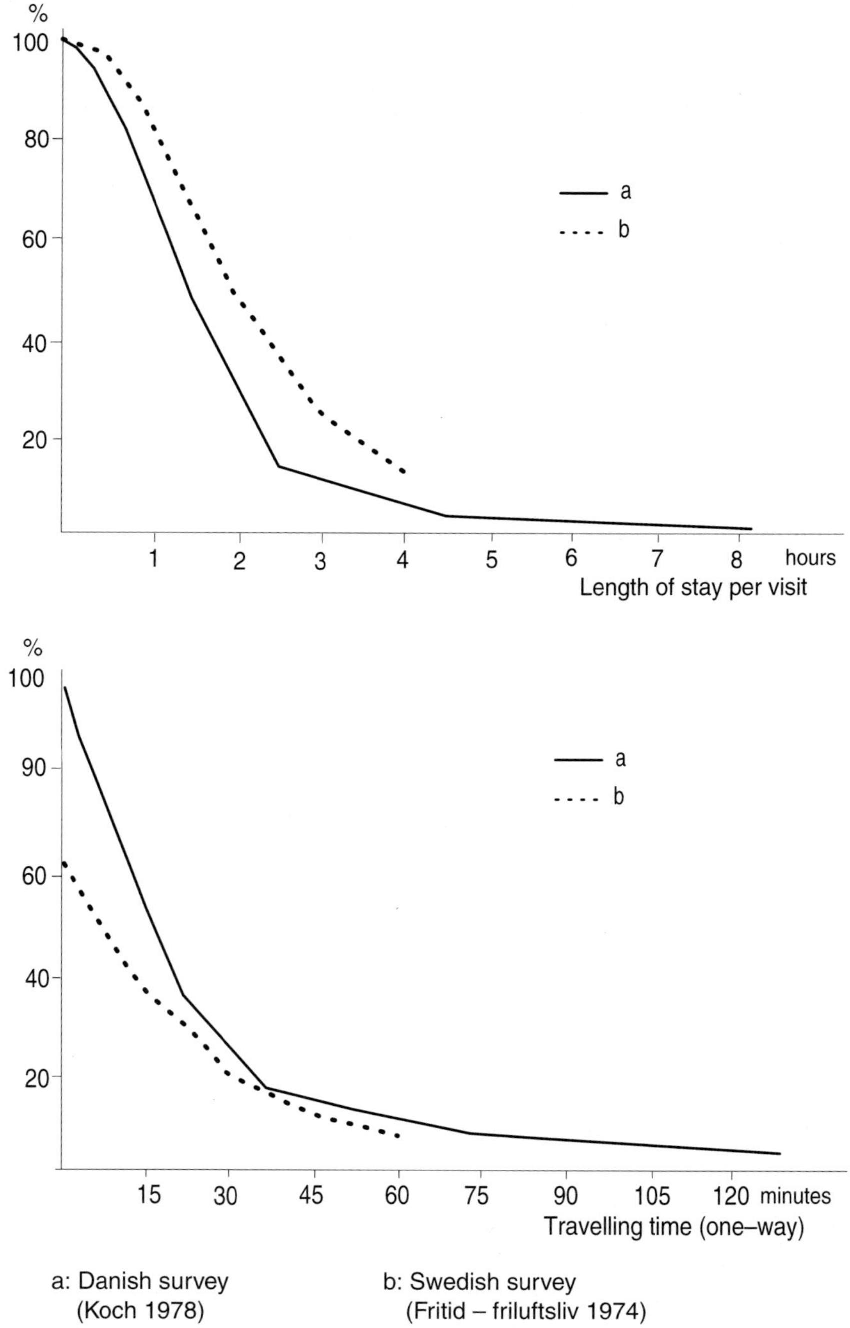

Figure 9.3 Comparisons of the cumulative distributions of two Nordic surveys according to length of stay and travelling time (Saastamoinen et al. 1984).

Time patterns

The Danish time patterns in forest recreation were studied in the 1980s (Koch 1984). It is quite natural that more visits are made at the weekends, and forests are more frequently visited in spring and summer than in autumn and winter. Very few visits occur in the late evening and night but there is a tendency to a small peak of visits around 7 o'clock in the morning (e.g. people jogging and walking the dog before the start of the working day).

This is a general Danish pattern, which will cover many situations, but of course there will be differences between the various forest areas – and the Nordic countries. Therefore, examples of differences and similarities in the visit variations through the week and the year are given in *Table 9.4*.

Variations through the week

In a Finnish survey by Saastamoinen (1979) and a Swedish survey by Kardell (1982) it is found that in areas very close to or within the city boundary, visits on Saturdays and Sundays account for a considerably smaller part of the total number of visits than in the Danish investigation. On the other hand, examples of areas in Norway and Sweden may be found where the number of visits during the weekend account for a larger part of the total number of visits than is the case in the Danish study (Haakenstad 1975, Kardell 1972).

Saastamoinen (1979) has contributed to an explanation of the tendency demonstrated in *Table 9.4*, that the number of visits per day is somewhat larger on Tuesdays and Wednesdays than on the rest of weekdays, because he finds that visits to a skiing trail in a forest in Rovaniemi by schools, companies, and other organizations are especially intensive on those days.

In northern Scandinavian forests, where the conditions are favourable for skiing, visit fluctuations through the year differ considerably from the results found in Denmark. February and March are among the most visited months, while May is among the least visited, and the range of variation is relatively large (Haakenstad 1975, Saastamoinen 1979, Saastamoinen & Sievänen 1981). It is evident that the length of the period with good snow for skiing influences the visit variations through the year in these areas, which is also demonstrated clearly by Saastamoinen and Sievänen (1981) as they investigated the variation through the year in forests partly in North Finland (around Rovaniemi), and partly in South Finland (around Kerava). This survey shows that it is possible to affect the variation through the year considerably; an illuminated cross–country skiing trail in the forest near Rovaniemi resulted in a visit variation through the year on Wednesday evenings almost similar to the variation through the year for Sundays at midday. The variation through the year on Wednesday evenings around Kerava, where there is *no* illuminated trail, is completely different with close to zero visits during the period of November–January.

Most of the investigations compared have in common that visits are especially intensive in July and August (cf. *Table 9.4*). Presumably, a contributing factor is the tourist visits culminating in these months. In addition, May has a large number of forest visits as this month is a pronounced foliation month (Kardell 1982, Koch 1984). Concerning the berry picking in autumn, there is no doubt that for some areas this will be the most intensive season.

Geographical use distribution within the forest

Just to know the number of visits in a certain forest is not always sufficient information when planning and managing for recreation. It is of interest to know how the visitors are distributed within a single forest area. In Denmark, a technique has been used with success by Koch (1984) and Jensen (1992). *Figure 9.4* shows a result of the use of this technique: when making interviews in the particular forest area, the respondents have been asked to draw the route of their forest visit on a detailed map (practically all respondents found this little exercise very interesting and amusing – and not difficult). As shown in the figure, it is obvious that the eastern part of the forest is the most popular. It is close to the city nearby and the two major parking lots.

A considerable variation in the geographical use intensity within the single forest area is found in several surveys, for example, in Norway (Oslomarka) by Haakenstad (1975), in Sweden (Bogesundslandet and Linköping) by Kardell (1972 & 1982), and in Finland (Sievänen 1989). Kardell (1972) finds the largest geographical use distribution in autumn. This condition is explained by the fact that visitors are picking berries and mushrooms at this time of the year. This observation is probably valid for many forest areas in Norway and Finland as well, as in both countries it is a tradition to pick berries and mushrooms. This activity has just started to become popular among the Danes.

It is also important to know how many visitors stick to the forest road and trail system and how many do not. Two Danish surveys by Koch (1984) and Jensen (1992) show that approximately 15–20 % of the visitors are walking outside the roads and trails (more men than women and less single visitors).

A similar percentage is found in a survey from Iceland in a forest area near Akureyri (Blöndal 1991). In Sweden, Kardell (1982) found that 5 % did not stick to the forest road and trail system in two forest areas. In Norway, more younger people and more people during the summer are found walking outside the road and trail system (Haakenstad 1972). Finally, a Finnish survey from two areas close to Helsinki concludes that: "...the recreationists visit terrain without trails nearly every time they visit a recreation area" (Kellomäki 1977).

Reference	Unit of measurement	Area(s)	Mon	Tue	Wed	Thu	Fri	Sat	Sun	Dec	Jan	Feb	Mar	Apr	May	Jun	Jul	Aug	Sep	Oct	Nov
NORWAY																					
Haakenstad			1	1	2	2	3	59	100	7	41	61	100	45	28	46	46	74	51	48	7
1975	One visit	Frogneseter /	13	14	25	16	20	30	100												
		Sognsvann	2	2	3	3	4	53	100												
Haakenstad			4	10	10	10	14	62	100	53	56	91	99	93	49	87	100	94	67	62	37
1975	One visit by car	Ringkollen	43	63	51	55	79	100	60												
			5	10	9	10	30	91	100												
SWEDEN																					
			______	______	12	______	______	52	100	(	,	)	(	,	)	(	,	)	(	,	)
Kardell			______	______	10	______	______	45	100		26			33			100			31	
1972	One car visitor hour	Bogesundslandet	______	______	4	______	______	31	100												
	and one visit		______	______	3	______	______	34	100												
			______	______	8	______	______	41	100												
Kardell	One visit	Lindköping S.	______	______	81	______	______	57	100	(	,	)	(	,	)	(	,	)	(	,	)
1982											11			99			100			53	
FINLAND																					
Saastamoinen	One visit	Rovaniemi																			
1979		(skiing route)	65	73	71	51	48	76	100	30	78	100	47	27	4						16
Saastamoinen &	One visit	Rovaniemi	35	35	38	41	41	68	100	55	90	100	100	52	13	19	16	19	19	42	55
Sievänen 1981		Kerava	32	32	35	33	61	68	100	22	100	100	65	33	26	15	17	22	20	37	35
DENMARK																					
Koch	One car visit	Århus	57	59	59	58	56	75	100	28	25	31	44	63	85	93	100	78	53	46	36
1984		Store Dyrehave	33	35	31	29	28	61	100	100	19	16	20	25	38	30	23	32	29	24	22
		Århus	28	27	27	25	24	55	100	36	41	37	48	75	100	70	82	61	48	63	43
		Århus	38	39	38	35	33	64	100	14	17	21	28	53	96	83	100	71	38	33	20
		Åtte Bjerge	18	16	18	16	11	45	100	24	56	69	23	66	100	94	98	91	68	44	24

Note: a): Index 100 = The highest mean number of visits per day of the week.
b): Index 100 = The highest mean number of visits per day per month.

Table 9.4 Comparisons between selected Nordic investigations on the variation through the week[a] and the year[b] in the relative number of visits per day (table based on Koch 1984).

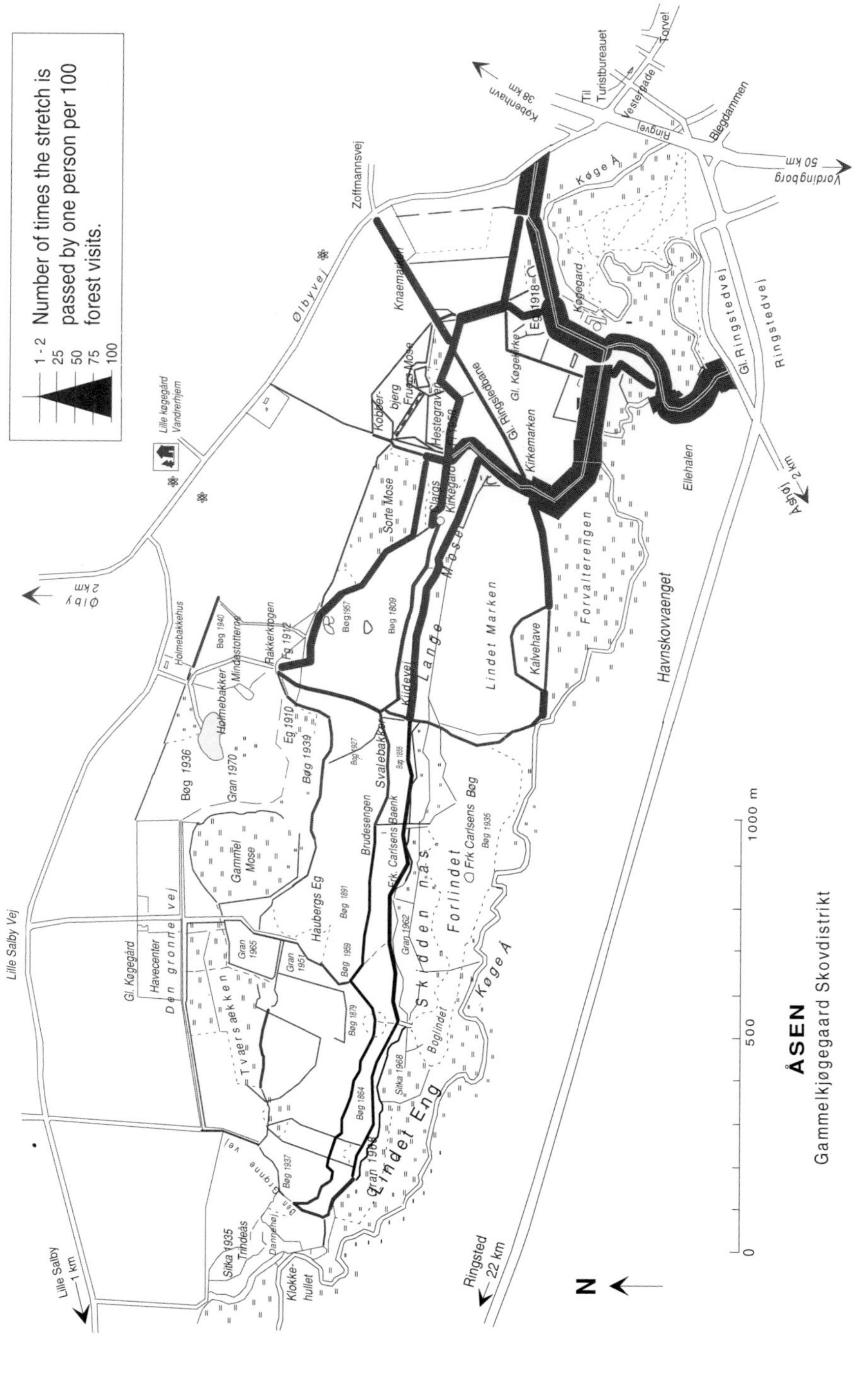

Figure 9.4 The geographical distribution of the recreational use of the Danish forest "Åsen" (Koch 1984).

Figure 9.5 Distribution of the Danish forest visitors according to activities pursued during the latest visit to the forest (Koch 1978).

Main activities

What kind of activities do visitors pursue during their visits to the forests? This question is of great importance when managing for recreation in a multiple–use context.

Figure 9.5 shows the distribution of Danish forest visitors. It is interesting that walking, and enjoying and studying nature are the most frequent activities, while the more "consuming" activities like fishing, hunting and making (legal) use of products from the forest, for example, picking berries and mushrooms, collecting nuts, moss and lichen are the least frequent in Denmark (a total of less than 5 %).

When comparing activities in the Nordic countries, it is found that activities like walking, cycling and exercising in general are relatively popular in all countries. Activities where we really find differences are berry and mushroom picking: participation rates in Finland are more than 70 % (Ulkoilututkimus 1979), and in some specific areas in Norway and Sweden rates up to 50 % have been observed (Kardell & Pehrson 1978, Kardell & Holmer 1985, Aasetre 1993). These are rates far above what are seen in Denmark.

Denmark and the other Nordic countries are different also when it comes to winter activities like cross–country skiing. Mainly due to lack of snow, skiing is very seldom possible in Denmark, while in some areas in the other Nordic countries it is clearly the main activity during the winter time. For example, Saastamoinen and Sievänen (1981) conclude: "The recreational use of urban forests has a special importance during the skiing period in Finland". Undoubtedly, this conclusion applies to Norway and Sweden as well (as a point of interest, there is actually only a couple of illuminated cross–country skiing trail in Denmark!).

Group size and transportation to the forest

From *Table 9.5* it can be seen that the Danes walk alone in the forest a little less frequently than the Swedes, the Norwegians, and especially the Finns. Perhaps fear of the forest is greater in Denmark, a densely populated agricultural country, than in the other Nordic countries? The Finns seem to walk alone much more often than all the other Scandinavians (here it is important to be aware of the fact that the Finnish survey is restricted to urban forests).

The figures in *Table 9.6* also reflect some minor differences in recreational behaviour in the Nordic countries. In this case, in connection with the choice of the means of transportation to the forest, the number of visitors who walk to the forest seems to be largest in Norway, while the number of visitors using a car is largest in Denmark. Common to all the countries is the fact that less than 10 % use public transportation.

Table 9.5 Percentage share of the forest visits with a group size of one person. Results from four selected Nordic surveys.

Sweden: (Fritid–friluftsliv 1974)	16 %
Norway: (Friluftsliv –undersøkelse 1975)	18 %
Finland: (Sievänen 1992a) [a]	67 %
Denmark: (Koch 1978)	13 %

a) Urban forests

Table 9.6 Distribution according to the means of transportation used to get to the forest. Results from four Nordic surveys.

	Norway[a]	**Sweden**[b]	**Finland**[c]	**Denmark**[d]
On foot/ski	59	43	33	27
Horse	-	-	-	1
Bicycle	2	4	15	7
Moped, motorbike, scooter	1	0	-	2
Car	33	47	43	55
Bus, train	3	4	5	6
Other	2	0	3	0
Total	100	100	100	100

Sources: a) Friluftsliv –undersøkelse 1975.
b) Fritid – friluftsliv 1974.
c) Ulkoilututkimus 1979.
d) Koch 1978.

9.3 Preferences of the population

If an estimate of the importance of forests for outdoor recreation is wanted, a *quantitative* survey of how and how much forests are used for outdoor recreation is not sufficient. It is essential to try to form an estimate of the *quality* of outdoor recreation experiences, which leads to a study of wishes/preferences of the population and forest visitors as regards planning and management of the forest.

What kind of forest do the Nordic forest visitors prefer? This question will be covered in more detail in the chapters by *Christina Axelsson Lindgren* and *Minna Komulainen*. Here I will only discuss preferences in relation to environmental, management and recreational aspects.

Environmental aspects refer to the influences the landscape manager does not control (e.g. climate and noise from aeroplanes). In a Danish survey by Koch and Jensen (1988), these kind of influences have only been tested by one verbal stimulus: "silence". It is notable that this influence reaches the highest average estimate of all (100) verbal stimuli used. Haakenstad (1972) finds similar results in his survey in the Oslo region in Norway. However, the interpretation of this result is rather complex. It is hardly the physical silence alone which is so important. Supposedly, bird song, for example, is compatible with "silence"? Perhaps this result really reveals more about an important reason for forest visits than it would be possible to uncover by a number of more direct questions. But, presumably, the impression of the word "silence" varies from person to person, place to place, and from one time to another. (In addition,

Haakenstad found that the weather conditions had an influence on the decision for taking a forest walk.)

Management aspects refer to the influences the landscape manager can adjust directly. Here only *facilities for forest recreation* will be discussed. A general problem attached to the method of measuring preferences for facilities is that the estimate of these arrangements is often based more on an aesthetic criterion than on a functional one.

In Denmark, forest roads and paths are considered a relatively positive element in the forest environment, and it can be concluded that the less specialized and "unnatural" a path is, the higher preference it has with the public. The order of precedence of exercise path, riding path or cycling path, corresponds to the public assessment of meeting the three kinds of forest visitors – joggers, riders, and cyclists (Koch & Jensen 1988). Also, Hultman (1983) finds that forest roads and paths reach a relatively high assessment in Sweden. This is by far the case in countries where quite a few new forest roads are established in connection with large cuttings – and wilderness areas. Aasetre (1992) lists quite many surveys where the results indicate a rather mixed feeling for forest roads and paths among Norwegian forest visitors.

Recreational aspects refer to the other forest visitors' influence on the quality of a forest visit. Not surprisingly, it is found that the less other forest visitors a Danish forest visitor meets on his way, the higher quality is attributed to the forest visit in general (*Figure 9.6*). Similar results have been found in Norway (Haakenstad 1975).

According to Koch & Jensen (1988), the forest visitors can be divided into three popularity groups according to the opinion of the Danish population: (1) the most popular group consists of families, riders, and joggers, (2) the second group consists of cyclists and hunters, and (3) the least popular group consists of motorists and moped riders (*Figure 9.6*). In Norway, Haakenstad (1975) has also found a certain aversion to motoring in the forest. Not surprisingly, great differences are found in the preferences of different segments of the population, as, in general, there is most sympathy for the type of visitors to whom the individual feels most related to, and considerably less sympathy for the rest.

Concluding remarks

In the above, primarily *mean* estimates are referred to. However, in conclusion it is important to notice that if a landscape manager seeks to consider the interests of the "average" person only, he will only satisfy a *few* people; whereas a *varied* supply of opportunities that is in harmony with the surroundings and the *different* wishes of the visitors, can provide for the interests of a lot more people. In addition, surveys have shown that the managers' own preferences, or their perception of the preferences of the population, do not always correspond to the actual preferences of the population, for example, in connection with recreation facilities (Jensen 1990).

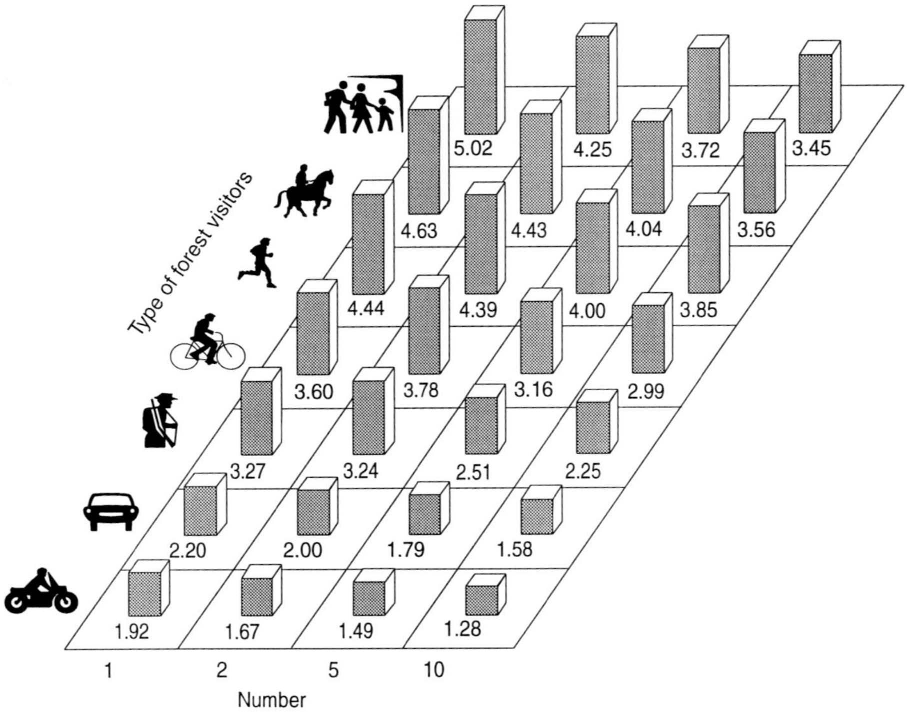

Figure 9.6 The Danish population's mean assessment of meeting various types of forest visitors, depending on their number (7: highest score, 1: lowest score) (Koch & Jensen 1988).

It is also interesting to make a note of a classification of the kind of comments Danish respondents in the survey by Koch and Jensen (1988) have put forward on the postal questionnaires. Of a total of 1,075 comments, *31 %* were related to recreation facilities including recreational use–related influences, while, for example, the silvicultural system plus the choice of tree species plus the tending of stands only got a total of *4 %* of the comments! This result indicates the need for detailed and comprehensive preference surveys. Insufficient knowledge of the different preferences of individual forest visitors might easily result in the landscape manager primarily seeking to comply with the wishes of the more articulated groups, which does not necessarily provide the best solutions. This is expressed by Gregory (1972) as landscape management in conformity with "the squeaking wheel principle".

Finally, Reunala (1984) points out that preference estimations can be useful for some purposes, but maybe their usefulness is more limited than generally believed. This might be due to forest experience being much wider than it appears, for example, it can include also symbolic values such as the forest as a mother symbol, and as an archetype.

9.4 Nature interpretation and nature schools

Before a description of the conditions in the individual Nordic countries (based on Naturvägledning... 1990), a short and more general discussion of the subject will be given.

In all Nordic countries nature interpretation is considered to be a very important part of both recreation and tourist policies. Furthermore, in many countries all over the world nature interpretation is considered to be one of the means of carrying out a modern nature protection policy. Such a policy does not only endeavour to protect strategic areas and formulate restrictions on those using nature, but also to increase understanding of conservation of the environment and species.

The Outdoor Life Project Group of the Nordic Council of Ministers has agreed on a common Nordic definition of nature interpretation (Naturvägledning... 1990):

"Nature interpretation is dissemination of feelings and knowledge about nature. Nature interpretation aims at increasing understanding of the basic ecological and cultural relations and of the part people play in nature. This way, the possibilities of positive experiences in nature and the possibilities of increasing concern for the environment of the individual and of society as a whole are improved".

The same project group has set up seven important subaims for nature interpretation. Nature interpretation can contribute to:

1) Encouraging simple outdoor recreation on the conditions of nature and in conformity with Nordic traditions.
2) Counteracting disturbances and damages to sensitive parts of nature.
3) Increasing the understanding of the necessity of different kinds of environmental and nature protection.
4) Contributing to a mutual understanding between landowners and the different groups of recreational forest visitors.
5) Increasing the understanding of how people use and influence nature seen in the perspective of cultural heritage.
6) Disseminating knowledge of how human activities influence the ecological system.
7) Encouraging development in harmony with nature and natural resources.

Denmark

Nature interpretation is not a new concept in Denmark. In the 1950s, nature interpretation was already included in information on nature, through signs, leaflets, exhibitions, guided trips, movies, etc. This work is still in progress. The National Forest and Nature Agency alone has produced about 100 leaflets on walking tours. Some of these have been translated into English and German.

In the 1960s, the first Danish nature school was formed outside Copenhagen. Today, there are about 25 nature schools which are run by the National Forest and Nature Agency and the local municipalities together, and other 25 nature schools which are run by the municipalities alone. During a one day visit, the class follows a program devised by the principal of the nature school. The aim of the visit is to give the child an understanding of nature through direct experiences. The forestry organizations have also produced information materials. One example is the teaching material "Forests in Schools, Schools in Forests" (*Figure 9.7*).

Another important aspect is the establishment of "forest/nature kindergartens". These are defined as kindergartens where the children, all year round and without facilities, are out in a forest area all the day. The interest in this kind of day–care has increased dramatically in Denmark: in 1991, 66 "nature kindergartens" were registered and in 1993 the number is estimated to be 3–4 times higher. Normally it is the municipality which organizes this type of day–care in cooperation with the local forest administrations.

An analysis in the 1980s showed that non–governmental organizations (ornithologists, scouts, anglers, etc.) were doing comprehensive and valuable work for nature interpretation, but the different interests were not coordinated and the work was done by unpaid volunteers. Therefore, the Ministry of the

Figure 9.7 Children participating in the Danish program "Forest in Schools, Schools in Forests". Photo: O. Vorre.

Environment decided to arrange nature interpretation more formally on a national level. Nature interpretation met with a favourable reception from the Danish population. During 1987–88, more than 50,000 people had contact with nature interpreters and their activities.

Special training in nature interpretation does not exist in Denmark. People with quite different educations are used as nature interpreters, and the National Forest and Nature Agency has devised a compulsory further training programme for the nature interpreters. Until today (1993), about 120 persons have taken part in this education.

Norway

The Norwegian parliament has recently taken a number of initiatives to strengthen outdoor recreation, among other things through "education, instruction and information". In an account about national parks in Norway from the mid–80s, a more direct attempt to deliver information and interpretation in cooperation with the national park management is proposed.

Activities and interpretation of outdoor recreation have been taken care of by non–governmental organizations and the educational system. Tourist associations, hunting and fishing associations, orienteering, skiing, and scout organizations have all played a part in further stimulating developments.

In recent years, the educational system has become more active as regards outdoor recreation. One of its aims is that all pupils, at least once, participate in a "camp school". The forest authorities also contribute through the project "Learn with the Forest" to learning in the recreational areas near the school.

Several nature interpretation centres have been established in connection with nature protection projects (*Figure 9.8*). There are leaflets which provide information on nature reserves, other nature conservation areas, and attractive outdoor recreation areas. The latest hiking maps are a combination of a map and an information brochure, where nature conditions are also described. Many municipalities hand out maps of outdoor recreation areas and opportunities for activities; forest or "nature" trails have also been established over a number of years throughout the country. For further description of nature interpretation in Norway, see Vistad (1993).

Sweden

Within the Swedish compulsory school system, nature schools are established as an extension of education on the environment and nature. In this way, "the child's curiosity and love of adventure are stimulated through direct experiences". The first nature school was established in 1982. Today there are about 25 schools. In recent years, a variant has been established, the so called "nature and culture school". To some extent, the nature schools also offer

Figure 9.8 A nature interpretation centre in Jostedalsbreen National Park, Norway. Photo: R. Bierach.

their services to the adult part of the population in the form of guided walks, excursions, lectures, etc.

As mentioned earlier, the forest organizations in Denmark and Norway have prepared teaching material. Actually, Sweden is the pioneer country in Scandinavia through its programme "Learn with the forest".

The establishment of visitor centres spread over the country has been for the benefit of both outdoor recreation and tourism. These information centres have been established in connection with nature reserves and other places that have a great number of tourist visits. At present, about 25 centres are spread over the country.

Different kinds of nature trails are becoming more common. At a rough estimate, there are 200 established nature trails in the country. Several different ways of disseminating information are used. The most usual way is through signs, but also leaflets, audio–tapes and guides are used. Finally it is worth mentioning that Sweden has put a lot of effort in information about the Right of Public Access (*Figure 9.9*).

Finland

The utilization of national parks and other conservation areas is increasing in Finland. This has led to the establishment of visitor centres in some of these areas. The aim is to comply with increased interest from the public and tourism through raising the level of service.

Right and Wrong in the countryside

About the Right of Public Access

In Sweden, the Right of Public Access allows you to roam about freely or to go camping in the countryside. You are entitled to enjoy the fragrance of the flowercovered meadows, the singing birds and the peaceful silence of the deep forest. But, please...

DON'T DISTURB AND DESTROY!

If you keep the following rules in mind you will be sure of keeping on the right side of Swedish laws and customs.

Figure 9.9 Swedish information leaflet on "the Right of Public Access".

Nature interpretation is also carried out at a local level. The bigger towns have for several years arranged free guided walking tours in the green open spaces. These tours are often guided by students and have a specific theme. Like in the rest of the Nordic countries, the non–governmental organizations arrange nature excursions, camps, etc. all over the country for children as well as adults.

Since 1969, mushroom consultants have been trained, and from the mid 1980s herb consultants have also been trained. The main objective at the courses has been to supply the students with sufficient knowledge to know and to distinguish the different mushrooms and herbs and to treat and prepare them correctly.

The first courses of "Wilderness guides" were carried out at the end of the 1970s. The main elements in the training have been wilderness skills, ecology, and customer–minded service. A certain interest in employing the trained wilderness guides can be found in the tourist associations and the hotel chains, and many of the trained wilderness guides have started their own businesses.

In addition, a lot of leaflets and maps are handed out or are on sale in order to promote nature interpretation and information about attractive outdoor recreation areas. Most of these exist in a Finnish as well as in a Swedish version. Some can also be found in English and German.

Iceland

With the increase in leisure time of the Icelanders, and the growing number of foreign tourists, the need for nature interpretation in Iceland has increased. As well as the other Nordic countries, Iceland has produced information about nature areas in the form of, for example, signs and leaflets. A relatively large part of the leaflets is translated into English, German, and French, and a part even into another Nordic language.

The Icelandic tourist association was founded in 1927. It offers various kinds of trips and walking tours. The guides working with the tourist groups are trained in nature, culture and language. This training was initiated in 1965.

In some national parks, daily nature interpretation is offered during the tourist season in the form of guided walking tours. Special interpretation for children has been carried out with great success. Guides are also found along hiking trails in the parks, where they concentrate on explaining about the vulnerability of nature.

In preparatory school, a part of the teaching is carried out in nature. It is considered to be very important that the teachers should have increased possibilities of taking the pupils outdoors in nature. In 1985, the first and till now only nature school was founded within the framework of a non–governmental organization. The school is situated near the capital Reykjavik. School classes can stay overnight but most classes come on one day visits. The teachers have to arrange for instruction themselves as there is no "nature teacher" at the nature school.

Closing remarks

Common to all Nordic countries is the aim to educate children and young people through information about nature and environment. These countries also have in common that to a great extent they make use of special training of nature interpreters. It appears to be very impressive when the visitors meet qualified and well–trained people in nature; perhaps even more impressive than other media such as exhibitions, leaflets, movies, etc. However, these methods are very valuable, too. In all Nordic countries, a large number of leaflets and maps are published with information about attractive outdoor recreation areas.

As regards development in tourism, nature interpretation is of fundamental importance. It helps the tourists to enjoy their experience of nature and contributes to their knowledge of vulnerable ecosystems.

Basic and common to the Nordic countries is an increasing demand and need for nature interpretation. There are several reasons. One is improved knowledge of the necessity of a rich and varied environment. Another is the intense competition between, on the one hand, the landowners' interests, and on the other hand, the users of outdoor recreation opportunities. A third reason for an increasing need for information is the growing number of tourists from other countries who emphasize outdoor recreation. The methods used in the field of nature interpretation in the Nordic countries are more or less similar, and work with the concept has been intensified in all the countries since the 1980s.

9.5 Urban forests and green spaces

Recreation in – or close to – urban settings is a relative conception, which has different meanings according to the situation, and it is to some extent interpreted differently in the Nordic countries. The following short discussion, mainly based on two sources (Nærrekreation – friluftsliv... 1987, Naturvägledning... 1990), gives a general overview of the importance of recreation possibilities in an urban setting. A more detailed discussion of outdoor recreation in urban settings is given by, for example, Gåsdal (1993).

Great importance is attached to good possibilities of outdoor recreation in everyday life in all these countries. In this connection, it should be mentioned that the absolutely most important reason for the afforestation program in Iceland is to increase the outdoor recreation possibilities of the population in urban forest settings. In the ambitious Danish afforestation program (where the objective is to double the Danish forest area during the next approximately 100 years) urban afforestation is also included as a very important element.

A trend is that increasing traffic, new traffic barriers, and expanding towns have increased the distance to "nature" and made access to the urban open spaces more difficult. In all Nordic countries, the municipality has the main responsibility for management of recreation interests in urban settings. Therefore, an important part of local planning is formed by physical planning and management – in which outdoor recreation possibilities have to be taken into account.

The groups with low mobility especially need good recreation areas in urban settings. Many senior citizens and walking–impaired citizens cannot choose their recreation area themselves. The same thing applies to certain handicapped groups (*Figure 9.10*).

Although many Nordic towns have relatively close contact with nature when seen in general perspective of the Western World, it is evident that a growing segment of Nordic people will live in towns with little contact with nature. This is unfortunate, not only because of the values of outdoor recreation, but also because a lack of experiences in nature in childhood *might* create an adult population without any (or with a feigned) interest in managing and preserving the natural resources of society.

Figure 9.10 It is important for handicapped groups to have the opportunity to enjoy recreation activities in nature. Photo: J. Holt.

9.6 Future

General perspectives in forest recreation

"It is difficult to foretell – especially about the future" a Danish jack–of–all–trades once said. This is very much to the point for the contents of this section. To give an estimate of future outdoor recreation applying to all Nordic countries will only be an informed guess.

Teigland (1990) expresses some ideas about outdoor recreation of the 1990s in Norway. Kaltenborn (1993) sets up some hypotheses on how changing society might influence outdoor recreation in the future. Based on my own and my two Norwegian colleagues' conceptions, the following section will try to look at some tendencies in the Nordic countries with the reservation in mind that there probably will be some differences in the trends in each individual country.

If we concentrate on the most popular and traditional sides of outdoor recreation like, for example, walking and bicycling in the forest, and on fishing and cross–country skiing, it appears that the level of activity has been rather constant during the last 20–25 years. However, it is not just a Nordic phenomenon that the *traditional* outdoor recreational habits have been rather steady. A survey from the USA has shown that only very few activities have changed in a dramatic way, even during a period of 20 years (Bevins & Wilcox 1980). Presumably, the reason for this is to be found in the fact that Nordic people as well as US citizens do not easily change their traditional hobbies, especially if they furthermore have invested a lot of time and money in developing skills and buying equipment. Such investments have taken place on a relatively large scale during the last couple of decades. However, the relatively steady outdoor recreation pattern does not exclude the occurrence of certain expected changes – and in some cases they are already in evidence:

- Increased unorganized outdoor recreation in forest/nature (e.g. the jogging/fitness movement, waterbased recreation and a greater interest/awareness of nature in general).
- Increased organized outdoor recreation in forest/nature (e.g. the scout movement).
- Increase in the types of sport that use forest as sports ground (e.g. orienteering).
- Increased risk of conflicts because of the specialized use of nature as recreative area (e.g. snowmobiling vs. cross–country skiing).
- A wish for "thrill–seeking" activities (e.g. mountainbikes, paintball/survival games).

And finally there seems to be trends towards:

- A wish for increased quality of leisure time as well as an increase in tourism.

If the above mentioned tendencies appear to be correct, these processes represent a big challenge for all people involved in forest recreation and forest management.

In this connection, it is interesting that according to foreign surveys, the norms and values that people are introduced to in childhood and youth form the basis for their future interests (Sofranko & Nolan 1972). In addition, a survey from 1991 (Recreation today... 1991) shows that a majority of Americans (59 %) began their favorite recreational activity in their childhood.

If these results apply to the Nordic countries too, the result *might* be a decline in recruiting to what could be referred to as more "traditional" Nordic forest recreation activities (such as walking in the forest, berry and mushroom picking and cross–country skiing). The earlier generations who were taught to appreciate these activities will be replaced by generations with a different or at least a more diversified basic attitude. Also, increasing urbanization might influence participation in the traditional "harvesting" –activities.

A demographic alteration which can be expected is the tendency to a higher level of education of the Nordic people. It is, of course, difficult to define what effects this will have on forest recreation, but it might lead to an increase in activities which at present are practised mainly by segments of the population with a relatively high level of education (e.g. wilderness backpacking and orienteering).

Finally, it could be discussed how probable changes in societies, like large (and permanent?) unemployment, a larger and more mobile and wealthy group of retired pensioners, relatively small youth cohorts, and a change in the ethnic composition of the Nordic population will influence trends in recreational patterns. As scientists, we probably just ought to realize that our knowledge of the facts and the alteration processes is not sufficient to form a basis for precise predictions about developments in the long term (*Figure 9.11*).

Area competition

When working with management of nature and environmental qualities and the resulting opportunities for experiences, it is important to distinguish between two main types of areas. One type is areas like national parks, conservation areas, and outdoor recreation areas which society has set aside for the sake of nature protection and/or recreational purposes. Even if the protection interests, through a considerable effort (and a presumable increase in support from the population), would be able to extend such areas in the years to come, they would hardly account for more than a very small percentage of the total area of the Nordic countries. In the rest of the area (maybe 90–95 %), outdoor recreation interests have to compete with other user interests – such as town, industrial and traffic extensions. In this respect, multiple–use planning will become of essential importance so that *all* interests are assessed in connection with the future development of an area.

Figure 9.11 What will the future bring in forest recreation? Photos: F. S. Jensen, O. Vorre, E. Oksanen.

As regards the areas which are set aside for nature protection and/or outdoor recreation, the main problem will be to maintain the various nature and environment qualities simultaneously with the improvement of facilities for different users. This is caused by a variety of wishes and demands from different user groups which cannot be met simultaneously everywhere.

Presumably, the future will show the necessity of increased user influence in the planning phase, just as there will be a demand for a clear overview of the types of experiences/qualities that are offered in a special area – whether it is qualities like solitude, unspoilt nature or improved recreation facilities and services.

To what extent a considerable improvement in facilities is wanted in connection with outdoor recreation cannot be described unambiguously. In the long term, it might prove more reasonable in terms of economy as well as ecology to teach visitors to use the nature areas in the most careful way and in a way that needs minimum improvement and maintenance of facilities. If such a strategy is chosen, the years to come will be a considerable challenge to governmental as well as non–governmental organizations to develop effective education and nature interpretation programs. Such programs should also include the increasing number of tourists from the more densely populated countries in Central Europe that are expected in northern Europe.

User fees

The tradition of relatively free access still prevails, but significant increases in user fees for some of the more capital–intensive opportunities have occurred since 1980 all over the Western World. These trends are likely to continue and be expanded to other opportunities in most countries.

It seems certain that the days of widespread relatively free publicly provided outdoor recreation opportunities are coming to an end in some countries, especially in Canada and the United States. In the Nordic countries, the practice of making land, especially state–owned land, readily available at low costs to the users, and the principle of Everyman's Right (Right of Public Access), have modulated use of many of the rationales for fees. These cultural traditions will probably persist in the near future and limit fee increases. The largest increases will probably be for activities requiring special capital investments, for example horse riding, fishing, etc. (Driver & Koch 1986).

What is happening in the United States?

It is not an unknown phenomenon that the trends seen in the U.S. sooner or later appear in one way or another in Europe. (In this context, comparative studies like the workshop proceedings on long–distance trails, edited by Sievänen (1992b), is of great value). In conclusion, it might be interesting briefly to call attention to some of the U.S. trends related to forest recreation (Cordell 1992):

- – The public is more concerned and informed than ever before about the environment.
- – Traditional resource management priorities are being challenged.
- – Agency priorities are tending towards amenities management.
- – Congress is shifting budgets to fund amenities management.
- – New partnerships and expectations are being forged.
- – The public has greater awareness and more avenues for involvement and influence.
- – Decision–making has become more complex and more scrutinized.
- – Recreation management must be factually, not emotionally, based in order to compete with other resources and to meet intense scrutiny.
- – Decline in available leisure time.
- – Increasing immigration.
- – Increasing number of "DINK's" (Dual Income No Kids).
- – Increasing growth in metro areas.
- – 1970s trend of rural growth reversed.
- – Increasing growth of the non–white population.

And what does it all mean? It *might,* for example, mean:

- – More urban people with less understanding of natural systems, but a growing interest in how the agencies and owners manage them.
- – A greater diversity of people with less time and a growing demand for a diversity of high quality recreation opportunities.
- – Greater place attachment and interest in learning and seeing (non–consumptive activities such as bird watching versus hunting; more fishing will change to catch and release).
- – Management emphasis needs to focus on accommodations, appearance, information, education, and being in touch.
- – A greater need for useful data in relation to outdoor recreation.

Perspectives in forest recreation research

The Nordic forest recreation research has been fluctuating from around 1970 until today. It seems that the individual countries have all had their "vigorous" periods as regards research within this field.

Sweden and Norway have been the pioneering countries, while it seems that Finland has had the most steady research over the years. The same relative steadiness is found in Denmark, but with considerably less resources available. In recent years, Norway has contributed positively by establishing an outdoor recreation research group as a subdivision of NINA (Norwegian Institute for Nature Research). This research group has published a competent monograph, describing outdoor recreation research in Norway (Kaltenborn &

Vorkinn 1993). In Iceland, it has not yet been possible to establish a research group for outdoor recreation.

Research in outdoor recreation in the Nordic countries up till now can be characterized by the need for descriptive studies (and mainly of the adult population). The following examples are a sample of future research topics given by Kaltenborn (1993), which more or less seems to fit all the Nordic countries:

- Repeated surveys to identify trends.
- Motives, attitudes and barriers.
- Obtained benefits, both individual and social.
- Effects of management interventions, such as nature interpretation.
- Retired pensioners and children seen in the perspective of outdoor recreation.
- The "close–to–home" outdoor recreational life.

Besides outlining some of the tasks for the future, the list above also clearly reveals the challenge involved in the discipline, *outdoor recreation research*, being situated somewhere between natural and social science.

As regards future research, the following items seem to be of essential importance in order to complete research of high quality in the relatively small research environments in the individual Nordic countries:

- Outdoor recreation research groups in the individual countries are secured a stable economic funding to enable *continuity* of the work.
- *Coordination* of research between the individual countries in order to secure optimal exploitation of the limited resources.
- *Liaison* projects are initiated to make use of larger external sources of knowledge and funding, like, for example, the European Union (EU).
- *Exchange* of scientists.
- Common research *seminars*.
- Construction of *interdisciplinary* research groups.

In conclusion, *cooperation* seems to be one of the keywords for future progress in Nordic outdoor recreation research.

References

Aasetre, J. 1992. Friluftsliv og skogbruk: en litteraturstudie. (Outdoor recreation and forestry: literature review.) Norsk institutt for naturforskning, Utredning 34. 52 pp. (In Norwegian.)

Aasetre, J. 1993. Friluftsliv i skog. (Outdoor recreation in forest.) In: Kaltenborn, B.P. & Vorkinn, M. (eds.). Vårt friluftsliv: aktiviteter, miljøkrav og forvaltningsbehov. (Our outdoor recreation: activities, environmental aspects and administration requirements.) Norsk institutt for naturforskning, Temahefte 3: 25–32. (In Norwegian.)

Bevins, M.I. & Wilcox, D.P. 1980. Outdoor recreation participation: analysis of national surveys, 1959–1978. University of Vermont, Vermont Agricultural Experimental Station Bulletin 686.

Blöndal, S. 1991. Forest recreation in Iceland. Unpublished. 19 pp.

Cordell, K. 1992. Keeping recreation management on top in an information age. Papers for the Utah State University Recreation Shortcourse. Southeastern Forest Experiment Station, Athens GA. Unpublished.

Driver, B.L. & Koch, N.E. 1986. Cross–cultural trends in user fees charged at national outdoor recreation areas. Proceedings from 18th IUFRO World Congress, Division 6, Ljubljana. p. 370–385.

Fridberg, T. 1989. Danskerne og kulturen: de 16–74 –åriges fritidsaktiviteter i 1987, sammenlignet med 1975 og 1964. (The Danes and culture: the leisure activities of the 16–74 year old people in 1987, compared to the situation in 1975 and 1964.) Socialforskningsinstituttet, Rapport 89:8. 181 pp. (In Danish.)

Friluftsliv –undersøkelse 1974. Summary: Outdoor life study 1994. 1975. Statistisk Sentralbyrå, Norges Offisielle Statistikk A 725. 133 pp.

Fritid – friluftsliv. En undersökning av vanor och önskemål hos den vuxna tätortsbefolkningen 1973. (Leisure time – outdoor recreation. A study of the habits and preferences of the adult population in urban areas in 1973.) 1974. Statens Naturvårdsverk 1974:20. (In Swedish.)

Gåsdal, O. 1993. Uteliv i byen. (Outdoor recreation in towns.) In: Kaltenborn, B.P. & Vorkinn, M. (eds.). Vårt friluftsliv: aktiviteter, miljøkrav og forvaltningsbehov. (Our outdoor recreation: Activities, environmental aspects and administration requirements.) Norsk institutt for naturforskning, Temahefte 3: 33–41. (In Norwegian.)

Gregory, G.R. 1972. Forest resource economics. The Ronald Press Company, New York. 548 pp.

Haakenstad, H. 1972. Skogbehandling i et utfartsområde: en opinionsundersøkelse om Oslomarka. Summary: Forest management in an area of outdoor life: an investigation of public opinion about Oslomarka. Meldinger fra Norges Landbrukshøgskole 51(16): 1–79.

Haakenstad, H. 1975. Skogskjøtsel i rekreasjonsområder: skog og friluftsliv i to modellområder i Oslomarka. Summary: Forestry in recreation areas: forest and outdoor recreation in two model areas in Oslomarka. Norges Landbrukshøgskole, Institutt for Skogskjøtsel, Ås. 174 pp.

Heglebäck, T. 1978. Rörligt friluftsliv i tre rekreationsområden i Stockholmstrakten: Nackareservatet, Järvafältet och Lovön. (Motion recreation in three recreation areas in Stockholm region: Nackareservatet, Järvafältet and Lovön.) Sveriges lantbruksuniversitet, Avdelningen för landskapsvård, Rapport 10. 84 pp. (In Swedish.)

Hultman, S.–G. 1983. Allmänhetens bedömning av skogsmiljöers lämplighet för friluftsliv. 2. En rikstäckande enkät. Summary: Public judgement of forest environments as recreation areas. 2. A national survey. Sveriges lantbruksuniversitet, Avdelningen för landskapsvård, Rapport 28. 91 pp.

Jaatinen, E. 1974. Recreational utilization of Helsinki's forests. Folia Forestalia 186. 35 pp.

Jensen, F.S. 1990. Landscape managers' and politicians' perception of the forest and landscape preferences of the population. Proceedings from XIX IUFRO World Congress, Division 6, Montreal. p. 61–72.

Jensen, F.S. 1992. Befolkningens anvendelse af Vestamager til friluftsliv, 1985–88. (The utilization of Vestamager for recreation, 1985–88.) Forskningscentret for Skov & Landskab, Lyngby. Unpublished. 138 pp. (In Danish.)

Kaltenborn, B.P. 1993. Hva så? – Noen utfordringer i forskningen på og forvaltningen av friluftsliv. (What then? – A few challenges to research on the administration of outdoor recreation.) In: Kaltenborn, B.P. & Vorkinn, M. (eds.). Vårt friluftsliv: aktiviteter, miljøkrav og forvaltningsbehov. (Our outdoor recreation: activities, environmental aspects and administration requirements.) Norsk institutt for naturforskning, Temahefte 3: 136–139. (In Norwegian.)

Kaltenborn, B.P. & Vorkinn, M. (eds.). 1993. Vårt friluftsliv: aktiviteter, miljøkrav og forvaltningsbehov. (Our outdoor recreation: activities, environmental aspects and administration requirements.) Norsk institutt for naturforskning, Temahefte 3. 141 pp. (In Norwegian.)

Kardell, L. 1972. Bogesundslandet som rekreationskälla: en studie av ett skogsområdes utnyttjande till rörligt friluftsliv. Summary: Bogesundslandet as a source of recreation: a study of the utilization of a forest area for motion recreation. Forskningsstiftelsen Skogsarbeten, Redogörelse 4. 40 pp.

Kardell, L. 1979. Farfars friluftsliv – om han hade något. (Grandfather's outdoor recreation – if he had any.) Sveriges Skogsvårdsförbunds Tidskrift 1: 6–21. (In Swedish.)

Kardell, L. 1982. Hur Linköpingsborna utnyttjar sina stadsnära skogar? (How the people of Linköping utilize the surrounding forests?) Sveriges lantbruksuniversitet, Avdelningen för landskapsvård, Rapport 23. 86 pp. (In Swedish.)

Kardell, L. & Holmer, M. 1985. Friluftslivets förändringar på Bogesundslandet 1969–1982. (The changes in outdoor recreation in Bogesundslandet, 1969–1982.) Sveriges lantbruksuniversitet, Avdelningen för landskapsvård, Rapport 33. 117 pp. (In Swedish.)

Kardell, L. & Pehrson, K. 1978. Stockholmarnas friluftsliv: vanor och önskemål. En enkät- och intervjustudie. Summary: Stockholmers outdoors: use of nature areas. A mail questionnaire and a home interview study. Sveriges lantbruksuniversitet, Avdelningen för landskapsvård, Rapport 13. 112 pp.

Kellomäki, S. 1977. Polut ulkoilun kanavoinnissa. Summary: Potential of trails in guiding recreational activity. Silva Fennica 11(4): 263–268.

Koch, N.E. 1978. Skovenes friluftsfunktion i Danmark. I del. Befolkningens anvendelse af landets skove. Summary: Forest recreation in Denmark. Part I: The use of the country's forests by the population. Det Forstlige Forsøgsvæsen i Danmark, vol. 35: 285–451.

Koch, N.E. 1980. Skovenes friluftsfunktion i Danmark. II del. Anvendelsen af skovene, regionalt betragtet. Summary: Forest recreation in Denmark. Part II: The use of forests considered regionally. Det Forstlige Forsøgsvæsen i Danmark, vol. 37: 73–383.

Koch, N.E. 1984. Skovenes friluftsfunktion i Danmark. III del. Anvendelsen af skovene, lokalt betragtet. Summary: Forest recreation in Denmark. Part III: The use of forests considered locally. Det Forstlige Forsøgsvæsen i Danmark, vol. 39: 121–362.

Koch, N.E. & Jensen, F.S. 1988. Skovenes friluftsfunktion i Danmark. IV del. Befolkningens ønsker til skovenes og det åbne lands udformning. Summary:

Forest recreation in Denmark. Part IV: Preferences of the population. Det Forstlige Forsøgsvæsen i Danmark, vol. 41: 243–516.

Levnadsförhållanden 1982–83: Fritid. (Living conditions 1982–83: Leisure time.) Sveriges officielle statistik, Statistiska centralbyrån, Rapport 56. (In Swedish.)

Mattsson, L. & Sødal, D.P. (eds.). 1989. Multiple use of forests: economics and policy. Proceedings of the Conference held in Oslo, Norway, May 1988. Scandinavian Forest Economics 30. 184 pp.

Nærrekreation: friluftsliv og fysisk planlægning i Norden. (Recreation close to urban areas: outdoor recreation and physical planning in the Nordic countries.) 1987. Nordisk ministerråd, Nord 1987:15/Miljørapport 1987:4. 164 pp. (In Danish.)

Nærrekreation. (Recreation close to urban areas.) 1990. Nordisk Ministerråd, København. 16 pp. (In Danish.)

Naturvägledning i Norden. (Nature interpretation in the Nordic countries.) 1990. Nordisk ministerråd, Nord 1990:52. 66 pp. (In Swedish.)

Niemi, I., Pääkkönen, H., Rajaniemi, V., Laaksonen, S. & Lauri, J. 1991. Vuotuinen ajankäyttö. Ajankäyttötutkimuksen 1987–88 taulukot. (Time use annually. The tables of the time use studies 1987–88.) (In Finnish.)

Pouta, E. 1990. Ulkoilualueen virkistyshyötyjen taloudellinen arviointi. (The economic valuation of the recreational benefits of a recreation area.) Pro gradu – tutkielma. Helsingin yliopisto, Kansantaloudellisen metsäekonomian laitos. (In Finnish.)

Recreation today: trends and statistics. 1991. Recreation Roundtable, Washington, D.C. Unpublished. 5 pp.

Reunala, A. 1984. Forest as symbolic environment. In: Saastamoinen, O., Hultman, S.–G., Koch, N.E. & Mattsson, L. (eds.). 1984. Multiple–use forestry in the Scandinavian countries. Communicationes Instituti Forestalis Fenniae 120: 81–85.

Saastamoinen, O. 1979. Valaistun hiihtoreitin käytön ajallinen vaihtelu. Summary: Time patterns in the use of an urban skiing route. Silva Fennica 13(1): 101–106.

Saastamoinen, O. & Sievänen, T. 1981. Keravan ja Rovaniemen lähimetsien ulkoilukäytön ajallinen vaihtelu. Summary: Time patterns of recreation in urban forests in two Finnish towns. Folia Forestalia 473. 24 pp.

Saastamoinen, O., Hultman, S.–G., Koch, N.E. & Mattsson, L. (eds.). 1984. Multiple–use forestry in the Scandinavian countries. Communicationes Instituti Forestalis Fenniae 120. 142 pp.

Sievänen, T. 1989. The trail inventory and forest recreation. In: Mattsson, L. & Sødal, D.P. (eds.). Multiple use of forests: economics and Policy. Proceedings of the Conference held in Oslo, Norway, May 1988. Scandinavian Forest Economics 30:89–105.

Sievänen, T. 1992a. Aulangon ja Ahveniston ulkoilualueiden käyttö ja kävijät. (The users and utilization of the recreation areas Aulanko and Ahvenisto.) Metsäntutkimuslaitoksen tiedonantoja 415. 70 pp. (In Finnish.)

Sievänen, T. (ed.). 1992b. Nordic Outdoor Recreation: international Comparative Studies. Proceedings of the Workshop held in Siuntio, Finland, 9–10.9.1992. Finnish Forest Research Institute, Research Papers 439. 144 pp.

Sievänen, T. 1993. Kaupunkiväestön ulkoilukäyttäytyminen ja ulkoilualueiden käyttö: Hämeenlinnan ulkoilututkimus. Summary: Outdoor recreation household survey in the city of Hämeenlinna. Folia Forestalia 824. 62 pp.

Sofranko, A.J. & Nolan, F.M. 1972. Early life experiences and adult sports participation. Journal of Leisure Research 4(1).

Teigland, J. 1990. Veivalg for norsk friluftsliv i 1990 –årene. Om nordmenns friluftsliv i 1970 og 1980 –årene, og perspektivene framover. (Choosing the way for Norwegian outdoor recreation in the 1990s. Norwegians' outdoor recreation in the 1970s and 1980s and future perpectives.) Lands–konferansen Friluftsliv, juni 1990, Bodø. Unpublished. 11 pp. (In Norwegian.)

Tveit, K.J. 1979. Fritid, kultur og kino. Arbeidsrapport fra en kinobruksundersøkelse i Stavanger. (Leisure, culture and movies. Working report of a study on the movies visiting activity in Stavanger.) Universitetet i Bergen, Geografisk Institutt. 56 pp. (In Norwegian.)

Ulkoilututkimus 1979. (Outdoor recreation study 1979.) 1980. Sisäasianministeriön ympäristönsuojeluosaston julkaisu C 1, n:o 5. 45 pp. (In Finnish.)

Vistad, O.I. 1993. Losing, læring eller sjekking? Om naturrettleing som forvaltningstiltak. (Guiding, teaching or checking? Nature interpretation as a management tool.) In: Kaltenborn, B.P. & Vorkinn, M. (eds.). Vårt friluftsliv: aktiviteter, miljøkrav og forvaltningsbehov. (Our outdoor recreation: activities, environmental aspects and administration requirements). Norsk institutt for naturforskning, Temahefte 3: 121–127. (In Norwegian.)

Vuolle, P., Telema, R. & Laakso, L. 1986. Näin suomalaiset liikkuvat. (This is the way how the Finns move.) Liikunnan ja kansanterveyden julkaisuja 50. Valtion painatuskeskus, Helsinki. (In Finnish.)

10 Forest aesthetics

Christina Axelsson Lindgren[1]

Abstract

Attention has been paid to forest aesthetics mostly in urban and recreational areas, and in areas with exceptional landscape qualities. Due to the Right of Public Access to private land, forest aesthetics is also essential in the countryside. The visual appearance of forests is of major concern in many Nordic works of literature, music and painting. Since the 1970s, research on forest aesthetics in the Nordic countries has mainly focused of forest interiors. The aesthetics of forests as part of broader landscapes has also been studied to some extent. Physical characteristics of importance to visual appearance seem to be tree species and ages, and to what extent they are mixed, as well as the density and structure of the stands. Experiments in management of the visual qualities of forest have been carried out, for example, in Järavallen and in Furulunds fure in Sweden. Today, computer graphics in combination with geographical information systems and simulation techniques make it possible to create visual scenarios of past and future landscapes.

Keywords: aesthetics, landscape, forest structure, perception, preferences, landscape management.

---

[1] Nordic Institute for Societal Planning
P.O.Box 1658, S–111 86 Stockholm, Sweden

10.1 Introduction

In Finland, Norway and Sweden, forests are larger than any other of the landscape elements. Landscapes are often covered by forests, and only fields, lakes and moors help to open up the scenery. In Denmark and Iceland, the areas covered with forests are much smaller. In these countries, forests are spectacular elements in the open landscapes.

Forests play a considerable role in landscape aesthetics in the Nordic countries, but general landscape aesthetics is not the scope of this chapter. In connection with multiple–use forestry, it is most relevant to consider aesthetics that may be influenced by forest management, or by avoiding forestry. This chapter thus deals with forest aesthetics, mainly concerning forest interiors, as these aspects have been most studied in the Nordic countries. To some extent, aesthetics of forests as a part of broader landscapes are also touched upon.

In forest aesthetics, the interest is focused on visual evaluation of forests. Attention to forest aesthetics has been paid mostly in urban and recreational areas, and in areas with exceptional visual landscape qualities. Due to the Right of Public Access to private land, forest aesthetics is also essential in countryside forest areas. The chapter concentrates on describing research on forest aesthetics in the Nordic countries from 1970 onwards. Some management essays with great originality are also presented and discussed.

10.2 Cultural importance of forest aesthetics

In forestry debate in the Nordic countries, great attention has been paid to forest aesthetics. In several countries there are works analyzing descriptions of forests, above all in literature, but also in art and music. The visual appearance of forests is of major concern in these descriptions. The complexity of the expectations Norwegian people may attach to a forest area is described in analyses of literature, sketches and photographs especially from Oslomarka (Oraug et al. 1974). Recently, a seminar on nature experiences in Oslo illustrated this theme in psychology, forestry, landscape architecture and art (Naturopplevelse i skog... 1992). The Swedish anthology "Sinnenas skog" alternates between literary descriptions of experiences of forests and descriptions in popular science of conditions for forest production (Sörlin 1983). A deep cultural approach to the importance of forest in Finland was conducted in the seminar and publication "The forest as a Finnish cultural entity" (Reunala & Virtanen 1987). The importance of forests was discussed in this multidisciplinary meeting from many viewpoints: philosophical, silvicultural, semiotical, aesthetic, etc. A summary of Danish literature on forest aesthetics from around 1900 and onwards has been compiled by Borup (1991).

Aesthetic experience of his surroundings has always been important to man. Hjort (1983) states in his dissertation that, in purely agricultural societies, aesthetic considerations were a part of the daily cultivation of forest and open land. The skilled craftsmanship in these cultivations has brought about many of the cultural landscapes that today are still appreciated the most. By household multiple–use cultivation, natural vegetation has been formed into meadows, enclosed pastures, dense forests, heaths, burn–beaten land, and so on.

At the realization of industrial society in the 1930–1950s, when functionalism had its era of prosperity in housing architecture, functionalistic forest aesthetics became all the more dominating (Frivold 1991). In the transition from craftsmanship to mechanical and industrial production, matters of design were closely associated with technical development and with the possibilities of mass production. This was the case also in housing architecture. "Functionalistic forest aesthetics" (see Frivold 1991) has its roots in the 19th century, when the forest industry was established. The view of forest as a producer of raw material became all the more dominant and effective even–aged monocultures became increasingly á la mode. Recently, post–modern trends, that in the last decade have highlighted the complexity of design issues in housing architecture, have also been noticed within forestry in the Nordic countries. Considerations emphasized in connection with multiple–use forestry and discussions of a multitude of forestry methods and ideals of forests have thus gained renewed relevance.

10.3 Perception of forest stands

In the Nordic countries, there is a concentration towards research on perception of forest stands. Research on perception of broad–scale forest landscapes is not so usual. Perhaps this is due to the good possibilities in the Nordic countries to wander in the forests. The studies often deal with preferences of forest stands, i.e. which stands are preferred least or most. In environmental psychology, dimensions of experience are used to further express the visual similarities and differences between, for example, forest stands. This approach to environmental perception contains possibilities of enhancing the use of knowledge on visual aspects of forests in multiple–use planning and design situations.

In Norway and Finland, studies on people's perceptions of forest stands, mainly concerning preferences, were carried out already in the early 1970s. The Norwegian studies (Haakenstad 1972, Lind et al. 1974) show that mixed forests with both coniferous and deciduous trees of different ages, and with glades that vary the density of the stands, are preferred. The greater the share of old trees in forest stands, the more appreciated they are (Loven 1973, Kellomäki 1973, Mikola 1973). Similar results are found by Hultman (1983), Kardell (1990) and Savolainen and Kellomäki (1981). In a Finnish study (Pukkala et al. 1988), a pure, old and rather sparsely stocked birch stand got the highest preferences for beauty, while pure and old pine stands were the most preferred for recreation (see also Loven 1973, Kellomäki 1973, Savolainen & Kellomäki 1981). In spruce stands, the mixture of other species constantly increased the beauty and recreation evaluations. The effect of the tree species depended on the age and size of the trees. In a very young stand, pines and birches were not better liked than spruces (Pukkala et al. 1988). In young coniferous stands, though, the scenic value increased when the proportion of intermixed birch exceeded 20 % (Savolainen & Kellomäki 1981). The beneficial effects of pines and birches increased as the stands got older. In connection with the adjective sum method used by Pukkala et al. (1988) and several other Finnish studies, it should be kept in mind that the adjective sum does not measure exactly scenic beauty or suitability of the stand for recreation. It is simply a sum of different stand parameters, such as variability and originality, which do not always increase amenity (Pukkala et al. 1988).

Rather different results are given in a study by Koch and Jensen (1988), where the photos to a large extent show broadleaved deciduous forest common in Denmark. Studies conducted in other Nordic countries have very few photos of broadleaved deciduous forest. The Danish subjects like deciduous forest better than coniferous forest, especially in summertime. Beech forest is most preferred. High preferences for beech forest are also found in Sweden (Hultman 1983). Deciduous forest gets more popular with increasing age of the trees (Koch & Jensen 1988). This is not the case with coniferous forest, as age has no effect on its popularity. Mixed deciduous forest is almost as preferred as the old, even–aged beech forest, one of the most beloved forests of the Danes. Otherwise, forest stands with mixed ages generally are liked better than even–aged stands, because stands with mixed ages have many layers. In wintertime, young stands of spruce are just as much liked as young beech stands. It is essential to notice that mixed coniferous forest with different ages, many old trees and a lot of deciduous trees – i.e. the forest type often preferred in other Nordic countries was not included in the study by Koch and Jensen (1988).

Most Norwegians did not want foreign tree species in the forest, when answering a verbal question (Haakenstad 1972). In Sweden, forest stands of *Pinus contorta* were perceived to be somewhat brighter and more unnatural than other middle–aged coniferous stands in the field (Kardell & Wallsten 1989). Otherwise, the stands did not appear different in pleasantness or in suitability for outdoor life.

Twigs are irritating and disturbing in all kinds of forests, according to a number of studies (Haakenstad 1972, Lind et al. 1974, Hultman 1983, Koch & Jensen 1988, Savolainen & Kellomäki 1981). The more twigs on clearcuttings, the more negative they are perceived to be. Clearcuttings are generally one of the most negative elements in the forest (Haakenstad 1972, Lind et al. 1974, Kardell 1990b, see also Aasetre 1992). Recently stump–pulled clearcuttings are perceived as very negative (Kardell 1989). Smaller clearcuttings may rather easily be accepted (Haakenstad 1972, Hultman 1983, Kardell 1990, Pukkala et al. 1988), while large clearcuttings are perceived as very negative by the great majority (see also Lind et al. 1974). Seed trees make the impression of clearcuttings less negative (Haakenstad 1972, Hultman 1983). The visual effect of seed trees remains also ten years after the cutting is done, as regeneration by seed trees is considered more natural (Kardell 1989).

The great amount of fallen trees make virgin forest unsuitable for outdoor recreation (Hultman 1983, Savolainen & Kellomäki 1981). Danes prefer beech forests without dead logs (Koch & Jensen 1988). Standing, winding old pines are perceived as positive elements in forests by Norwegians (Lind et al. 1974). Environmental protectionists and biologically educated people are positive to dead and fallen trees, while other people want such elements to be removed (Kardell 1990, see also Savolainen & Kellomäki 1981). However, these conditions do not prevent the mere consciousness of the existence and conservation of a virgin forest from meaning a lot to the individual. Rare visits can also lead to exceptional and valuable experiences (Frank S. Jensen, pers. com. 1993).

10.4 Combinations of forest stands

Physical characteristics of importance to visual variation seem to be tree species and ages and to what extent they are mixed, as well as the density and structure of the stands (Axelsson Lindgren 1990, see also Haakenstad 1972, Aasetre 1992, Savolainen & Kellomäki 1981). The study by Koch & Jensen (1988) indicates that Danes prefer forest areas with both deciduous and coniferous stands, especially if the majority of the area is covered by deciduous trees. Few Norwegians prefer pure deciduous forests (Haakenstad 1972, see also Aasetre 1992). Just as in the Danish study, a great majority of visitors to Oslomarka do wish, however, to see birch, rowan and other deciduous trees as components of the forest landscape (Lind et al. 1974). In Finland, Mikola (1982) states that broadleaves provide lots of lightness in a coniferous forest landscape, and *Prunus* and *Sorbus* add more colour.

So far, perception of forest interiors of single stands has been discussed the most. As Hultman (1983) suggests, variation between forest stands may well be just as important to the recreation experience. According to Kellomäki (1973), mixed forests are liked the most, as the richness in colours and shapes relaxes and gives new inspiration. Variability of forest stands may give more aesthetic and recreational experiences, as the attraction depends on the sequence of stands (Pukkala et al. 1988), and on the manner in which they are arranged.

Diversity of forest landscapes is appreciated, as for example changes between mixed forests and monocultures, variation in age, height and density, and variation in the degree of cultivation of forests (Loven 1973, Kellomäki

1975, Mikola 1982). High diversity in forest stands may, however, be perceived as monotonous when spread over large forest areas (Kellomäki 1973, Koch & Jensen 1988). Less appreciated forest types, such as clearcuttings and young tree cultivations, may contribute in a positive way to the variation if they are present in small amounts (Kellomäki 1975). Small clearcuttings may, in a positive way, open the closed canopy and the views (Pukkala et al. 1988, Kellomäki 1978). Diverse structure of age classes may thus be positive, as every age class may contribute with its kind of recreational value. Proportion and location of tree species are also important to perceived differences between forest stands.

A study on what people perceive as similarities and differences between forest stands (Axelsson Lindgren 1990) resulted in five similarity groups:

– Young deciduous stands.
– Forest with mixed density and mixes of species and age.
– Old, parklike deciduous forest.
– Coniferous forest with high stems.
– Young, very dense coniferous stands (*Figure 10.1*).

It is probable that the similarity groups represent extremes of visual forest types, so that spectras of visual forest types may be identified between similarity groups. The study showed extraordinary correspondence between expert and layman evaluations. This may indicate that visual variation is a quite fundamental aspect of landscape experience, where few different opinions may be found. At the same time, there are reasons to believe that you may find many individual differences in the perception of visual nuances.

For those who plan and design recreation forests, it is important to be able to evaluate the effects of combinations or sequences of forest environments. Field studies of strolls along forest trails show that an increased number of visual forest types along a trail increases the desire to practice different open–air activities (Axelsson Lindgren 1990, 1988a). Evaluations of the length of the trails and the time required to walk them were more accurate as the number of visual forest types increased.

In a study on both sides of Öresund, a strait dividing Sweden from Denmark, Swedish and Danish students from agricultural universities evaluated environments around Swedish and Danish forest paths (Axelsson Lindgren 1988a). On each side, one trail passed through few, and one through many, visual forest stands. Just as Koch & Jensen (1988) found, the Danes were very positive to all four of the forest trails described below. In this study, however, the Danes liked the uniform, uneven–aged mixed coniferous forest trail the best. They also found this trail to be the most special one. The trail through uniform beech forest was perceived as the most usual trail by the Danes, and this was also the trail least liked by them. The two varied trails were evaluated in between the uniform trails.

From left upper corner:

1) Young deciduous stands.
2) Coniferous forest with high stems.
3) Forests with varying density and mixture of species and age.
4) Old, park–like deciduous forest.
5) Young, very dense coniferous stands.

Figure 10.1 Similarities and differences between forest stands. The five photos each represent a similarity group of distinct visual forest types, according to a perception study (Axelsson Lindgren 1990).

The Swedes evaluated the pleasantness and originality of the four trails in completely the opposite way. Most appreciated by the Swedes was the trail through uniform beech forest, which they also considered most special. The two varied forest trails were evaluated as more pleasant than the uniform mixed coniferous forest trail by the Swedes, although earlier research (Haakenstad 1972, Lind et al. 1974, Hultman 1983) indicates that this should be the trail most appreciated. The uniform mixed coniferous trail was also the most usual one, according to the Swedes.

As this Swedish–Danish trail study indicates, the covariation between variation and preferences is probably not linear. An all too marked visual variation between forest stands may even have negative effects. The study also gives examples of how dimensions of experience such as pleasantness, variation, originality and unity can covariate with each other.

10.5 Design and management of visual variation

By identifying forest visual opportunity spectra it may be possible to create multiple–use alternatives for forests, where visual aspects as well as, for example, interests of environmental protection and forestry are taken into account. Each specific visual forest type gives specific possibilities and limits for forestry as well as for conservation aspects. Opportunities are thereby given to plan with patterns of possibility–combinations for the different aspects concerned (Axelsson Lindgren 1990).

A study of the possibilities of working practically with a forest visual opportunity spectrum was carried out in the forest of Järavallen (Axelsson Lindgren 1990). The experience of every visual forest type was described verbally and registered using Küller's method for semantic environmental description (1972). The method consists of eight dimensions of experience: pleasantness, complexity, unity, openness, social status, potency, affection and originality. This semantic instrument has been tested all over the world, and the dimensions are proved to be stable and usable in all kinds of cultures. It is based on people's experiences of built–up environments but was successfully used in forest stand interiors (Axelsson Lindgren 1990). Some adjustment to the method for semantic environmental description may, however, be needed to captivate the specific experiences fostered by vegetation, such as forests, which may activate other dimensions of experience than built–up environments.

Together, the dimensions of experience in the semantic instruments captivate the atmosphere in each visual forest type. The vegetation structure in each visual forest type was described by an inventory form, where aspects of presumed importance for identification of visual forest types were registered. Verbal scenarios of the spontaneous development of the vegetation in the 28 visual forest types were created. Further, suggestions for management of each of the visual forest types were made in order to maintain the existing visual character described by Küller's method. It may be possible to use similar techniques to develop the shaping of visual forest stands towards specific future aesthetic goals. In all, the forest area in the study was 225 hectares.

Another example of research for management of the visual qualities of forest is conducted in the "Furulunds fure" forest (17 hectares). From discussions of principles of design and architecture of forests and from the structural composition of the forest, management recommendations were made. The maintenance plan proposed restoration measures as well as the initiation and resumption of continuous maintenance measures. Six different main characteristics were presented, from well–defined patches of natural woodland which are left untouched, to cultivated "pillared halls" of pine and beech. In addition, rich–in–species and multilayered forest stands and the opening of more glades and meadows were proposed. An important suggestion was the planting of attractive woodland flowers to enhance the experience of spring and early summer (Gustavsson & Fransson 1991) (*Figure 10.2*).

In a smaller urban forest (around 2 hectares), Falck and Rydberg (1990) promoted twelve different forest types by specific management. The aim was to study the possibilities of steering the composition of forest stands towards goals other than pure forestry. Forest stands used were, for example, the rowan forest, the uneven–aged spruce forest and a glade with border zone.

The study by Falck and Rydberg (1990) actualizes the question of minimum sizes for visual forest types. In mixed forests, researchers on forest structure found a minimum size of forest stands to be around 0.25 hectares. For dense, young forest and shrubs, one might identify stands of half that size, i.e. around 30 x 30 m. More open forest stands, looking like pillared halls, should probably have a minimum size of at least 100 x 100 m (Gustavsson & Fransson 1991). It is plausible that a visual forest type ought to cover areas quite a bit larger than these minimum sizes, so that the visitor has time to experience the character of the stand before entering a new visual forest type.

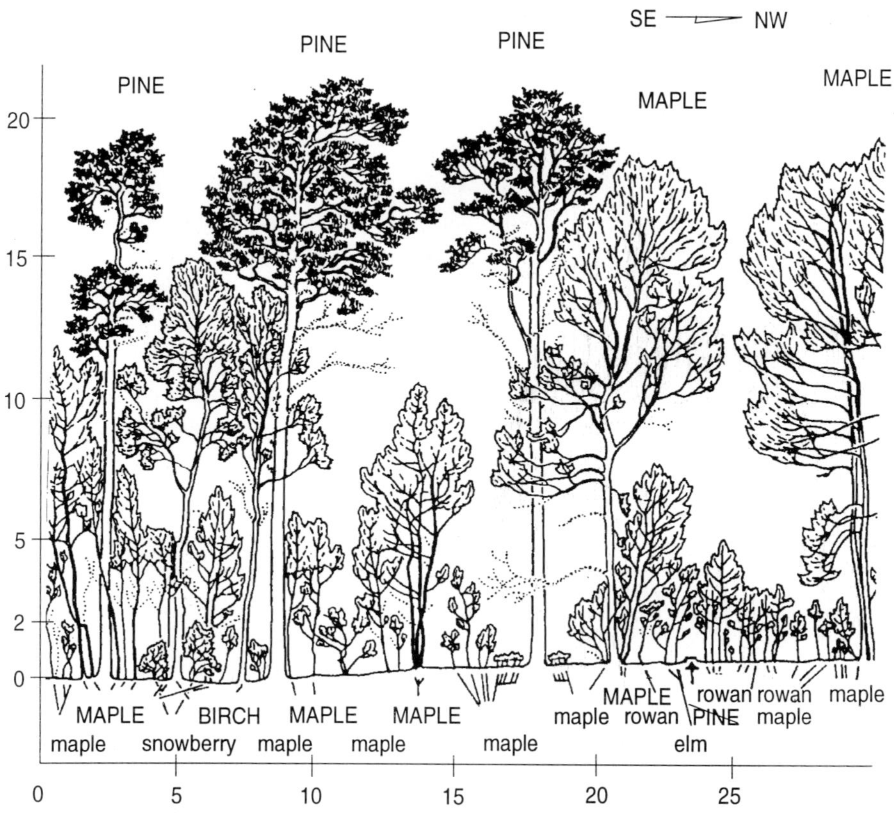

Figure 10.2 An example of a profile from Furulunds fure, where the structural composition of the forest is analyzed (Gustavsson & Fransson 1991).

Figure 10.3 Winter landscape in eastern Iceland with *Larix*, *Pinus contorta*, *Picea glauca* and indigenous *Betula*. Photo: Sigurdur Blöndal.

10.6 Forest and the landscape

Aesthetic management of forest as a landscape element has been debated above all in connection with large clearcuttings and afforestation of agricultural land. The national landscape of Koli in Finland represents an image of Finnish national identity which is important to preserve (described in chapter 11 by Minna Komulainen). In Iceland, foresters test tree species from suitable climates all over the world for the future afforestation (Blöndal 1991). In this work, the positive effects of variated forests are emphazised (*Figure 10.3*). Aesthetic argumentation along the entire range in between these two extreme examples may be found in all of the Nordic countries.

In open landscapes, the problems often concern the creation of new forest stands in concurrence with the environment. Choices of species are important, as also the connection of the forest to existing or created groves, hedges, shelter belts, and so on. The shapes of border zones are especially important for the experience of forests in open landscapes. The length of border zones in the landscape has been used as a way to measure the degree of variation in landscapes (Skärbäck 1980). In Denmark, forest landscape aesthetics is one of five aspects, apart from wood production, which have to be taken into account in the multiple use of forests as prescribed by the law (Koch & Kristiansen 1991). This is especially important in publicly owned forests. In the Danish project "Forest and People" (skov og folk), preferences for forests as parts of broad, open landscapes were studied, as well as preferences for forest interiors (Koch & Jensen 1988).

In connection with the on–going afforestation in Denmark, that could result in a forest area twice as large as the present total forest area in the country, Borup (1991) actualizes the aesthetic aspects of border zones. She furthers the handbook tradition from forest authorities in the USA and Great Britain in developing a practical method for aesthetic design of forest border zones. She mainly combines the Visual Resource Management System (VRMS) from the USA with a method which she calls "The Six Principles" from Great Britain. Other Danish reports (Koch & Canger 1987, Koch & Kristiansen 1991), give examples on how the VRMS criteria: line, texture, color and form, in combination with contrast, sequence, axis, convergence, co-dominance and enframement, may be used in aesthetic management of forests in the landscape. A source of inspiration for Danish reports is also the work of Dame Sylvia Crowe. She often let shifting in selfbred forests due to changes in soil, moisture, and height of the site inspire the design (Crowe 1966, 1978).

Gustavsson (1988) gives recommendations on afforestation of cultural land, inspired by the VRMS, in the course compendium "Forest on agricultural land". He also mentions the importance of distinctive elements, of developing the complexity of the landscape, of preserving the feeling of openness, and of sticking to older cultural patterns, as well as developing completely new ones. In this course compendium, a more thorough analysis on afforestation and landscape complexity based on existing research is also given (Axelsson Lindgren 1988b).

10.7 Discussion and concluding remarks

In design of forest landscapes, perception of relations between forest areas with different visual qualities becomes very essential. Evaluations of preferences are in a static way attached to singular, uniform areas. Working with specific views, as in many handbooks, is also static. Several authors, such as Borup (1991), and Pukkala et al. (1988), ask for methods to handle the changes in forest over time. The future challenges for aesthetics of forest in the landscape are to describe differences and similarities between different forest stands as well as between larger forest areas and their surroundings, in a nuanced way. This should concern both description of the geographical dispersement of visual qualities in present landscapes, and different kinds of aesthetic changes in forests in the past and in the future.

Work with aesthetics of forests as part of broader landscape in the Nordic countries has often been inspired by the kind of handbooks Borup (1991) has used. At best, these handbooks are based on professional experiences of aesthetic forest management. Very seldom are they based on research of people's perceptions and experiences of forests. Parts of the British method used by

Borup are clearly based on pictures of graphics and geometrical patterns such as are used in perception psychology. Researchers in this field firmly state that they have never said anything about people's perceptions of environment. Environmental psychology has grown up since around 1960, exactly because perception psychology is incapable of describing people's perceptions of environments.

Dimensions of experience used in environmental psychology (see Küller 1972) concern the moods people experience in specific environments. The VRMS to a greater extent concerns the qualities of forest volumes as part of a view, which shares similarities with the analysis of painted art. There are reasons to believe that descriptions of perceived landscapes become considerably different from each other, depending on whether the landscape is described as a picture, i.e. as an object, or as a three–dimensional room, i.e. as a place to experience different moods in. The criteria texture, color and form in the VRMS –system are actually found to be dimensions of experience for objects (Sorte1982). Other research has shown that the VRMS –criteria might rather be expressions of expert praxis, in work with landscape and forest views. The criteria are not so much expressions of lay people's experiences of landscape and forest (Grden 1979, Kopka & Ross 1984). How expert knowledge may be related to lay people's experiences of forests is discussed by Axelsson Lindgren (1990).

A positive aspect of Borup's method (1991) is the formulation of aesthetic goals for the design of forest border zones. Thereby, the designer explicitly chooses to make aesthetic decisions from a spectrum of possibilities. Comparisons between existing and future environments are difficult to make, though, as Borup does not use the same aesthetic documentation of existing environment for describing aesthetic goals for the future. The method Borup uses for spatial analysis of existing forest landscape is developed for city environments. From the results obtained by Borup, it is doubtful whether it is relevant to use such a method in spatial analyses of forest landscapes.

The semantic instrument used by Axelsson Lindgren (1990) for describing the experiences of forest interiors (Küller 1972) could probably be used for describing experiences of perceived forest landscape. The criteria: pleasantness, complexity, unity, openness, social status, potency, affection and originality seem to be similar to criteria used in other methods in several aspects. As this semantic instrument is developed for built–up environments, adjustments are most likely necessary before it can be conveniently used in forest environments.

Today computer graphics in combination with geographical information systems and simulation techniques make it possible to create database–related visual scenarios of past and future landscapes. There is, however, a need for careful control of what pictures made by computer graphics are actually showing in comparison with real field situations.

References

Aasetre, J. 1992. Friluftsliv og skogbruk: en litteraturstudie. (Outdoor recreation and forestry: a literature study.) Norsk Institutt for Naturforskning, Utredning 34. 52 pp. (In Norwegian.)

Antikainen, M. 1992a. Metsämaiseman suunnittelu Kolin kansallispuistossa. Summary: Forest landscape planning in Koli National Park. Metsäntutkimuslaitoksen tiedonantoja 456. 88 pp.

Antikainen, M. 1992b. Tammimetsien hoito. (Silviculture in oak forests.) Helsingin yliopiston metsäekologian laitoksen julkaisuja 1. 105 pp. (In Finnish.)

Axelsson Lindgren, C. 1988a. Danskars och svenskars upplevelse av variation i skog: en pilotstudie. (Danes' and Swedes' attitudes towards the variation in forest: a pilot study.) Sveriges lantbruksuniversitet, Institutionen för landskapsplanering, Stencil 88:12. 37 pp. (In Swedish.)

Axelsson Lindgren, C. 1988b. Skog på jordbruksmark: betydelsen av landskapets utseende för rekreationsutbytet. (Forest on agricultural land: the importance of the appearance of landscape for the supply of recreation.) In: Skog på jordbruksmark: ekonomi och miljö. (Forest on agricultural land: economy and environment.) Sveriges lantbruksuniversitet, fortbildningsenheten, Kurskompendium. 9 pp. (In Swedish.)

Axelsson Lindgren, C. 1990. Upplevda skillnader mellan skogsbestånd: rekreations– och planeringsaspekter. Summary: Perceived differences between forest stands: recreation and planning aspects. Dissertation. Stad & Land 87. 443 pp.

Blöndal, S. 1991. Socioeconomic importance of forests in Iceland. In: Alden, J., Mastrantonio, J.L. & Odum, S. (eds.). Forest development in cold climates. Plenum Press, New York. 13 pp.

Borup, A. 1991. Landskabelige hensyn og fremtidige skovbryn. (Landscape aspects and future forest edges.) Den Kgl. Veterinær– og Landbohøjskole, Frederiksberg. 181 pp. (In Danish.)

Crowe, S. 1966. Forestry in the Landscape. Forestry Commission Bulletin 18. HMSO, London. 32 pp.

Crowe, S. 1978. The Landscape of Forests and Woods. Forestry Commission Booklet 44. HMSO. London. 47 pp.

Falck, J. & Rydberg, D. 1990. Skogsskötselmodeller för tätorternas skogsklädda grönområden. (Silvicultural models for the urban forest areas.) Statens råd för byggnadsforskning, Stockholm. 51 pp. (In Swedish.)

Frivold, L.H. 1991. Synen på blandskog genom tiderna. (The way to look at mixed forests through the ages.) Skog & Forskning 2: 6–10. (In Swedish.)

Grden, B.G. 1979. Evaluation and recommendations concerning the visual resource inventory and evaluation systems used within the Forest Service and the Bureau of Land Management. Proceedings of Our National Landscape, 23–25 April 1979, Incline Village, Nevada. Pacific Southwest Forest and Range Experiment Station. p. 296–304.

Gustavsson, R. 1988. Ny skog i odlinglandskapet: landskapsvårdsaspekter. (New forest in agricultural landscape: landscape management aspects.) In: Skog på jordbruksmark: ekonomi och miljö. (Forest on agricultural land: economy and environment.) Sveriges lantbruksuniversitet, fortbildningsenheten, Kurskompendium. 13 pp. (In Swedish.)

Gustavsson, R. & Fransson, L. 1991. Furulunds fure: en skog i samhällets centrum. Summary: Furulunds fure: a forest in the centre of the town. Stad & Land 96. 131 pp. (In Swedish.)

Haakenstad, H. 1972. Skogbehandling i et utfartsområde. En opinionsundersøkelse om Oslomarka. Summary: Forest management in an area of outdoor life. An investigation of public opinion about Oslomarka. Meldinger fra Norges Landbrukshøgskole 51(16): 1–69.

Hjort, B. 1983. Var hör människan hemma? (Where does a person feel at home?) Dissertation. Stockholm. (In Swedish.)

Hultman, S.–G. 1983. Allmänhetens bedömning av skogsmiljöers lämplighet för friluftsliv. 2. En rikstäckande enkät. Summary: Public judgment of forest environments as recreation areas. 2. A national survey. Sveriges lantbruksuniversitet, Avdelningen för landskapsvård, Rapport 28.

Kardell, L. 1990. Skog och natur i Nordmaling: en attitydstudie 1986. (Forest and nature in Nordmaling: an attitude study 1986). Sveriges lantbruksuniversitet, Institutionen för skoglig landskapsvård, Rapport 45. 115 pp. (In Swedish.)

Kardell, L. 1991. Talltorpsmon i Åtvidaberg. 1. Förändringar i upplevelsen av skogen mellan 1978 och 1989. Sveriges lantbruksuniversitet, Institutionen för skoglig landskapsvård, Rapport 46. 103 pp. (In Swedish.)

Kardell, L. & Mård, H. 1989. Några gruppers attityder till stubbrytning 1976 och 1988. (A few groups' opinions about stump pulling in 1976 and 1988.) Sveriges lantbruksuniversitet, Avdelningen för landskapsvård, Rapport 41. 87 pp. (In Swedish.)

Kardell, L. & Wallsten, P. 1989. Några gruppers attityder till Pinus Contorta. (A few groups' opinions about Pinus Contorta.) Sveriges lantbruksuniversitet, Avdelningen för landskapsvård, Rapport 40. 55 pp. (In Swedish.)

Kellomäki, S. 1973. Recreational potential of a forest stand. Silva Fennica 12(3): 179–186.

Kellomäki, S. 1975. Forest stand preferences of recreationists. Acta Forestalia Fennica 146. 36 pp.

Kellomäki, S. & Savolainen, R. 1984. The scenic value of forest landscape as assessed in the field and the laboratory. Landscape Planning 11(2): 97–107.

Koch, N.E. & Canger, S. 1987. Skovopbygning til glæde for friluftslivet. (Forestry for joyful outdoor recreation.) Skov– og Naturstyrelsen, Hørsholm. 239 pp. (In Danish.)

Koch, N.E. & Jensen, F.S. 1988. Skovenes friluftsfunktion i Danmark. IV del. Befolkningens ønsker til skovenes og det åbne lands udformning. (Forest recreation in Denmark. Part IV: Preferences of the population as regards the development of forests and the open land.) Det Forstlige Forsøgsvæsen i Danmark, vol. 41: 243–516. (In Danish.)

Koch, N.E. & Kristiansen, L. 1991. Flersidigt skovbrug: et idékatalog. (Multiple–use forestry: a handbook of ideas.) Skov– og Naturstyrelsen, Hørsholm. 39 pp. (In Danish.)

Kopka, S. & Ross, M. 1984. A study of the reliability of the Bureau of Land Management visual resource assessment scheme. Landscape Planning 11: 161–166.

Küller, R. 1972. A semantic model for describing perceived environments. Statens råd för byggnadsforskning, Document D 12. Stockholm.

Lind, T., Oraug, J., Skjervold Rosenfeld, I. & Östensen, E. 1974. Friluftsliv i Oslomarka: analyse av intervjuundersøkelse om publikums bruk og krav til Oslomarka. (Outdoor recreation in Oslomarka: analysis of an interview study on the publics' use and expectations in Oslomarka.) Norsk Institutt for By–og Regionforskning, Arbeidsrapport 8. 96 pp. (In Norwegian.)

Loven, L. 1973. Metsäympäristön viihtyisyystekijät. Summary: Amenity factors in forest environment. Helsingin yliopisto, metsänarvioimistieteen laitos, tiedonantoja 3. 101 pp.

Mikola, P. 1973. Metsätalouden ympäristövaikutukset ja niiden merkitys metsien käytön suunnittelussa. (Environmental impacts of forestry and their significance in the planning of forestry.) Helsingin yliopisto, metsänhoitotieteen laitoksen tiedonantoja 9. 51 pp. (In Finnish.)

Mikola, P. 1982. Suomen metsämaiseman kehitys. Summary: Development of the Finnish forest landscape. Terra 94(1): 56–63.

Naturopplevelse i skog: referat fra et fagseminar. (Nature experiences in forest: a summary of a professional seminar.) 1992. Naturvernforbundet i Oslo og Akershus/Oslo og Omegn Turisforening. 80 pp. (In Norwegian.)

Oraug, J. Östensen, E. Lind, T. & Skjervold Rosenfeld, I. 1974. Forventninger til rekreasjonsmiljøet i Oslomarka. Delrapport fra prosjektet: Friluftsliv i Oslomarka. (Expectations concerning the recreation milieu in Oslomarka. A report from the project: Outdoor recreation in Oslomarka.) Norsk Institutt for By– og Regionforskning, Arbeidsrapport 10. (In Norwegian.)

Pukkala, T., Kellomäki, S. & Mustonen, E. 1988. Prediction of the amenity of a tree stand. Scandinavian Journal of Forest Research 3(4): 535–544.

Reunala, A. & Virtanen, P. (eds.). 1987. Metsä suomalaisten elämässä. Summary: The forest as a Finnish cultural entity. Silva Fennica 21(4): 317–480.

Savolainen, R. & Kellomäki, S. 1981. Metsän maisemallinen arvostus. Summary: Scenic value of forest landscape. Acta Forestalia Fennica 170. 74 pp.

Skärbäck, E. 1980. Öresundsprojektet: landskapsanalys. (Öresund –project: landscape analysis.) Metodrapport, SNV pm 1268. 209 pp. (In Swedish.)

Sorte, G. 1982. Visuellt urskiljbara egenskaper hos föremål i den byggda miljön. (Visually distinguishable properties of objects in built–up environment.) Statens råd för byggnadsforskning, Rapport R 5. 141 pp. (In Swedish.)

Sörlin, S. (ed.). 1983. Sinnenas skog: antologi. (The forest of the senses: an anthology). Falköping. (In Swedish.)

Personal communications

Frank Søndergaard Jensen, researcher, The Danish Forest and Landscape Research Institute.

11 Landscape management in forestry

Minna Komulainen[1]

Abstract

In the past, landscape experience was related to the use of forest; to grazing, hunting and collecting wood. Nowadays, easthetic feelings aroused by forests are often connected with recreation. During the period 1880–1950, many forest managers emphasized the importance of conserving and creating beauty in forests. The modern efficient timber production oriented forestry developed after the Second World War and aesthetic considerations became less important. In the 1980s, the rise in labour costs and changes in people's values have forced forestry to use more natural methods in fellings and other operations. Recent aesthetic trends pay more attention to ecological aspects. This has led to the development of the landscape ecology approach. The need for deeper knowledge of how to adapt forest operations to the aesthetic expectations of people opened the way for landscape research in the Nordic countries in the 1970s. Methods of forest management including cultural and aesthetic considerations have been studied, for example, in Ruissalo and in Koli National Park in Finland. The importance of aesthetics in forestry is still increasing because of tourism, urbanization and recreation.

Keywords: landscape planning, cultural heritage, recreation, forest management, landscape ecology, aesthetics.

[1] Finnish Forest Research Institute
Unioninkatu 40 A, FIN–00170 Helsinki, Finland

11.1 Introduction

Scandinavian landscape is typically small–scale forest landscape, where fragmented variation is created by lakes, fields and cutting (*Figure 11.1*). Forests surround also cultural landscapes. Forest management has a great impact on landscape, and the abrupt changes due to geometrical fellings have provoked strong criticism towards forest industry. For example, in autumn 1993 the stopping of clearcuttings was intensively demanded in Finland. Experience of the landscape is related to feelings, and thus it has become more important in forest management to adjust the needs of economic use of forests with the other forest values. This can be done with the help of landscape research, political strategies and silvicultural guidelines.

Both the landscape experience and the Nordic forestry have changed in the course of time (Geelmuyden 1989a). According to Miettinen (1993), consideration of beauty in silviculture in Finland can be divided into three time periods:

1) Time before the First World War (before 1917):

Forest aesthetics is developed following the central European models. Beauty is presented as part of idealistic education; the beauty of forest occurs when forests are managed.

2) Time from World Wars to 1965:

Silviculture has developed to an independent sector. Vividness of forest views is created through ordinary timber management, and through management of park forests, sceneries, forest edges, and also through protecting forest for nature conservation purposes.

3) Time from 1965:

Beauty is included in the new concept of multiple–use forestry. Landscape research is started.

11.2 Forest aesthetics

Forest scenery is a part of the everyday living environment for the Scandinavians. Aesthetic experiences are felt when moving in the landscape and when viewing the landscape from a distance. Human beings have a need for beauty, security and variation in the landscape where they live in (Landskapsplanering... 1992). Forest landscape is perceived as elements (colours, smells, sounds), visual qualities (space, shapes) and qualities like unity, contrasts and scale (Sepänmaa 1987, Horelli 1982). While observing the environment, one also judges its aesthetic quality (Sepänmaa 1986).

Aesthetic qualities have been studied by assessing the beauty of forests, regarding them as aesthetic objects (Sepänmaa 1986, Borup 1991). Forest aesthetics describe the aesthetic experience as a result of different senses in a cultural context. According to Sepänmaa (1987), the *Forstästhetik* can also mean principles of aesthetic forest management. This concept was first presented by a German, Heinrich von Salisch (1885).

Forest landscape contains many immaterial values, which vary according to cultural history, traditional land use and architectural style (Linkola 1983). Landscape values often reflect the cultural heritage of the country. The character of northern Scandinavian landscape is wild, natural and closed, due to the forests. It differs from the general landscape aesthetics in Germany and England, where cultivated landscape is more common.

The aesthetics of Nordic forests has been studied from landscape paintings, literary descriptions and folklore (Schulin 1949, Sepänmaa 1978). Perception studies measure the landscape values of modern people, but the roots of these values have been affected by cultural history. In a study of landscape heritage, forest aesthetics may explain why certain landscape types are respected. Some landscape values remind us of archetypes in people's minds, and some come from works of art (Reunala 1987).

11.3 History of forest aesthetics

In the 19th century, forest landscape was shaped by farmers and their traditional land use. The forest was a setting for their everyday life; it was used for grazing and shifting cultivation (Linkola 1983). Natural landscape was increasingly transformed into fields and meadows. In the past, landscape experience has been related to the use of forest landscape; to collecting wood, grazing and hunting. Nowadays, aesthetic feelings aroused by forests are often connected with recreation (Geelmuyden 1989a). Until the 19th century, the word "landscape" (Swedish and Norwegian: landskap, Danish: landskab, Finnish: maisema) in Scandinavian languages meant land, area, county and soil, while it nowadays is often considered as the scenery of the surrounding areas (Keisteri 1990).

At the end of the 19th century, untouched landscapes were seen from a romantic point of view (Borup 1991). This was related to the nations' birth and development. For example, in Norway and Finland, where economic development went hand in hand with strong cultural identity, the natural resources, including forests, were respected (Geelmuyden 1989a).

In Finland, the myth of landscape heritage generally applies to areas where natural elements of landscape are in powerful contrast (e.g. narrow eskers surrounded by water) and where human activity has had only slight influence (Antikainen 1993). The earliest idea of nature and landscape is found in the national epos, Kalevala. Its landscape aesthetics reflect the Finnish natural landscape, with forests and lakes, and also the use of forests for shifting cultivation (Sihvo 1984).

Appreciation of natural landscapes arose in the 19th century. The expansion of shifting cultivation moved the border of wilderness landscape to eastern Karelia and North Finland. Wild nature landscapes were appreciated in Finnish painting and literature. Scenes with lakes and forest–growing eskers from the Finnish lake district were represented as an ideal landscape. An ideal northern panoramic landscape was presented by the poet, J.L. Runeberg (Laitinen 1984). His descriptions of summer time scenery, small–scale variation of woods, lakes and fields became the archetypal landscape for Finns. Even nowadays, the most appreciated landscapes are areas where many landscape types meet.

Later this symbolic and ethical landscape became more recognised and located in a certain area (Klinge 1984). A Finnish writer, Zacharias Topelius, described the character of different landscape areas in his poems and writing. He appreciated the mild, cultivated landscape, but also described wild forest landscape. Topelius named the three main elements of Finnish landscape: granite rocks, coniferous woods and lakes (Suutala 1986). Also the contrasts in topography and the verdure of the shifting cultivated birch forests were important in his descriptions. Many landscapes presented in "Our Country", a famous book on Finnish geography, history and folklore, written by Topelius in 1887, are nowadays appreciated by Finns as landscape heritage. For example, the Punkaharju eskers and the Puijo hill offer famous views over blue lakes, seen from a high hilltop, and through crooked pines.

Figure 11.1 A typical landscape from Finland. Photo: M. Komulainen.

Figure 11.2 Artists have often depicted the Finnish landscape by means of forest bordered views over a lake, seen from a high hilltop. The Koli landscape is a prime example of the embodiment of these elements: a lake, a high place, crooked pines and verdure. Eero Järnefelt: "Koli". Gouache, 1927. (The Neste Collection, Espoo).

At the end of the last century, national landscapes started to reflect people's image of their home country. Artists like Axel Gallen–Kallela, Jean Sibelius, and other "Karelianists" travelled to areas of Kalevala's birthplace. In their works, the panoramic landscapes, for example, from the Koli ridge or from esker areas, represented the genuine Finland (*Figure 11.2*). Coniferous forests became symbols of wild, free and untouched nature, located in the hinterland or the border zones of culture. Groves and broadleaved trees symbolized the expansion of culture and civilization (Julkunen & Kuusamo 1987). A writer, Juhani Aho (1893) divides the landscape of Koli, one of the most famous national landscapes, into the wild, heathen eastern panoramic landscape and into the Christian, shifting cultivated landscapes dominated by deciduous trees.

Landscape experiences are related to cultural context. Appreciation of landscape types has become more diversified. In the beginning of the 19th century, only cultivated landscape was highly appreciated – virgin nature meant chaos and a threat. Romanticism brought appreciation of wilderness landscapes: cliffs, wind–broken trees and a stormy sea. But it was not until the beginning of the 20th century, with the rise of nature conservation, that a monotonous natural landscape, like a plain bog, and the outer archipelago, also started to represent the beauty of virgin, barren nature (Sepänmaa 1978).

11.4 Foresters as managers of nature and landscape 1880–1950

As the first foresters learned methods of silviculture from Germany in the 19th century, they were also influenced by the cultural era of National Romanticism (Geelmuyden 1989a). In Scandinavia, discussion of forest aesthetics started in the late 19th century when the Danish Opperman (1897) wrote about a Romantic point of view to nature in the book "Skoven, Skovbruget og Det Skönne" ("Forest, Forestry and Beauty"). He was also inspired by von Salisch's book *"Forstästhetik"* (1885), where aesthetics was connected to silvicultural methods. Opperman points out that the beauty of forest is born from contrasts in environment, and the forest should give an impression of nature, not culture. These principles of forest aesthetics were carried further by the director of State Forests, Lorenzen (1918), who said that a forester's duty was to conserve and create beauty (Borup 1991). This same thought was also emphasized by the Finnish silviculturist, P.W. Hannikainen, who wrote one hundred years ago of how foresters awoke negative feelings in the public when forest practices were carried out against naturalness and a sense of beauty. According to him, economic benefits and aesthetics exist side by side in forestry (Hannikainen 1893).

With knowledge of natural sciences and wide experience of the state of forest nature during the intensive shifting cultivation period, foresters also were active in the newly founded organizations for nature conservation (Reunala & Heikinheimo 1987). The idea of protecting the most sensitive and beautiful areas was suggested by the first Norwegian foresters, P. Chr. Asbjörnsen and J.B. Barth, in 1864 (Geelmuyden 1989a). In Finland, A.E. Nordenskiöld was the first to present the idea of establishing state–owned nature parks in 1881. Due to his famous suggestion, intensive discussion started in scientific societies. The Forest Research Association suggested the establishment of national and nature parks in Finland in 1906. In 1910, the National Forest Board started to protect valuable areas in state–owned forests (Cajander 1909).

In eastern Finland, an intensive shifting cultivation culture developed in connection with the settlement of wilderness areas. By the end of the 19th century, it was estimated that there were around 4 million hectares of shifting cultivated areas, 20 % of which were without forest cover. The large burnt areas and signs of tar production in the Ostrobothnian region, in western Finland, provoked a fear of deforestation. Due to intensive shifting cultivation and selection cuttings by measure, the first forest law was issued to protect forests in 1886. It was emphasized that the "forests should not be devastated". This also formed the main principle of later forest laws (Reunala & Heikinheimo 1987).

In writings about the Finnish landscape, intensive human acts against the landscape were considered to cause destruction of aesthetic values (Inha 1924, Kalliola 1949, Sepänmaa 1978). A hundred years later, traditional land use and natural succession have made the shifting cultivated landscapes more harmonious and appreciable, due to the fact that these areas are now covered by beautiful birch forests.

The concepts of scenic beauty and ecological values were not especially emphasized in forestry at the beginning of this century, but they were included in silvicultural methods (Geelmuyden 1989a), when farmers were taught how to regenerate forests spoilt by shifting cultivation, tar production and by selection cuttings by measure. In Finland, the theory of natural forestry was developed by Erkki K. Kalela in the 1940s (Kalela 1949). According to Kalela, all management of forests should be done in such a way that man follows the natural processes.

In Finland, ideas of landscape aesthetics in forestry were promoted by Torsten Rancken and Reino Kalliola. They emphasized naturalness in landscape management, using biological laws as guiding principles of beauty. This differs from von Salich's *Forstästhetik*, where aesthetic management varies according to the styles being applied (Kalliola 1949). Later, Kalliola's aesthetic criteria were analyzed by Yrjö Sepänmaa (1978). In Sepänmaa's study, beauty of nature is classified with the help of the following concepts: harmony, contrast, richness, economy, simplicity, wildness, elaborateness, purposefulness, changeableness, general opinion and knowledge.

Torsten Rancken developed the basis for modern urban forestry. He wrote about how forests should be managed aesthetically when close to towns, villages and recreation areas. He described how to manage trees to develop a well–shaped crown, how to open views and how to enhance views with solitary trees (Rancken 1956, 1964). In Norway, the concept of park forest was created between garden parks and national parks, in 1945, to classify aesthetically important forests which were managed by applying silvicultural methods (Geelmuyden 1989a).

Reino Kalliola, a nature conservation inspector in Finland, described landscape management as aesthetic and social nature conservation, the goal of which is to conserve natural beauty, recreational opportunities and cultural values. Landscape management does not exclude the economic exploitation of an area, but tries to guide it so that human action affects the landscape harmoniously. According to Kalliola (1949), landscape management and nature conservation are separate issues because of their different goals. The goal of nature conservation is to protect the object from human impact, and it has an ethical basis for preserving flora and fauna (Kardell 1991).

Landscape aesthetics and forestry were not in conflict until the Second World War. In this period, silviculture symbolized responsible management of natural resources and national values. Forestry and aesthetics had parallel management goals (Geelmuyden 1989a, Reunala & Heikinheimo 1987).

11.5 Functional aesthetics and efficient forest management 1950–1970

A modern, rational forestry developed after the Second World War. It emphasized economic interests and technical skills, whereas ecological and aesthetic considerations became less important.

In the 1950s and 1960s, after an expansion of clearances even to the hinterlands, an efficient timber management began in Finland. Soon after the guidelines of natural silviculture had been presented by Erkki K. Kalela (1949), new technical skills made it possible to move from natural regeneration to cultivation–based forestry. Clearcutting and planting spread from Lapland to southern Finland. Other new forestry practices were also introduced: site treatment, ploughing, fertilization, and chemical weed and sprout control. Also, technology became more artificial, as machines started to replace man and horse in felling and in other forest work (Reunala & Heikinheimo 1987).

Clearcutting became a symbol of intensive forestry, alongside ploughing (Reunala & Heikinheimo 1987). Until the end of the 1960s, the size of clearcutting in Lapland could be thousands of hectares. When problems occurred in the reforestation of large open areas, strip felling became a common method of regenerating old spruce forests. The resulting rectangular shapes did not adapt to the landscape. Geometrical clearcutting and strips were already used in the 1920s and 1930s on a small scale. When these methods were used in large areas, the fellings were criticized (*Figure 11.3*).

The first guidelines for landscape and nature management were prepared for state owned forests by the National Board of Forestry, and also for private forest by the Central Forestry Board Tapio, in 1970. In Finland, thinning of forests is usually not a problem for landscape. Instead of tree selection, the natural character of landscape can be emphasized, for example, by increasing the amount of deciduous trees. The main problems occur in regeneration felling. Usually the natural methods for regeneration, shelterwood felling and seeding felling, have been favoured in aesthetically important places, because by using them the scenery will change gradually from dense forests to a stand formed by scattered trees. The impacts of regeneration have been softened by leaving solitary trees and tree groups in clearcutting areas (Savolainen 1990).

In Norway, a new forest law focusing on timber production was drafted in 1965. It promoted the strategy of large–scale clearcuttings, building of forest roads and drainage. This type of forestry soon ended up in conflict with nature conservation and aesthetic values of forests. At that time, the landscape experience became a topic in the public debate on forest methods for the first time (Geelmuyden 1989a). The building of forest roads and the large–scale fellings were considered disturbing by Norwegians, because these operations reduced the characteristics of the forest being a wilderness. Also the use of Norwegian spruce and Sitka spruce in afforestation was criticized, because these species

Figure 11.3 Strip felling in northern Finland. Fellings were done in long strips on slopes. The geometrical shapes do not adapt to the character of the rolling landscape. Photo: E. Oksanen.

are not natural in western Norway. Planting of these species has often been done in the middle of deciduous forest and on mountain slopes, where their geometrical shape and distinctive appearance can be seen from long distances.

Conflicts between timber production and recreational values were sharpened in Oslomarka, a forest area surrounding Oslo. This area became so essential, that the events there affected forestry and forest law drafting for the whole country.

Oslomarka has been an important recreation area for a long time. Criticism started in the 1960s, because of the increased building of forest roads. Later, clearcuttings were also criticized (Reunala & Heikinheimo 1987). Due to these conflicts over forest operations, a new nature protection law was passed. The law made it possible to establish special landscape areas to conserve aesthetic values. In Oslomarka, landscape values were given priority in determining management practices (Geelmuyden 1989a).

As a consequence of the conflicts in forestry, more information was spread and compromises in methods were made. In Norway, a new forest law on "forestry and forest protection" (Lov om skogbruk og skogvern) was passed in 1976. Under this law, the recreational and scenic importance of forests was stressed (Geelmuyden 1989a).

Also in Sweden, forestry was rationalized and forest owners concentrated in maximizing their economic benefits from forests in the 1960s. The clearcutting of large, over 100 hectare areas was started in Norrland. In ordinary silviculture, there was a tendency to develop pure conifer forests, and mixed forests were reduced (Reunala & Heikinheimo 1987). Criticism of clearcutting started in 1967. This process led to a regulation which required that proposed fellings should be announced in advance (Kardell 1991, Falk 1991).

A conflict was also created when large spruce sapling stands appeared in the cultural landscape of southern Sweden in the 1960s. Local people and nature conservationists started to complain about the dark spruce fields which replaced the flowering meadows and semi–open pasture forests which are highly appreciated by Swedes. In 1963, the Swedish National Board of Forestry established a working group to conserve cultural landscapes (Kardell 1991). Furthermore, during the 1970s, the amount of beech forests was reduced, because spruce forests were easier to cultivate. Due to the fear that southern Sweden would become more and more coniferous, a new law to protect broadleaved forests (Ädellövskogslagen) was passed in 1986. This law prescribes that all oak, beech, lime and other broadleaved forests which are felled, should be regenerated by broadleaved trees.

Since 1974, landscape aesthetics has been one factor in the policy of the Swedish National Board of Forestry, for example, in its new guidelines for nature– and landscape management (Kardell 1991, Natur och landskapsvård... 1974). The guidelines emphasize the importance of active counselling with forest owners. Reduction of the size of clearcuttings, avoidance of geometrically shaped edges in fellings and protection of beautiful solitary trees are recommended as well. Aesthetic considerations have also been featured in the new Swedish forest law (1993). It contains, for example, regulations for the size of clearcuttings and shapes of cutting areas.

Kardell (1991) makes the criticism that there has not been much landscape management action in practice. In the period 1965–1980, the dominating type of landscape management was opening the forests along the shores and leaving birches and alders evenly spaced. Often these operations were not aesthetically successful. In Finland this kind of clearance work was also for a long time a common management method implemented for aesthetic purposes.

In open landscapes, which are common in Denmark and Iceland, forests are appreciated for recreation but also for ecological reasons as wind and erosion shelters. Afforestation has often resulted in geometrical planting following private borders. New guidelines for adapting the afforestation better to the character of the rolling landscape are being worked out. In Iceland, the open areas, from which trees disappeared hundreds of years ago, are now being partly afforestated in order to prevent erosion and to improve the landscape (Blöndal 1987). These new forests have been highly appreciated by the Icelanders, but some criticism has been made of the cultivation of exotic species (e.g. Sitka spruce). Nowadays more birch is planted, as the amount of natural birch trees has decreased to 1 % of the original area (Arnalds 1987).

11.6 Ecological trends in landscape management in the 1980s

The rise in costs for manual labour and changes in values have forced forestry to use more natural methods in felling and other operations. Forest owners have become urbanized and for many of them timber production is not the most important management objective. A recent study in Finland has shown, that for more than 30 % of forest owners the main uses of their own forest are connected with recreation, i.e. to spending free time, or to affection for their home region (Ihalainen 1992).

Many conflicts between forestry and nature conservation can be caused by an emotional response to changing landscape. Landscape is connected to feelings, and abrupt changes in it cut personal connection to the past. The change in landscape either creates or destroys the feeling of security, depending on how we see or feel our environment.

Recent aesthetic trends show a more ecological approach than before (Sepänmaa 1986, Geelmuyden 1989a, Bramsnæs 1991). Functional and visual aesthetics have been reflected by the rise of landscape ecology (e.g. Forman & Gordon 1986). Sepänmaa (1986 & 1987) talks about ecological aesthetics, where the beauty of nature is not only beauty of forms of visual appearance, but also beauty of natural processes. He says that forest aesthetics should be based on a nature– and culture–ecological approach, where the appropriateness of nature is the criterion for beauty.

In Norway, the guidelines of landscape and nature values in forests highlight the importance of landscape in forestry because of increasing recreation (Landskapvern... 1978). In 1986, a guidebook "Multiple–use forestry" ("Flersidig skogbruk") was published. In this book, landscape was viewed both on a broad scale and at one stand level. Fellings and forest roads should follow the terrain's natural contours in order to keep broad landscape as natural as possible. In tree selection, deciduous trees should be favoured, and special attention should be paid to forest edges, for the sake of natural and scenic diversity (Geelmuyden 1989a).

In 1989, a new official strategy paper also called "Multiple–use forestry" (Flersidig skogbruk 1989) suggests that multiple use should be one condition of getting state support for forestry planning. Private forest organizations should use it as a starting point when making forest plans. Various elements of multi–purpose forestry are mapped and registered at forest stand level. These elements include edges, peatland forests, shrub layer, multi–layer forests, old broadleaves, elks' feeding places, nesting trees for birds, etc. In the report, it is mentioned that consideration and protection of different elements in forests result in diversity, variation, and with time, stable forest views without significant economic losses.

In the "Flersidig skogbruk" –report, cooperation with local authorities and forest owners has been suggested when planning afforestation. Attention should be paid to the selection of tree species when planting trees in old, cultural landscapes. In western Norway, broadleaved trees should be favoured instead of exotic spruce plantations.

The main problems with forestry and landscape are often connected to the shapes of clearances on summits or slopes, because they can be seen from long distances. Not much attention has been paid to the shaping of fellings on a broad scale, although it can be considered one of the most important landscape management practices. Kardell (1991) asks how kilometers long eskers in Skåne or river valleys in Norrland can be managed, when land is owned by many land owners. In Finland, fellings often follow private borders, which has been suggested as the reason for geometrically shaped clearances. In Norway, special orders are given to locate plantations and fellings according to the natural lines of the landscape (Flersidig skogbruk 1989). Cooperation across private borders is recommended to prevent plantations and fellings from appearing too geometrical (*Figure 11.4*).

In Denmark, the Forest Policy Committee established by the government in 1987 stressed that multiple use of forests should be supported and enhanced, with special consideration to immaterial values, such as aesthetics and recreation. Later, in the 1989 Forestry Act, landscape aesthetics is mentioned to be one of the five aspects to be taken into account in forestry in addition to timber production. According to the law, forests should be managed by considering landscape aesthetics, environmental conservation, recreation, and natural and cultural values. The law prescribes the promotion of conservation of broadleaved edges, diversification of landscape types and afforestation of broadleaved forests and edges (Koch & Kristiansen 1991).

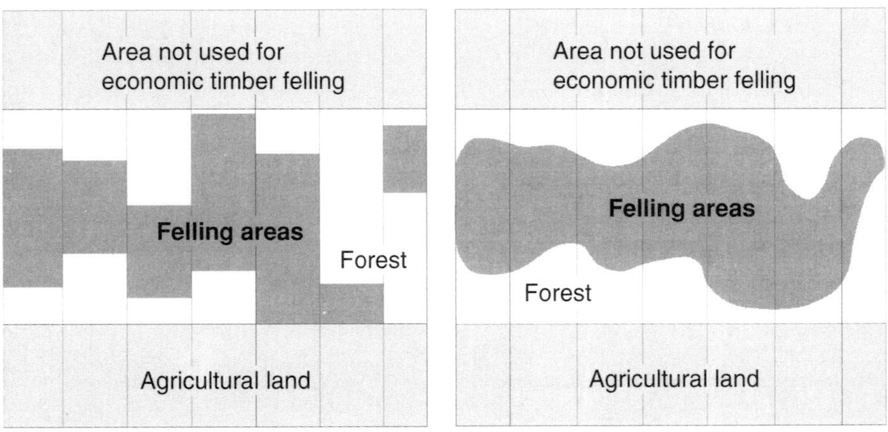

Figure 11.4 Principles of afforestation to adapt the operations accross private borders to the lines of landscape (Flersidig skogbruk 1989).

In Sweden, the government has appointed the National Board of Forestry, the Agricultural University and the State's nature conservation organization (Naturvårdsverket) to work out a strategy for landscape planning for Swedish forestry. The working group points out that landscape considerations should be included in all forest operations (Landskapsplanering... 1992). They divide the various elements of landscape planning into three main categories: 1) the elements to be regarded in a geographic area, 2) the qualities of landscape which should be emphasized and developed, and 3) the development of strategies for managing the forest.

In the report of the working group, landscape aesthetics has been put into a wide ecological and cultural context; for example, historical changes and hydrology should also be considered in landscape planning (*Figure 11.5*). Various reforms to improve landscape planning in forestry are also suggested including financial support by the state, and more research and practical models. Furthermore, landscape considerations should be included in new forest legislation.

Ecological approaches to silviculture are also presented in a project called "Ståndortsanpassat skogsbruk", which can be translated "forestry following natural site variation" (Lundmark 1988), and in a campaign called "Rikare skog" (Richer forest) (Rikare skog 1990), which shows new trends for the 1990s. Furthermore, a new education program, "Det nya landskapet" (New Landscape) starts in 1994 (Gustavsson & Ingelög 1994).

Recently, a new program for the protection of landscape heritage was started in Finland (Arvokkaat maisema–alueet 1992, Maisemanhoito 1992). In this program, regional types of landscapes have been classified and financial support has been suggested for the management of the most important landscape areas. These conservation plans mainly focus on cultural landscapes, but

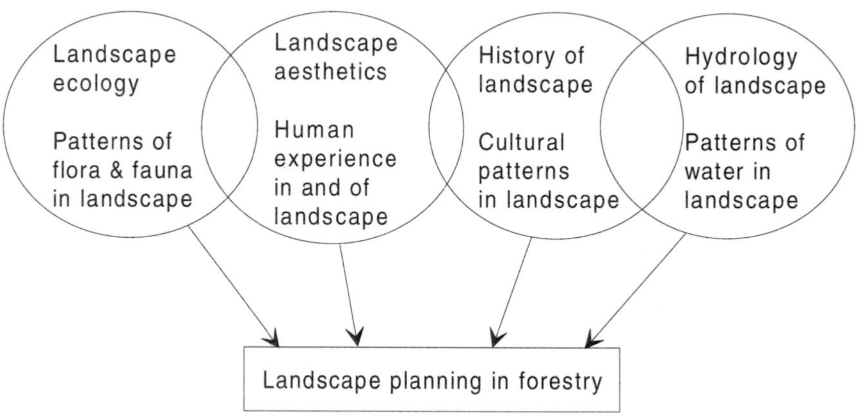

Figure 11.5 Landscape planning includes considerations of landscape ecology, aesthetics, landscape history and hydrology. Cultural patterns have also great importance for ecological and aesthetic layers of landscape (Landskapsplanering... 1992).

areas of forest scenery are also included. The importance of adapting forest operations harmoniously to the edges of cultural landscape is featured. Another program for the protection of nationally significant landscapes was presented in the autumn of 1993.

In Finland, forest organizations have responded to increasing criticism by creating programs such as "Green change" (Kalland & Pätilä 1993) and "Plus-forest". New strategies and guidelines for considering ecological and aesthetic values of forests have been prepared by the forest industry and the state supported forestry administration organizations. The manager of state forests, the former National Board of Forestry, changed its name to the Forest and Park Service, which also reflects its growing importance as the manager of national parks and recreation areas. Because of criticism over forestry operations in state forests, the Finnish Forest and Park Service has prepared an education program and a guidebook about environmental forestry. The book contains practical management methods regarding nature protection, landscape and recreation (Metsätalouden ympäristöopas 1993)

The organization for the promotion of private forestry, Forest Central Tapio, is renewing its silvicultural guidelines in order to increase the diversity of species and landscapes, and to protect water resources in forest operations. In addition, mixed forests are recommended instead of developing pure coniferous forests. In site treatment, light methods are preferred instead of heavy ploughing.

In these new multiple–use strategies, forestry organizations are revising their ethical responsibility in the light of ecological and landscape values. Because of critical arguments about the impacts of forestry operations on landscape and nature, and also because of the threat campaigns in the paper buying European countries, the interests of forest industry and environmental activists are beginning to meet. Reasons for the agreement can also be found among the economic expectations towards cooperation in the European Union. For example, landscapes are essential assets for tourism, which could generate income for local communities (Geelmuyden 1989a).

11.7 Research on forest landscape planning

Due to the need for deeper knowledge on adapting forest operations to landscape, landscape research started in the 1970s in the Nordic countries. Forest landscape planning and management have been studied for different goals and by different methods. In northern Scandinavia, where timber production and wilderness areas dominate, preserving ecological aspects in forest operations is given more consideration (Mikola 1973, Antikainen 1993). In southern Scandinavia, efforts have focused on creating rich environments for recreation (Axelson Lindgren 1990, Gustavsson & Fransson 1991) and on afforestation of agricultural lands (Borup 1991).

Different landscape planning methods have been developed to analyze the visual factors of landscape, and the potential impacts of forest operations on landscape. Often the terminology and methods applied in the Nordic countries to forest landscape planning have been developed by the US Forest Service (Forest Landscape... 1972, National Forest... 1973, National Forest... 1974) and Forestry Commission in Britain (Crowe 1966, 1978, Forest Landscape... 1989, Lucas 1991).

According to the Visual Management System (VMS), the character and variation of landscape is formed by four dominating elements: form, line, texture and colour. They are organized by contrast, sequence, axis, convergence, codominance and enframement (Forest Landscape... 1972). Danish reports (Koch et al. 1987, 1991) give examples of how these criteria can be used in forest management. However, in VMS methods, the forest view is seen through its present visual picture from a certain viewpoint; ecological development and complexity of landscape are not considered. In Britain, site factors are used to inspire design. According to Crowe, the visual character of forest is affected by the shapes of the topography, variations in scale and vegetation types and patterns, texture and colours. Aesthetic qualities of forest landscape can be analyzed by different visual factors: shape, scale, visual force, diversity, unity, sensitivity and "spirit of place" (Crowe 1978, Forest Landscape... 1989, Lucas 1991, Bell 1993).

Landscape planning methods in the Nordic countries underscore more the functional unity of forest landscape, and changes due to ecological and cultural factors, compared with the methods developed in the USA and in Great Britain (*Table 11.1*). Different ecological and visual layers of landscape have been applied in historical landscape scenarios (e.g. Brusewitz & Emmelin 1985) and in a Danish analysis called "Seks slags landskabsanalyse" (Stahlschmidt 1983). In Finland, landscape is also seen as a product of geomorphological, ecological and cultural development (Rautamäki 1990, see also Holt–Jensen 1988). Computer graphics in combination with GIS and simulation techniques have made possible the visualization of scenarios of future forest landscapes (Pukkala 1988).

There have been few studies on landscape management to fulfil the needs of practical forestry. This has often resulted in guidelines for aesthetic forest management being developed first in practical forestry, by following one's prefences, or guidelines being borrowed from northern America. In the 1990s, the interest in research on landscape ecology seems to be increasing in the Nordic countries (e.g. Gustavsson 1986, 1993, Sarlöv–Herlin 1993, Antikainen 1993, Borup 1991, Bååth et al. 1993).

The structure of broadleaved forest landscape has been studied by Roland Gustavsson (1986). He has classified different forest and edge types and followed their structural development and visual appearance after different forms of management. In his study of Furulunds fure, a pine forest in southern Sweden, variation is created in a visually unified area by different management

Table 11.1 Fields of forest landscape research in the Nordic countries (Geelmuyden 1989b, Antikainen 1993).

Area of research	Forest aesthetics	Environmental psychology	Landscape architecture	Forestry planning
Goal	Aesthetic landscape experience	How milieu affects human beings, valuations	Information for milieu planning	Uniting measured forest data to landscape value
Study method	Analysis of paintings, literature, historical changes	Preferences from photographs, video or on field	Analysis of physical environment and its aesthetic valuation	Mathematical models and computer programs
Result	Theoretical background of landscape valuations	Measured knowledge on valuations	Practical planning methods	Numeric data from landscape
Problems	No concrete information for practice	Static landscape picture, problems of measuring	Dependent on valuations of expert	Restricted concept of landscape
Recent studies in the Nordic countries	Sepänmaa 1978	Koch & Jensen 1988, Kellomäki & Savolainen 1981	Rautamäki 1990, Gustavson & Fransson 1991, Borup 1991 Antikainen 1993	Pukkala 1988, GIS-systems

strategies changing the character of parts of the forest, for example, from a pillar–hall pine forest to dense mixed forest (Gustavsson & Fransson 1991). Per Stahlschmidt (1988) also presents in his article, "Forest seen as Architecture", different structural types of recreation forests and edges. Different types of plantations can be created by planting various tree species in different structures, for example, in points, strips or mosaics. By managing the canopy layers differently, variation in forest views can be developed.

A new approach to combining landscape aesthetics and landscape ecology in forestry is presented in the case area of Tranemåla in southern Sweden (Bååth et al. 1993, Gustavsson 1993). Landscape patterns were studied both on a broad and detailed scale. In visual analysis of Tranemåla, different cultural and aesthetic values were mapped, such as open spaces, landscape damage, edge zones, contrasting tree groups and historical roads. Experiments were also made to increase natural and cultural elements in the frames of economically efficent but close–to–nature silviculture. Corridors, borders, buffer zones and key areas were protected and also created by leaving some areas in their natural state. The changes in landscape were studied by pollen analysis and the development of meadows and the surrounding edges was also predicted.

In Denmark, connected with the on–going afforestation that aims at the doubling of the total forest area of the country, Borup (1991) actualizes the aesthetic aspects of border zones. She has developed a method for the aesthetic design of forest border zones, with the help of the Visual Management System (National Forest... 1973 & 1974) and "The Six Principals" from Great Britain (see Forest Landscape... 1989). Aesthetic goals for the design of border zones are formulated, and some examples are illustrated in the following paragraphs.

In Finland, a method for forest landscape planning has been developed through an analysis of landscape structure and aesthetic qualities (Antikainen 1993). Aesthetic views were related to forest factors and location in landscape in order to create alternative management models. According to the theory of landscape structure, the present appearance of forests is formed by the ecological structure of landscape (Rautamäki 1990). Visual factors like scale, harmony, shapes of space and diversity are born from the visual presence of physical factors of nature and culture .

Landscape structure analysis, visual analysis, and the history of land use were studied in case areas in the south–western coast (Ruissalo) (*Figure 11.6*) and the Karelian ridge area (Koli National Park). Ecological layers of landscape, for example, location of soil types, exposition, water and microclimate, were studied. A visual analysis of views, shapes of topography, landmarks and dominating elements was conducted in order to enhance the character and qualities of landscape and to avoid deterioration. The history of human land use was studied in order to understand the present view of the landscape, and to predict future development. In these analyses, the forest area was divided into landscape types according to their location in landscape structure. When the visual aspects of the forest were connected to their location in the landscape, models for aesthetic management were developed. In a forested ridge area, as in Koli, different forest types are: summit forests, slope forests, border zones and forests in valleys. For each type alternative management directions were given (*Table 11.2*).

According to these analyses, silvicultural management models were recommended, depending on the location of the stand in the landscape structure. Forests on summits, slopes and edges ought to have their own typical tree species and density. Also the regional type of landscape could be emphasized by silviculture. For example, in northern Karelia the summits of moraine ridges should be closed and covered by Norway spruce forest.

In southern Finland, in Ruissalo, rocky hills and eskers are typically covered with half–open Scots pine forest. By thinning the interior space of a forest, the stand can be shaped to suit the topography. In Ruissalo, the summit forests are, by their character, often light and spacious pine forests, due to poor soil quality. The topographical shapes and big stones may be featured in a canopy landscape, and management could in many cases aim at making views more open. On more fertile soils vegetation is richer, and thus forms a dominant visual factor. It brings vividness with its colours, variation in the structure of canopies, and changes in light and shadow. Thus, a forest in hollows can be kept more dense and mixed forests can be maintained (Antikainen 1993).

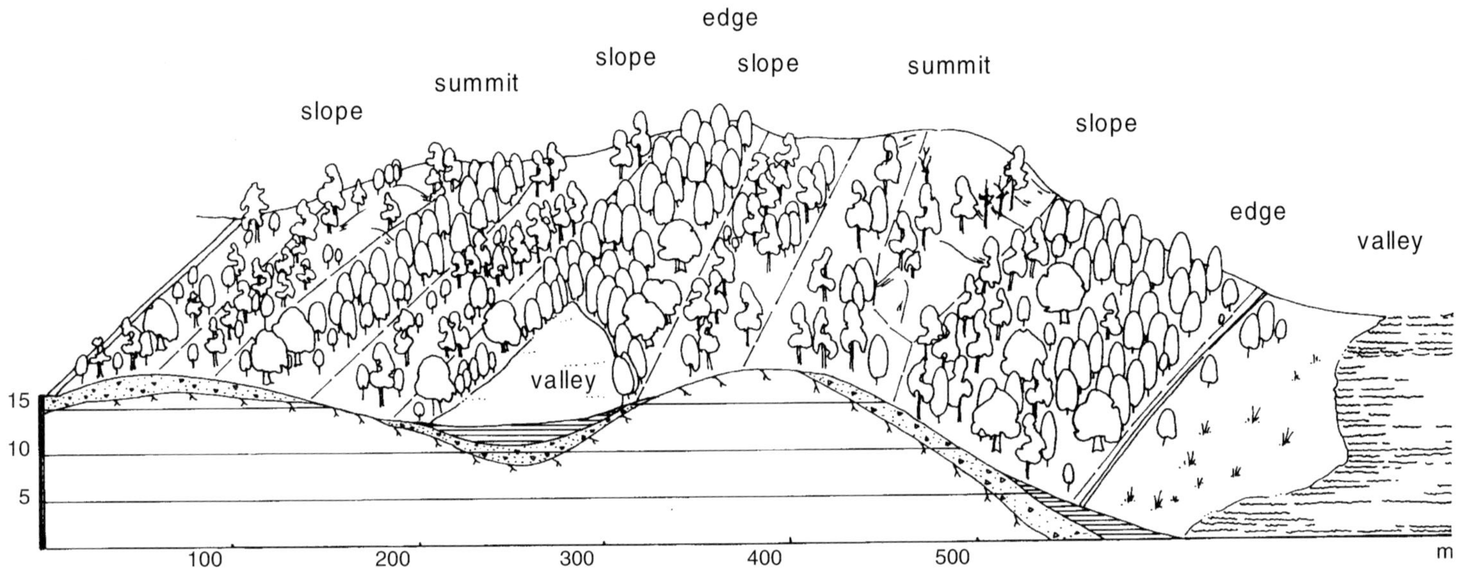

Figure 11.6 A landscape profile from Ruissalo, southwest coast of Finland. The summits of rocky hills are typically half–open pine forests. In valleys and in edges, forests are more dense and richer (Antikainen 1991).

Table 11.2 Classification of forest area into landscape types in Koli National Park (Antikainen 1993).

Landscape type	Location	Goal of forest view	Aesthetic problems	Landscape management
Summit forest	– on summits of hills, ridges, eskers	– closed in broad landscape – skyline – harmonious texture	– broken skyline by clearcutting or seeding felling	– saved closed by horizontal fellings, narrow clearcutting or dense shelter–wood fellings
Forest on slope	– on slopes	– naturalness	– geometrical and vertical fellings	– shape and scale of felling – emphazising topography by different tree species
Edges	– between open space and forest	– density	– monotonous edge – clearcutting – removing all undervegetation	– to enrich edge by different tree species
Valley forest	– in valley or flat land	– richness of foreground landscape – unity of small woods	– deforestation of fields	– on flat land invisible, behind edges – to enrich foreground

In the study of Koli National Park, descriptions of Finnish landscape heritage were also studied in order to find criteria for aesthetic forest management. National landscapes can give criteria of cultural and natural beauty for present landscape management, for example, to emphasize the contrasts of natural elements, to conserve diversity and to adapt human operations to landscape character (Antikainen 1993).

11.8 Future aesthetic trends in multi–value forestry

The importance of aesthetics in forestry has increased because of tourism, urbanization and recreation. During the present era of internationalization, it has become more important to preserve the identity and local character of landscape (Algreen–Ussing 1992). The increasing amount of threatened species also forces forestry to consider non–material values more than before. According to the Swedish government report (Landskapsplanering... 1992), landscape planning is an important tool for protecting the forest's biological diversity and cultural landscape.

New trends in forestry seem to be emphasizing more the whole landscape instead of separate stands (Landskapsplanering... 1992). The trends in forestry seem to lead to a more dynamic landscape, where nature protection is complemented by active and creative management of natural and cultural heritage.

Table 11.3 New trends in forest landscape planning according to the working group of landscape planning in Swedish forestry (Landskapsplanering ... 1992)

Until today	Tomorrow
Conserving	Dynamic, developing new
Systematical forestry	Multi–value forestry
Large operation units	Small operation stands, details in large stands
Flora, fauna	Aesthetics, cultural values
Planning in sectors	Cooperation between plans
Specialists	Synergists
Separate stand	Landscape perspective
One interest	Totality

Instead of managing large areas with one method, for example, clearing a clearcut completely, there is a tendency to increase small–scale variation of stands, and the amount of details in large stands. More attention is also paid to ecological, historical and aesthetic factors of landscape (*Table 11.3*).

Forest laws prescribe "the consideration of landscape in forest operations", but often practical methods for landscape management are not given. There is a need for practical models in landscape planning and management, with illustrations to visualize the impact of each practical action on the landscape. More research is needed to analyze the regional characters of landscape and to develop suitable management models.

Landscape planning has often been considered to be based on intuition, the idea being that aesthetic values can not be studied. The reason for this maybe a lack of suitable concepts for describing landscape (Bell 1993). As a result of this, only the feelings experienced when seeing fellings can be expressed, but deeper planning or justifications are not discussed by foresters. For better design and a landscape ecological approach, suitable concepts and criteria for landscape management have to be developed for use in the management of commercial forests. These concepts and criteria have to be produced for the Nordic landscapes, and can not be borrowed straight from international research, because the landscape ecological patterns and cultural way of reacting emotionally to landscape varies a lot in different Nordic countries and regionally. Several regional classifications of landscape and nature types have been prepared in projects supported by the Nordic Council of Ministers (Representativa naturtyper... 1983, Terrängformer 1984, Vegetationstyper 1984). These classifications can be used to analyze the character of a region, and for developing strategies to adapt fellings and plantations to the local landscape. New computer programs for predicting visibility and impacts of operations should also be developed to facilitate the visualization of landscape design.

Criticism of modern forestry is often caused by radical changes in landscape. There are no adequate methods to estimate what would be an acceptable change. It has been easier to describe the change by methods of biodiversity research. How can timber production be harmoniously adapted to the shapes and scale of the landscape? What kind, what amount and what speed of change is acceptable in a forest area or cultural landscape as a result of forest operations and afforestation of fields?

The fear of increasing costs is one reason for the negative attitudes to landscape management in forestry. The impact of landscape management on income and costs of silviculture should be studied. As there is a demand for harmonious landscape in society, the state should support landscape management through lower forest taxes or environmental subsidies if a forest owner is required to reduce his fellings for the sake of the landscape.

References

Aho, J. 1893. Kauniita näköaloja Suomessa. 1. Kolin vaara. (Beautiful landscapes in Finland. 1. The Koli Ridge.) In: Oinonen–Eden, E. (ed.). 1984. Kolin taiteilijakareliaanit. Unpublished Report. (In Finnish.)

Algreen–Ussing, G. 1992. Bevaringsværdi: hvad er det. (Protection value: what is it.) Arkitekten 15. (In Danish.)

Antikainen, M 1991. Ruissalon metsäsuunnitelma 1991–2000. (Forestry plan for Ruissalo 1991–2000.) Turun kaupunki, Ympäristönsuojelutoimisto, Julkaisu 7/1991. 23 pp. (In Finnish.)

Antikainen, M. 1992. Tammimetsien hoito. (Silviculture in oak forests.) Helsingin yliopiston metsäekologian laitoksen julkaisuja 1. 105 pp. (In Finnish.)

Antikainen, M. 1993. Metsämaiseman suunnittelu Kolin kansallispuistossa. Summary: Forest landscape planning in Koli National Park. Metsäntutkimuslaitoksen tiedonantoja 456. 88 pp.

Arnalds, A. 1987. Ecosystem disturbance in Iceland. Arctic and Alpine Research 19(4): 508–513.

Arnalds, O., Aradottir, A. & Thorsteinsson, I. 1987. The nature and restoration of denuded areas in Iceland. Arctic and Alpine Research 19(4): 518–525.

Arvokkaat maisema–alueet. Maisema–aluetyöryhmän mietintö II. Summary: Important landscape areas; Report II of the working group on landscape areas. 1992. Ministry of the Environment, Environmental Protection Department, Working group report 66/1992. 204 pp.

Axelsson Lindgren, C. 1990. Upplevda skilnader mellan skogsbestånd: rekreations–och planeringsaspekter. Summary: Perceived differences between forest stands: recreation and planning aspects. Dissertation. Stad & Land 87. 443 pp.

Bååth, M., Piga, C. & Säfström, Å. 1993. Tranemåla: exempel på landskapsekologi och landskapsestetik i skogsbruket. Tranemåla: an example of landscape ecology and landscape aesthetics in forestry.) Sveriges Lantbruksuniversitet, Institutionen för landskapsplanering, Stencil 2. 49 pp. (In Swedish.)

Bell, S. 1993. Elements of visual design in the landscape. E & FN Spon, London. 212 pp.

Blöndal, S. 1987. Afforestation and reforestation in Iceland. Arctic and Alpine Research 19(4): 526–529.

Borup, A. 1991. Landskabelige hensyn og fremtidige skovbryn. (Landscape aspects and future forest edges.) Den Kgl. Veterinær– og Landbohøjskole, Frederiksberg. 181 pp. (In Danish.)

Bramsnaes, A. 1991. Ecology in planning. The social and cultural choice. Proceedings of the European IALE –seminar on Practical Landscape Ecology. Vol. IV. Roskilde University Centre.

Brusewitz, G. & Emmelin, L. 1985. Det föränderliga landskapet: utveckling och framtidsbilder. (The changing landscape: development and future views.) LTs förlag, Stockholm. 127 pp. (In Swedish.)

Cajander, A.K. 1909. Metsätieteellinen tutkimustoiminta ulkomailla ja ehdotus sen järjestämiseksi Suomessa. (Forestry research abroad and a proposal for the organization of it in Finland). Helsinki. 138 p. (In Finnish.)

Crowe, S. 1978. The landscape of forests and woods. Forestry Commission Booklet 44. HMSO, London.

Falk, B. 1991. Rätten att formulera skogsbrukets problem. (The right to define the problems in forestry.). Skog & Forskning 1: 6–8. (In Swedish.)

Flersidig skogbruk: skogbrukets forhold til naturmiljø og friluftsliv. (Multiple–use forestry: the relationship of forestry with natural environment and outdoor recreation.) 1989. Norges Offentlige Utredninger, NOU 1989:10. Forvaltningstjeneste, Oslo. 139 pp. (In Norwegian.)

Flersidig skogbruk: veiledende retningslinjer for det praktiske skogbruk. (Multiple–use forestry: guidelines for practical forestry.) 1986. Norsk Skogbruk 5: 1–40. (In Norwegian.)

Forest landscape management. Vol. 1. 1972. USDA Forest Service, Northern Region.

Forest landscape design. Guidelines. 1989. Forestry Commission.

Forman, R. & Gordon, M. 1986. Landscape ecology. John Wiley & Sons, New York. 619 pp.

Frivold, L.H. 1993. Tendenser i skogbehandlingen. (Tendencies in forest management.) Paper presented in a Nordic Conference in Trondheim 27–30. 1993. (In Norwegian.)

Geelmuyden, A.K. 1989a. Landskapsopplevelse og landskap: ideologi eller ideologikritikk? Et essay om de teoretiske vilkårene for vurdering av landskap i arealplanleggingen. (Landscape experience and landscape: ideology or ideology critique? An essay on the theoretical prerequisites for valuing landscape in land–use planning.) Norges Landsbrukshögskole, Institutt for landskapsarkitektur. (In Norwegian.)

Geelmuyden, A.K. 1989b. Landskapsopplevelse: en planfaktor i norsk skogbruk gjennom tidene. (Landscape experience: a consideration in Norwegian forestry through the ages.) Unpublished. (In Norwegian.)

Gustavsson, R. 1986. Struktur i lövskoglandskap. (Structure in broadleaved forest landscape.) Stad & Land 48. 470 pp. (In Swedish.)

Gustavsson, R. 1993. Tranemålamodellen: ideer och grundläggande kunskaper bakom en forskning om landskapsplanering, skötselmetoder och skogsmiljöns förändring i anknytning till Tranemåla. (Tranemåla –modell: ideas and basic information for the research of landscape planning, management methods and the change of forest environment in Tranemåla.) Manuscript. (In Swedish.)

Gustavsson, R. & Fransson, L. 1991. Furulunds fure: en skog i samhällets centrum. Summary: Furulunds fure: a forest in the centre of the town.) Stad & Land 96. 131 pp. (In Swedish.)

Gustavsson, R. & Ingelög, T. 1994. Det nya landskapet: kunskaper och idéer om naturvård, skogsodling och planering i kulturbygd. (The new landscape: knowledge and ideas of nature management, afforestation, and planning in cultural areas.) Skogsstyrelsen, Jönköping. 360 pp. (In Swedish.)

Hannikainen, P.W. 1893. Kauneuden aisti metsänhoidossa. (The sense of beauty in silviculture.) Suomen metsänhoitolehti 2. (In Finnish.)

Holt–Jensen, A. 1988. Geography: history and concepts. 2. edition. Chapman, London. 186 pp.

Horelli, L. 1981. Ympäristöpsykologia. (Environmental psychology.) Weilin & Göös, Espoo. 250 pp. (In Finnish.)

Ihalainen, R. 1992. Yksityismetsänomistuksen rakenne 1990. (The structure of private forest ownership 1990.) Metsäntutkimuslaitoksen tiedonantoja 405. 41 pp. (In Finnish.)

Inha, I.K. 1925. Suomen maisemia. (Landscapes in Finland.) 2. painos. WSOY, Porvoo. 498 pp. (In Finnish.)

Julkunen, E. & Kuusamo, A. 1987. Kansallisomaisuus. Metsän mielikuvat isänmaallisissa lauluissa ja metsämainoksissa. Summary: Images of forests in patriotic songs and forest advertisements. In: Reunala, A. & Virtanen, P. (eds.). Metsä suomalaisten elämässä. Summary: The forest as a Finnish cultural entity. Silva Fennica 21(4): 351–361.

Kalela, E.K. 1949. Luonnonmukainen metsien käsittely. (Nature–like silvicuture.) Tapio, Helsinki. (In Finnish.)

Kalland, F. & Pätilä, A. 1993. The green change. Finnish Forest Industries Fereration, Helsinki. 29 pp.

Kalliola, R. 1949. Metsätalous ja luonnonsuojelu. (Forestry and nature conservation.) Suuri metsäkirja. Vol. 1. Metsänhoito. WSOY, Helsinki. (In Finnish.)

Kardell, L. 1991. Skogsbruket och landskapsvården. (Forestry and landscape management.) Skog & Forsking 3:13–20. (In Swedish.)

Keisteri, T. 1990. The study of changes in cultural landscapes. Fennia 168(1): 31–115.

Klinge, M. 1984. Suomalainen maisema. (Finnish landscape.) In: Löytöretki maisemaan, suomalaisuus kuvataiteessa 1700 –luvulta nykypäivään. (Excursion into landscape, Finnishness is paintings from the 18th century until today.) Tampere. (In Finnish.)

Koch, N.E. & Canger, S. 1987. Skovopbygning til glæde for friluftslivet. (Forestry for joyful outdoor recreation.) Miljøministeriets projektundersøgelser, Teknikerrapport 8. 239 pp. (In Danish.)

Koch N.E. & Kristiansen, L. 1991. Flersidigt Skovbrug: et idékatalog. (Multiple–use forestry: a handbook of ideas.) Skov– og Naturstyrelsen, Hørsholm. 39 pp. (In Danish.)

Laitinen, K. 1984. Metsästä kaupunkiin: esseitä ja tukielmia kirjallisuudesta. (From forest to town: essays and studies from literature.) Otava, Helsinki. 333 pp. (In Finnish.)

Landskapsplanering i svenskt skogbruk. (Landscape planning in Swedish forestry.) 1992. Rapport till Regeringen och den Skogspolitiska Kommitté. (A report to the government and the forest policy committee.) Sveriges Lantbruksuniversitet, Institutionen för Landskapsplanering, Stencil 92: 8. (In Swedish.)

Landskapsvern og naturvern i skogen: veiledende retningslinjer i skogen. (Landscape and nature protection in forest: guidelines for forestry.) 1978. Norsk Skogbruk 2: 3–43. (In Norwegian.)

Linkola, M. 1983. Suomalainen kulttuurimaisema. (Finnish cultural landscape.) In: Kinnunen, A. & Sepänmaa, Y. (eds.). Ympäristöestetiikka. 2. painos. (Environmental aesthetics.) Gaudeamus, Helsinki. p. 118–149. (In Finnish.)

Lorenzen, P. 1918. Fra Forstlig Diskussionsforening. (From forest discussion association.) Dansk Skovforenings Tidsskrift 6. (In Danish.)

Lucas, O. 1991. The design of forest landscapes. Oxford University Press, Oxford. 381 pp.

Lundmark, J.–E. 1988. Skogsmarkens ekologi: ståndortanpassat skogsbruk Vol 2. Tillämpning. (Ecology of forest soil: site adapted forestry. Vol 2. Application.) Skogsstyrelsen, Jönköping. 320 pp. (In Swedish.)

Maisemanhoito. Maisema–aluetyöryhmän mietintö I. Summary: Landscape management; Report I of the working group on landscape areas. 1992. Ministry of the environment, Environmental Protection Department, Working group report 66/1992. 199 pp. (In Finnish.)

Metsäluonnon hoito ja suojelu yksityismetsissä. (Management and protection of nature in private forests. 1986. Metsäkeskus Tapio, Helsinki. (In Finnish.)

Metsätalouden ympäristöopas. (Environmental guidelines for forestry.) 1993. Metsähallitus, Vantaa. 112 pp. (In Finnish.)

Miettinen, J. 1993. Luonnon kauneuden huomioon ottaminen metsänhoidossa. (Paying attention to the beauty of nature in silviculture.) Helsingin yliopiston metsäekologian laitoksen julkaisuja 7. 158 pp. (In Finnish.)

Mikola, P. 1973. Metsätalouden ympäristövaikutukset ja niiden merkitys metsien käytön suunnittelussa. (Environmental impacts of forestry and their significance in the planning of forestry.) Helsingin yliopisto, Metsänhoitotieteen laitoksen tiedonantoja 9. 51 pp. (In Finnish.)

National forest landscape management. Vol. 1. 1973. USDA Forest Service, Agric. Handbook 434, Washington D.C.

National forest landscape management. Vol. 2. 1974. The Visual Management System. USDA Forest Service, Agric. Handbook 462, Washington D.C.

Natur– och landskapsvård. (Nature and landscape management.) 1974. Skogsstyrelssen, Jönköping. (In Swedish.)

Oppermann, A. 1887. Skoven, Skovbruget og Det Skønne. (Forest, forestry and the beauty.) Tilskueren 14. København. (In Danish.)

Pukkala, T. 1988. Methods to incorporate the amenity of landscape into forest management planning. Silva Fennica 22(2): 135–146.

Rancken, T. 1956. Puistometsien hoito. (Management of park forests.) Metsäkäsikirja, 1 osa. Rauma. (In Finnish.)

Rancken, T. 1964. Träden i park och landskap. (Trees in parks and landscape.) Frenckellska tryckeri AB, Helsinki. (In Swedish.)

Rautamäki, M. 1990. Maakunnallinen maisemaselvitys: Varsinais–Suomi. (Provincial landscape investigation in South–West Finland.) Varsinais–Suomen seutukaavaliitto, Ympäristöministeriö. 108 pp. (In Finnish.)

Rautamäki–Paunila, M. 1982. Maisemamaakunnat, valtakunnallinen viheraluejärjestelmä. (Landscape provinces, national system for greenbelts.) Teknillinen korkeakoulu, arkkitehtiosasto, maisemalaboratorio, julkaisu 3. Otapaino, Espoo. 135 pp. (In Finnish.)

Representativa naturtyper i Norden. (Representative types of nature in the Nordic countries.) 1983. Nordiska rådet, NU 1983:2. 139 pp. (In Swedish.)

Reunala, A. & Heikinheimo, M. 1987. Taistelu metsistä: voimaperäinen metsätalous Suomessa ja muissa maissa. (The fight for the forests: intensive forestry in Finland and other countries.) Kirjayhtymä, Helsinki. 188 pp. (In Finnish.)

Reunala, A. & Virtanen, P. (eds.). 1986. Metsä suomalaisten elämässä. Summary: The forest as a Finnish cultural entity. Silva Fennica 21(4): 317–480.

Rikare skog: 90 –talets kunskaper om naturvård och ekologi. (Richer forest: the knowledge of nature protection and ecology in the 90s). 1990. Skogsstyrelsen, Jönköping. 133 pp. (In Swedish.)

von Salisch, H. 1885. Forstästhetik. Julius Springer, Berlin.

Sarlöv–Herlin, I. 1983. Woodland edges in the agricultural landscape. Studies of structural characteristics as a base for management, new construction and conservation. Paper presented in a meeting in Denmark 24.5.1993. Unpublished.

Savolainen, R. 1990. Metsä maisemassa. (Forest in landscape.) Teollisuuden metsäviesti 2. (In Finnish.)

Schulin, V.P. 1949. Skovdyrkningens forhold til det skovbesøgende publikum. (The relationship of forest management and the people visiting forests.) Dansk Skovforenings Tidsskrift. (In Danish.)

Sepänmaa, Y. 1978. Ympäristön esteettinen kuvaus, tulkinta ja arvotus: periaatteet ja analyysiesimerkki. (The aesthetic description, interpretation and valuation of environment: principles and an analysis example.) Helsingin yliopiston yleisen kirjallisuustieteen ja teatteritutkimuksen laitoksen moniste 5. (In Finnish.)

Sepänmaa, Y. 1983. Tarkoituksenmukaisuus kauneuden kriteerinä. (Appropriateness as a criterion for beauty.) In: Kinnunen, A. & Sepänmaa, Y. (eds.). Ympäristöestetiikka. (Environmental aesthetics.) 2. painos. Gaudeamus, Helsinki. p. 199–232. (In Finnish.)

Sepänmaa, Y. 1986. The Beauty of Environment: a general model for environmental aesthetics. Annales Academiae Scientiarum Fennicae, ser B 234.

Sepänmaa, Y. 1987. Metsäestetiikka ja metsän estetiikka. Summary: Forstästhetik and forest aesthetics. In: Reunala, A. & Virtanen, P. (eds.). Metsä suomalaisten elämässä. Summary: The forest as a Finnish cultural entity. Silva Fennica 21(4): 374–385.

Sihvo, H. 1984. Suomalaista maisemaa sanamaalarien kuvaamana. (Finnish landscape described by verbal painters.) In: Löytöretki maisemaan, suomalaisuus kuvataiteessa 1700-luvulta nykypäivään. (Excursion into landscape, Finnishness is paintings from the 18th century until today.) Tampere. (In Finnish.)

Stahlschmidt, P. 1983. Seks slags landskabsanalyse. (Six types of landscape analysis.) Den Kgl. Veterinær– og Landbohøjskole, Institut for have og landskap, (In Danish.)

Stahlschmidt, P. 1988. Skoven set som arkitektur. (Forest seen as architecture.) Grönt miljö 5. (In Danish.)

Suutala, M. 1986. Luonto ja kansallinen itsekäsitys. (Nature and national identity.) In: Manninen, J. & Patoluoto, I. (eds.). Hyöty, sivistys, kansakunta. (Benefit, civilization, nation.) Pohjoinen, Oulu. (In Finnish.)

Terrängformer i Norden. (Forms of terrain in the Nordic countries.) 1984. Nordisk ministerråd.

Topelius, Z. 1887. Maamme –kirja (Our land –book.) WSOY, Porvoo. (In Finnish.)

Vegetationstyper i Norden. (Vegetation types in the Nordic countries.) 1984. Nordisk ministerråd, Oslo. 539 pp.

12 Cultural heritage in multiple–use forestry

Katrine Højring[1]

Abstract

It is important that the marks left on the landscape by history be considered on an equal footing with productional, biological and recreational interest in multiple–use forestry. Legal protection of historical remains is quite similar throughout the Nordic countries. Historical sites and monuments protected by heritage preservation laws may not be excavated, covered, altered, damaged or removed, and a protective zone is defined around them. Despite these laws, many historical monuments and sites are still damaged and even ruined. This is usually done by people exploiting the forests for production purposes. A special problem in relation to heritage preservation is the preservation of living remains of past utilization forms. A historical view of nature necessitates a change in the understanding of what heritage preservation is. From mostly being concerned with monuments and buildings and other traces of settlements, the concept has been enlarged to include traces of, for example, old field systems and pastures. The most ambitious level of heritage preservation is the wish to create whole, coherent, historical landscapes.

Keywords: historical remains, forests, culture, legislation, administration, historical landscapes.

[1] Planetvej 5, DK–3100 Hornbæk, Denmark

12.1 Introduction

The Nordic nature and landscape is not only a product of geological, meteorological and biological processes, but also a product of cultural history. Hardly any part, however remote it may be, has through the centuries remained untouched by human activities. This is commonly known, but nevertheless most people in today's urbanized society think of the areas outside human settlements as "nature", implying something wild, unspoilt and original.

The idea of nature as wild has its origin in the romantic period around 1800, when the ideal of nature shifted from what was "cultivated" to what was "wild/uncultivated". To the romanticist, nature was not a place where you "produced", but a place, where you "reproduced" – a divine creation, untouched by civilization, designed to restore the mind from work and worry – something to be looked upon with admiration, and not to be soiled by human civilization (Hansen 1987, Reunala 1989). Barrows and dolmens –"giants' graves" and "pagan altars" – were mysterious signs of mankind's savage and primitive past.

The production landscape was submitted to rational human exploitation and as a consequence, was not covered by the romanticist idea of nature. Landscape was divided into a recreative, natural part, and a productive, denaturalized part.

In the light of the romantic view of landscape, a conflict in forest management becomes immediately visible. On the one hand, the forest produces immaterial goods, as sensuous experiences, necessary for mental reproduction. On the other hand, it produces material goods of great economic importance. It is this fundamental antagonism, caused by the romantic view of nature, which multiple–use forestry tries to overcome through a more holistic view, regarding all interests as integrated parts of a coherent landscape (Koch 1990, Koch & Kristiansen 1991). As a result, the multiple–use discussion has primarily focused on recreative and biological interests, as opposed to wood production.

Traces of cultural activities have only to a very small degree been drawn seriously and constructively into discussions of multiple–use forestry. In most cases, they have acted as a sort of appendix, listed with other aspects of multiple–use forestry, to show the variety and multitude of this approach to forest management. Foresters have always been aware of cultural remains in forests, but only as small "historical sanctuaries" – a grave mound, a reindeer pitfall, a prehistoric stronghold – viewed as isolated objects of interest and preservation.

Concurrently with the attempt to establish a holistic view of landscape and nature not only in forestry but in landscape management in general, it becomes relevant to consider the integration of historical interests in such a new nature concept. Cultural history offers the possibility of eliminating the conceptual antagonism between a "production landscape" and a romantic "natural landscape". Our landscape, with both its "natural" and its "cultural" elements, is the result and evidence of a historical process. Therefore, cultural history is the process which connects the two elements, and explains their coexistence, mutual connection and dependence. On this basis, it is important that the marks left on the landscape by history be considered on an equal footing with productional, biological and recreative interests when weighing the interests in multiple–use forestry.

12.2 Land conquest and heritage preservation

Today the general tendency in heritage preservation is directing attention towards the individual monument, instead of the monument as an element in an integrated cultural landscape. Heritage preservation has been treated as "protection of species" instead of "protection of structures and processes". Cultural remains are regarded as isolated retreats, where traces of past events can be studied.

This view of cultural remains is in many ways the natural consequence of the circumstances under which heritage preservation came into being. The interest in and awareness of the need to protect cultural remains from destruction grew in the first half of the last century. Big improvements were made in agriculture, the landscape being exploited in a much more rational and efficient way than before. New areas formerly uncultivated were taken into cultivation, and historical monuments and sites, previously lying utterly undisturbed out in the wilderness, would suddenly appear to lie in the middle of cultivated fields. To modern, rational land use historical monuments were partly a sheer inconvenience, partly a raw material source – both aspects causing their rapid demolition.

Heritage preservation rose from the growing awareness of the necessity to save the remaining traces of history before they were destroyed. The first antiquarian laws protecting ancient monuments were formulated in most of the Nordic countries at the turn of the century. A far–seeing exception was Sweden, which passed a heritage preservation law already around 1600 (Lund 1988, Nielsen, I. 1987, Nielsen, V. 1987).

As a direct consequence of the context in which the problem of heritage preservation arose, the first monuments to be taken under protection were the kinds that were most common in open agrarian landscape. These monuments were typically barrows, dolmens, passage graves, monoliths, prehistoric fortifications and other conspicuous remains of the past (Laursen 1988). Most of these types of monuments were common in southern Scandinavia, because this was where agricultural and urban development was most comprehensive. These were the regions with the closest contacts to European markets, and therefore the best chances to profit from industrial development, and the need for agrarian products in the new metropolises.

12.3 The forest as a heritage preserver

Outside the forests many monuments have been destroyed, not only because of land reclamation, but also because of the need for raw materials. Monuments built of stone have been most severely endangered. Stones have been taken from dolmens to build churches, houses and bridges, from cairns and other stone structures to build fences and roads. Earthen monuments have, because of their contents of humus, been used for soil improvement, particularly in areas with poor soils.

Most of the monuments in the forests were protected from this, because they were too out of the way and the trees made transportation difficult. Also forestry, compared to agriculture, was a much more extensive sort of land use. The soil was hardly tilled, and all planting was done by hand.

As a consequence, antiquarian concern for historical monuments in forests was limited. In Denmark, a systematic registration of historical sites and monuments was begun in the second half of the last century. Principally it covered the whole country and all landscapes, but in praxis registrations in forests were not done as thoroughly as out in the open (Laursen 1988). This was partly of course for practical reasons. Cultural traces in forests are much more difficult to register, because trees and undergrowth restrict sight. But strategic reasons were just as important. Remains in forests were not considered to be as seriously threatened as the ones in the agrarian landscape.

This idea of historical monuments being very well protected in forests has been held until quite recently. Whereas almost all historical sites in the open have been registered in the Danish Land Registry, only a random selection of sites in forests have been registered. Growing mechanization in forestry has changed this in recent years, and the administration has found itself forced to register sites in forests as well as in the open to protect them from big modern forestry machines.

12.4 Forest management after the Second World War

The increasing mechanization and rationalization of forest management since the Second World War has gradually eliminated the protective effect of the forest. Big and heavy machines are causing considerable traffic damage to historical monuments. New roads change the traditional layout of the forest landscape around the historical monuments and sites. In some cases, they even cross and damage them. Much more drastic forms of felling and clearing

Figure 12.1 A cairn ruined in forest work. Photo: Christian Meschke.

and much more comprehensive soil treatment than before definitely raise the risk of damaging historical remains (Reunala 1989).

With some of the machines used in modern wood production, only one case of thorough cultivation of an area is enough to do serious damage to a historical monument (*Figure 12.1*). After two cases, only very big structures will still be recognizable. Traces of historical fields and field boundaries will not be distinguishable any more, neither will small barrows (Højring 1991). The traditional, lenient forestry methods were the ones that ensured the preservation of these monuments.

The requirement for productivity and yield has led to the planting of forest on marginal forest grounds, for example, through draining wet or humid areas (Reunala 1989). Draining and planting cause significant changes in the landscape. In relation to historical remains, this may be very unfortunate. Living traces of old cultivation forms will in most cases be destroyed, and historical monuments will loose their natural scenic context. In the case of a medieval castle, for instance, draining and planting of the surrounding swamps will obliterate a landscape element which is crucial to the understanding of the defensive situation of the castle. In the case of historic settlements, draining and planting will make it difficult to understand what caused a settlement at that particular place, namely, the possibility of exploiting several different biotopes at the same time.

Another serious consequence of draining is that invaluable scientific source material such as organic remains of prehistoric settlements may be destroyed because of the altered chemical conditions. Through the years some of the most valuable finds have been made in bogs, because of the special preserving faculties of peat.

Seen from a heritage preservation point of view, what has been happening in forests in the period since the Second World War is analogous to what happened in open agrarian landscape 100 years ago.

12.5 Protection of historical remains today

Today a great variety of site types and monuments have been set under legal protection. Not only those traditionally known from the intensively cultivated areas, but also less spectacular remains and remains from the less densely inhabited parts of the Nordic countries are required to be protected.

Legal protection of historical sites and monuments is quite similar throughout the Nordic countries. Historical sites and monuments protected by heritage preservation laws may not be excavated, covered, altered, damaged or removed. Only the definition of the types of monuments covered by the law differs somewhat from country to country (Denmark: Lov om naturbeskyttelse, Lov nr. 9 af 3 januar 1992; Finland: Lag om fornminnen, N:o 295, 17. juni 1963, § 1; Iceland: Ólafson 1991, pers. com.; Norway: Lov av 9. juni 1978 nr. 50 om kulturminner, § 3; Sweden: Lag om kulturminnen m.m. SFS 1988:950, 2. kap. Fornminnen, § 6).

In Norway, only historical remains predating 1537 AD and Sámi sites more than 100 years old are included in the law. The Sámi sites were included in the law in 1978 as an acknowledgement of the role heritage plays in the reproduction of modern Sámi culture (Storm in Myklebust 1988). Sites younger than the above mentioned have to be considered individually, whether they qualify for protection (Lov av 9. juni 1978 nr. 50 om kulturminner, §§ 4 & 5).

In Sweden, sites and monuments are protected if they have been definitively abandoned from use, which means that Swedish law gives no precise age limit for protected sites and monuments (Lag om kulturminnen m.m. SFS 1988:950, 2. kap. Fornminnen, § 1).

In Iceland, all cultural remains more than 100 years old are protected by the Heritage Preservation Law (Ólafson 1991, pers. com.). Iceland was probably not populated until after 800 AD, and therefore lacks the traces of early human activities common in the other Nordic countries.

Finnish and Danish laws mention no age limit, but in Denmark the limit has through administrative practice been defined as sites and monuments more than 100 years old.

12.6 Protection of the surroundings

All the Nordic heritage preservation laws define a protective zone around sites and monuments, covered by the law (Finland: Lag om fornminnen, N:o 295, 17. juni 1963, §§ 4 and 5; Norway: Lov av 9. juni 1978 nr. 50 om kulturminner, § 6; Iceland: Ólafson 1991, pers. com.; Denmark: Lov om naturbeskyttelse, Lov nr. 9 af 3 januar 1992, §§13 and 18; Sweden: Lag om kulturminnen m.m. SFS 1988:950, 2. kap. Fornminnen, § 2). The size of the zone differs from 2 m in Finland, over 5 m in Norway, and 20 m in Iceland, to 100 m in Denmark. The Danish 100 m zone only applies to monuments outside forests – protecting the visual experience of the site from being obscured by buildings, constructions and plantations. Inside existing forests the protective zone is 2 m. In Iceland, there is no special rule concerning monuments in forests, as there is relatively little forest and therefore very few possibilities of conflicting interests arising.

According to Swedish law, the protective zone around a historical monument is to be determined individually, depending on the single monument's need for protection. This is also the case in Finland and Norway, but in these two countries the monuments also have a fixed protective zone in force until an individual zone is determined. The possibility of establishing individual protective zones opens up opportunities for safeguarding the surroundings of a protected monument, for example, the fields around an abandoned, medieval settlement, or the coastline of the Stone Age sea, explaining the situation of a Stone Age settlement site (Fornlämningar och... 1988). The individual fixation of the protective zone only happens on request, and as Sweden has about 200,000 registered historical sites and Norway has somewhere between 100,000 and 200,000 registered historical sites, the administrative task of determining individual protective zones for each monument is more or less unmanageable (Mikkelsen in Myklebust 1988, Riksantikvarieämbetet, Fornvårdsenheten 1991, pers. com.). In practice, the zones are in most cases determined, when foresters report felling plans for forest compartments or in connection with applications to the municipalities for building permits or the like.

12.7 Unprotected remains

A special problem in relationship to heritage preservation is the preservation of living remains of past cultivation forms, for example, old stands of trees which have been pollarded for fodder or fencing material, forest stands created by livestock grazing, vegetational patterns created by shifting cultivation, and forest paths used for centuries when bringing cattle from one grazing area to another (*Figure 12.2*). These cultural traces are not protected by heritage preservation laws. If they happen to be set under protection, it is mostly done on biological grounds, i.e. when a certain flora and fauna in a forest stand is found to be threatened, because the traditional kind of land management has been given up (Worsøe 1988). The preservation of these endangered species has in many cases been carried through as status quo preservation of their habitats, but as these habitats are a result of a certain historical land management, they are not secured in the long run, because of the

Figure 12.2 Cattle on the way to summer pasture in Norway. Photo: Jens Nytoft Rasmussen.

gradual regeneration of the forest. Neither are, of course, the traces of historical cultivation forms.

In the nature conservation acts of Norway, Denmark and Sweden, there exist implicit possibilities to create nature parks or reserves, not only on biological and geological grounds, but also on cultural grounds (Norway: Lov om naturvern, 19. juni. Nr. 63, 1970, §§ 5 and 11; Denmark: Lov om naturbeskyttelse, Lov nr. 9 af 3. januar 1992, § 33; Sweden: Naturvårdslagen, 1964:822, §§ 7, 18c, 19 and 20). The reasoning, though, has almost solely been based on natural–historical arguments (Emanuelsson & Johansson 1989).

Also the area planning laws of the different countries open up certain possibilities of protecting cultural landscapes, as they can be implemented in determining in which way a particular area may be utilized, and which conditions must be fulfilled (Norway: Plan– og bygningslov 14. juni. Nr. 77, 1985, § 25. 6.; Denmark: Lov om planlægning, Lov nr. 388 af 6. juni 1991, §§ 1, 6 and 15).

12.8 The organizational context of heritage preservation

In Sweden and Finland, heritage protection respectively belongs to the Ministry of Culture and of Education, whereas in Norway and Denmark it belongs to the Ministry of Environment. This organizational relationship indicates the different views of what historical sites and monuments represent. From an educational point of view, they are considered as historical sources and historical symbols. From an environmental point of view, they are rather considered as landscape elements and signs of the historical development of the cultural landscape.

The environmental point of view is carried quite far in Denmark, where heritage preservation is part of the assignment of the National Forestry and Nature Agency (Skov– og Naturstyrelsen), and where the heritage preservation rules are formulated in the Nature Conservation Act (Naturbeskyttelsesloven. Lov nr. 9 af 3. januar 1992). Heritage preservation and forestry were united into one organisation in 1987. The organisational joining of the two administrative branches has in recent years led to a growing awareness of the conflicts between heritage preservation and timber production, and also to a growing effort to try to solve these problems. One of the first visible results is that the new Danish Forestry Act (Skovloven) cites the care of cultural–historical remains as an aspect which should be included in multiple–use forestry (Skovloven, Lov nr. 383 af 7. juni 1989, kap. 1). This does not only concern remains already protected by the Nature Conservation Act (Naturbeskyttelsesloven), but also other traces of human activities, such as culturally and historically interesting forest stands, or traces of ancient fields (*Figure 12.3*).

Also in Finland consideration for historical remains has lately been included in the Provision on Forest Administration (Förordning om forststyrelsen), but in this case only as a reference to the current law on heritage preservation.

Figure 12.3 Participants of an excursion on a former field in Denmark. Photo: Niels Hørlück Jessen.

The provision does not mention additional historical traces, which are not covered by the law (Förordning om forststyrelsen, Nr. 373. 22. februari 1991, § 5).

12.9 Wood production and heritage preservation

In spite of the fact that heritage has been protected legally for more than 50 years in all the Nordic countries, many historical monuments and sites are still damaged and even ruined. According to antiquarian experience, damage to historical monuments and traces is rarely caused by forest guests visiting the forest for recreational purposes. Only if a monument is very popular, it may be subject to certain wear and tear.

Damage to historical remains is mostly caused by people exploiting the forest for productional purposes, using heavy machinery on soft ground, planting and constructing new roads. In most cases the damage occurs because the forest workers and administrators are negligent of the existence of historical monuments in forests and of the care they require. In some cases damage is done purposely, because the land owner is not willing to accept the limitations the heritage preservation law puts on his right of property and his possibilities to obtain maximum yield.

The extensive use of big machines in modern forest management often leads to the creation of uniform and uninteresting "forestscapes", which do not support the visual experience of historical sites and monuments. In many cases, old culture forest disappears and is replaced by forest which is more suited for rational forest management, able to meet society's demands for wood and fibres.

Seen from a heritage point of view, this is the dominant conflict in forest management today – the conflict between wood production interests and heritage protection interests. In practice, the conflict is a territorial battle, concerning whether areas should be defined as historical remains and thus be withdrawn from forestry administration and from the possibility of economic exploitation. The solution to the conflict is probably not only of a practical nature, but also ideological. It presupposes a general change in the conception of what historical traces in forests are. Today they are mainly seen as small, random refuges in a landscape which is organized on the basis of a non–historical nature view and with completely different interests in mind. Historical remains are not seen as natural co–products of the process which also creates the economic and the natural values of the forest.

In line with this separation of historical interests from the other interests in multiple–use forestry, historical interests have not been subject to discussions about cost and benefit in the same way as have biological and recreational interests. We do not speak of cultural resource economics as we speak of natural resource economics. To antiquarians, historical remains are generally considered a common boon, whose unquestionability is established by the law. To foresters, historical monuments are, on the other hand, an expensive disturbance in the employment of customary working processes and planning procedures.

Preliminary research indicates that only in very few and exceptional cases will considerations for historical remains be very costly for the forest owner. In most cases, it is a matter of integrating considerations in the short–term organizing of working processes, and long–term planning of forest layout and production (Højring 1991, 1992).

12.10 Altering perspectives

A practical attempt to implement a different perspective on the role of cultural heritage in the landscape has in recent years been made by Swedish antiquarians and foresters. They have been working on a general strategy for how Swedish forests should be managed, if the traces of traditional forms of land use are to be saved from demolition. The author of the group's recommendations stresses that protection of cultural remains is not only a question of protecting the monument or the site itself. The ways in which the water, the forest and the vegetation have been exploited are important parts of cultural heritage, as are the constructions and physical remains themselves. Therefore entireties and correlations between structures should be preserved as well as the monuments (Gårdö 1993). The group suggests a regional division of Swedish forests, based primarily on traditions for cultural exploitation, secondly on geological and climate conditions. Each region has its own guidelines for the ideal kind of forest management, seen from a heritage preservation point of view (*Figure 12.4*).

The Swedish strategy for forest management creates a framework for the weighing of historical interests in multiple–use forestry. It considers the landscape in general as a result of a historical process, where all elements have significance for the understanding of the entirety. The Swedish strategy is one of the first practical results of the growing wish amongst Nordic landscape managers to implement a wider and more process–oriented heritage concept (Kristiansen 1990).

The Swedish considerations on the protection of cultural landscapes in forests may bode a change of approach to the administration of heritage preservation. But one must be aware that the adoption of a historical nature view necessitates a change in the understanding of what heritage preservation is. If the landscape as a whole is a historical product, it will not be possible to continue the traditional way of preserving heritage by withdrawing areas with historical remains from productional exploitation.

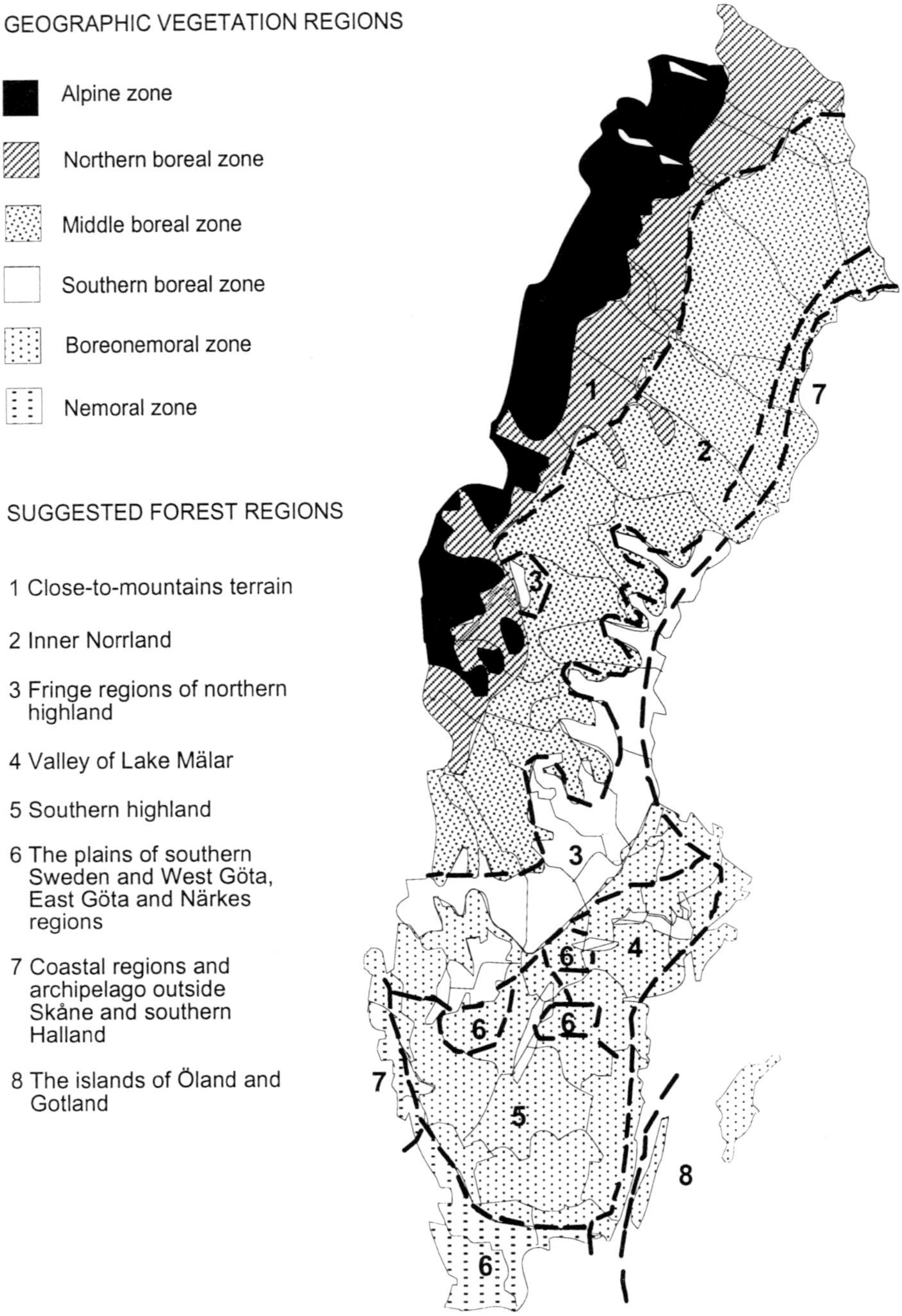

Figure 12.4 The suggested regional division of Swedish forests in accordance with cultural-historical landscape interests. Source: Gårdö & Jönsson (1991).

12.11 A widening of the concept

The exclusiveness of the traditional heritage concept is an expanding problem for antiquarian administration because of general widening of awareness of what types of historical remains should be preserved. Still more and bigger areas are subject to heritage interests. From mostly being concerned with monuments and buildings and other traces of settlements, the concept has been enlarged to include traces of old field systems and pastures. Modern antiquarians feel the growing pressure of the necessity to weigh up historical interests.

Awareness was first awakened towards the most ancient traces of old fields, those from the iron age. Later, awareness was raised towards the protection of high–backed fields from the middle ages, which were very often found in forests. In recent years, even traces of land–use forms from more recent centuries have been the object of registration and preservation interests.

Also the forest itself as a cultural landscape has become an object of interest. There is a growing wish to know how forest resources have been used through the centuries as pastures, hunting grounds and a raw material source for building materials and tar– and charcoal–burning – and how the first industrial exploitation of the forests for timber and firewood for the mining industries was organized. In Sweden and Finland, because of the big areas covered with forest, the exploitation of the forest as a resource is of particular historical interest and cultural–historical significance.

12.12 Biological interests

Generally historical interests are not contrary to other "soft" interests in multiple–use forestry. In most cases, historical monuments are an asset in the recreational exploitation of the forest, because they add to the span of possible experiences and to the variation and mysteriousness of the landscape. Biologically interesting species of plants and animals are often dependent on particular types of monuments and traditional forms of land use for their survival. In specific cases there may be conflicts between recreational and biological interests and heritage preservation, for example, when visitors' behaviour causes the demolition of a monument, or when attention to certain biological interests necessitates the modification or removal of historical remains. Seen in the light of a historical nature view, the conflict between biological and historical interests is a question of whether the remains from this or that historical period should have priority.

The floral and faunistic interests in heritage preservation are most conspicuous in relation to living traces of old cultivation methods, for example, particular forms of forest management, abandoned pastures and old stands of pollarded trees. Different ways of exploiting the natural environment have all created special biotopes, characterized by certain plants and certain animals, especially different kinds of insects.

Particular types of monuments may also be habitat for special plants and animals. For instance, in Denmark and southern Sweden, where accumulations of stone are rather rare, monuments built of stones, such as fences or cairns, attract species which are fond of this sort of environment (Thorsen 1984).

Draining is another field with merging interests between the biological and the cultural–historical aspects of multiple–use forestry. Both ecologist and cultural historians consider the conservation of bogs, moors and lakes important. Their preservation helps ensure biological diversity, which is important both from a scientific point of view and out of common curiosity. And they may represent a part of the landscape which is essential to the understanding of past activities. They might even hide structures and objects, preserved only because of the humid environment.

12.13 Livestock husbandry

The possibility of letting livestock graze and pigs find mast in the forest is very attractive from a historical point of view. This utilization of the forest has been an important factor in the creation of the forest landscape in all the Nordic countries.

Over large areas, on the outskirts of the farming settlements, pasture created light, park–like forests with picturesque trees. Pollarding trees for winter fodder and for fencing material left open, garden–like forest stands around the villages. Herding livestock gave rise to a dense network of forest paths leading from pasture to pasture and from village to village (Reunala 1989, Gårdö 1993, Worsøe 1988). Because of legal regulations, it is no longer possible to maintain this kind of forestry in all the Nordic countries.

In Denmark, economic utilization of the forests led to serious deterioration several hundred years ago. Around 1800, a series of initiatives were taken to protect the little remaining forest (less than 2 % of the land) from complete extermination. Forest areas were defined, and it was forbidden to let domestic animals forage in these areas (61. Forordning om Skovenes Udskiftning, Vedligeholdelse og Fredning i Kongeriget Danmark, 1805). This rule has now, almost 200 years later, caused some difficulty in preserving some very characteristic and biologically and historically valuable forest stands. The stands are in some cases set under protection, because of these particular values, but only in quite exceptional cases will you be permitted to carry on the kind of livestock husbandry that has created them and that would maintain them. If you want to preserve these landscapes, you will in most cases have to imitate the patterns artificially, which is in most cases much more laborious and expensive.

In Finland, livestock grazing in forests is still permitted, but general changes in agricultural production – not only in Finland, but in almost all the western countries – have decreased the demand for forest grazing areas. As the old forest pastures have become superfluous, Finnish forestry has turned them into modern production forests (Reunala 1989).

12.14 Recreational exploitation of forest

The social value of historical traces is not only their value as scientific source material for cultural historians. To see and understand historical sites and monuments is of great importance to average people as well. It gives the possibility of experiencing cultural and landscape continuity, and it adds to the variation of the landscape. Also historical traces arouse curiosity and give a sense of mystery.

This means that care of historical remains does not imply only the physical protection of monuments and sites from destruction. It also implies the question of how to give people access to this common good – how to make people see, experience and even understand the monuments. Presenting cultural heritage involves considerations of what sort of image the site should present, how its surroundings should look, how easily it should be picked out in the forest, and last, but not least, what sort of labour one is willing and able to invest in the effort. Heritage presentation can be extremely costly and labour demanding, but can also be practised effectively on a less ambitious level (Gårdö 1993, Højring 1992).

Most sites and monuments have to be made visible. Only few visitors will search for and recognize historical traces if they are covered by scrub and brushwood. A further improvement of the experience would be the creation of a suitable context around the historical monument. For example in the case of a medieval fortification, it greatly furthers the understanding of the structure if it is possible to see its natural surroundings, or to see all the barrows in a group, indicating a prehistoric cemetery, or to see a bridge crossing a stream as well as the sunken roads leading down the slopes towards it. In some cases the selection of indigenous trees may be valuable for the presentation of historical traces, because they supplement the experience of the historical landscape. The choice of indigenous trees may also support economic interests in forestry, because these trees are generally better adapted to the natural environment, and therefore less exposed to the risk of crop failure and vermin.

The wish amongst foresters to create a varied forest picture and to take into consideration the scenic context of the forest gives way to the integration of considerations for the traditional cultural landscape in forest planning. For example, it would be possible to let forest compartment boundaries comply with old field boundaries in the forest. In this way old fences could be saved from damage or from being demolished, and the traditional pattern of boundaries between properties or between areas, utilized in various ways, would be preserved (Højring 1991). Such considerations in forest planning would serve the preservation of historical remains and would reduce conflicts between historical and wood production interests.

As it is, the aesthetic considerations in forestry are mainly based on a traditional nature ideal, aiming at creating organic lines in the landscape (Forest landscape... 1989). The organic lines may hide or even obliterate historically caused boundary lines, and so remove historical information from the landscape. Seen from this point of view there may be agreement between the interests of traditional forestry and heritage preservation, because the traditional, geometrical border lines are practical in a rational organization of working processes.

12.15 "Nature" parks and ecomuseums

The most ambitious level of heritage preservation is the wish to create whole, coherent historical landscapes. There seems to be a growing interest these years in creating nature parks, which include cultural aspects in the landscape, and where the aim is to preserve a genuine picture of past forms of agricultural and industrial exploitation, and the derived flora and fauna (Emanuelsson & Johansson 1989).

An offshoot of these ideas of landscape preservation is the "ecomuseum" concept, which has been gaining a foothold in the Nordic countries these past few years (Ormio 1988). The basic intention with ecomuseums is to preserve economic units of historical significance in their entirety as living elements in the landscape – as a sort of open air museums, which not only include buildings, but also the landscape that makes up the economic basis of a settlement (*Figure 12.5*).

This kind of preservation of cultural landscapes is very valuable seen both from a scientific point of view and from a heritage presentation point of view. It offers the opportunity to study biological species, nutrient economy and human subsistence in a certain ecosystem, possibly suggesting environmentally sounder alternatives for modern agriculture and forestry. It also gives the public an opportunity to visualize a way of life in a certain historical period (Emanuelsson & Johansson 1989, Emanuelsson 1990).

The establishment of ecomuseums is still within the framework of a traditional nature concept, dividing the landscape into different sectors. The ecomuseum concept does not offer any immediate and operational methods for integrating historical interests in multiple–use forestry. Also the maintenance of an ecomuseum is extremely resource demanding and will probably only be possible in very few cases, particularly on state–owned land.

Figure 12.5 Settlement surrounded by swidden areas, open pastures, forest pastures and broadleaved stands utilized for winter fodder at the beginning of the 19th century. Photo: W. W. Wilkman. (Museiverket, Helsinki).

12.16 Conclusions

Until now cultural history has only been given very slight attention in the weighing of interests in multiple–use forestry. Discussions have been concentrated around biological, recreational and productional interests. The reason for this is probably to be found in the traditional nature ideal, which distinguishes wild nature and the production landscape. Historical remains are considered random, strange phenomena, which have survived from our cultural childhood.

Multiple–use forestry wishes to abolish the sectorization of the landscape that follows from the traditional view of nature, and wishes instead to create landscapes where all interests are equal parts of the entirety. The solution of this task is not only a practical problem, but also a mental one. It presupposes a change in our view of nature, which includes productional, recreative, biological and historical exploitation. If the landscape as a whole is seen as a product of the historical process, history will fulfil the purpose of explaining the merging in time and space of elements which are in a traditional nature view considered mutually exclusive.

The conflicts between historical interest and other aspects of multiple–use forestry have not been subject to systematic analyses, and therefore we lack systematized knowledge of the character and dimensions of the conflicts, the costs of paying regard to historical interests in forest management, and the possibilities of practical solutions to the conflicts. Today's preservation of historical remains in forests is solely based on the practical everyday experience of foresters and antiquarians engaged in their protection.

Heritage administration rests on the same sectorized view of the landscape as other landscape administration. Historical remains are seen as small areas, reserved for the purpose of attending to historical interests. A change in our view of nature, based on a historical understanding of the landscape, therefore necessitates a change in our attitude towards the management of historical interests in the landscape, and a change in administrative practice. At the moment, most antiquarians do not consider themselves to be landscape managers, but a sort of custodians, guarding cultural heritage against attacks from the outside. This understanding of heritage preservation is a natural consequence of the history of heritage preservation. If this role is to change, it is necessary to offer the antiquarian side of administration opportunities to be involved in landscape management right from the planning phases.

References

Bramsnæs, A. 1987. Landskabsarkitektens natursyn. (The landscape architect's way to see nature.) In: Bang, S. et al. (eds.). Naturen stopper ikke ved bygrænsen: om natursyn og naturforvaltning. (Nature does not end by the town border: the way to see nature and nature management.) Miljøministeriet, Miljøskrift 4. (In Danish.)

Damgaard, H. 1988. Kulturhistorie i Fredningsplanlægningen: metode og resultater 1978–1986. Summary: Cultural history in protection legislation: methods and results 1978–1986. In: Fortidsminder og kulturhistorie. Miljøministeriet/ Skov– og Naturstyrelsen, Antikvariske Studier 9.

Damgaard, H. 1988. Landskabsbevaring i historisk perspektiv. Summary: Landscape protection in historical perspective. In: Fortidsminder og kulturhistorie. Miljøministeriet/Skov– og Naturstyrelsen, Antikvariske Studier 9.

Emanuelsson, U. 1990. Växtnäringsämnenas roll vid övergången från traditionellt till modernt jord– och skogsbruk. Summary: The role of fertilizers in the transition from traditional to modern agriculture and forestry. In: Bang, C.H. (ed.). Kultur og miljø: en baredygtig udvikling. (Culture and environment: sustainable development.) Fortidsminderådet og Miljøministeriet/Skov– og Naturstyrelsen.

Emanuelsson, U. & Johansson, C.E. (eds.). 1989. Rekommendationer för kulturlandskapet. (Recommendations for cultural landscape.) Nordiska ministerrådet. (In Swedish.)

Forest Landscape Design. Guidelines. 1989. Forestry Commission, Edinburgh. 32 pp.

Fornlämningar och skogsbruk. (Historical remains and forestry.) 1988. Riksantikvarieämbetet. (In Swedish.)

Förordning om forststyrelsen. (Statutes concerning the Forest Service.) Nr. 373. 22. februari 1991. (In Swedish.)

Fredning av fornlämningar. (Protection of historical remains.) Museiverket, Helsingfors. Brochure. (In Swedish.)

Gårdö, M.B. & Jönsson, B. 1991. Bevarande av kulturmiljøer vid skogsbruk. (Preservation of cultural milieus and forestry.) Riksantikvarieämbetet. Unpublished. (In Swedish.)

Grue, U.D. 1990. Flerbruk: norsk jordbruk er mer enn mat. (Multiple use: Norwegian agriculture is more than food.) Statens fagtjeneste for landbruket. (In Norwegian.)

Grue, U.D. 1990. Gårdsmiljø: tilpassing av gammelt og nytt. (Farm milieu: adaptation of the old and new.) Statens fagtjeneste for landbruket. (In Norwegian.)

Grue, U.D. 1990. Skjøtsel av kulturlandskap: praktisk vejleder. (Management of cultural landscape: practical quidelines.) Statens fagtjeneste for landbruket. (In Norwegian.)

Hansen, J.S. 1987. Natursyn og naturforståelse. (Nature perception and understanding nature.) In: Bramsnæs, A. et al. (ed.). Sådan ligger landet. Dansk Byplanlaboratorium. (In Danish.)

Højring, K. 1991. Kulturhistoriske interesser og skovdrift. (Cultural and historical interests and forestry.) Skov– og Naturstyrelsen/Miljøministeriet. Unpublished. (In Danish.)

Højring, K. 1992. Skovdrift og fortidsminder. (Forestry and historical remains.) Skov–info 7. (In Danish.)

Kristiansen, K. 1990. Fra romantik over antikvarisme til historisk natursyn. (From romanticism via antiquarianism to historical nature perception.) In: Landet og Loven. Skov– og Naturstyrelsen, Hørsholm. p. 57–65. (In Danish.)

Koch, N.E. 1990. Flersidigt skovbrug i går, i dag og i morgen. (Multiple–use forestry yesterday, today and tomorrow.) In: Landet og Loven. Skov– og Naturstyrelsen, Hørsholm. p. 109–113. (In Danish.)

Koch, N.E. & Kristiansen, L. 1991. Flersidigt skovbrug: et idékatalog. (Multiple–use forestry: a handbook of ideas.) Skov– og Naturstyrelsen, Hørsholm. 39 pp. (In Danish.)

Kulturminnesvårdsprogram för Skåne. Del: Malmöhus Län. (A program for the preservation of cultural heritage in Skåne in Malmöhus Län.) 1984. Länsstyrelsen i Malmöhus län. (In Swedish.)

Lag om fornminnen. (Law concerning historical remains.) N:o 295. 17. juni. 1963. Finland. (In Swedish.)

Lag om kulturminnen m.m. (Law concerning historial remains.) 1988:950. (In Swedish.)

Laursen, J. 1988. Skove og fortidsminder. Summary: Forest and historical remains. In: Fortidsminder og kulturhistorie. Miljøministeriet/Skov– og Naturstyrelsen, Antikvariske Studier 9.

Linkola, M. 1988. Skogen som finländskt kulturlandskap. Summary: The forest as cultural landscape. Nord nytt 33/34: 71–80.

Lov av 9. juni 1978 nr. 50 om kulturminner. (Law concerning cultural remains.) Miljøverndepartementet, rundskriv T–5/79. (In Norwegian.).

Lov om naturbeskyttelse. (Law concerning nature protection.) Lov nr. 9 af 3. januar 1992. (In Danish.)

Lov om naturvern. (Law concerning nature protection.) 19. juni. Nr. 63, 1970. (In Norwegian.)

Lov om planlægning. (Law concerning planning.) Lov nr. 388 af 6. juni 1991. (In Danish.)

Miettinen, M. 1990. Skogens kulturhistoria i Finland: arkeologiska och historiska aspekter. (The cultural history of forest in Finland: archaelogical and historical aspects.) Nordisk Bygd 4: 24–26. (In Swedish.)

Museiverket. (National Board of Antiguites.) Museiverket, Helsingfors. Brochure. (In Swedish.)

Museoviraston toimintakertomus 1990. (Annual report of the National Board of Antiguites 1990.) 1991. Museiverket, Helsingfors. 32 pp. (In Finnish.)

Myklebust, D. (ed.). 1988. Kulturarv og vern: bevaring af kulturminner i Norge. (Cultural heritage and protection: preservation of cultural remains in Norway.) Universitetsforlaget. (In Norwegian.)

Naturvårdslagen. (Law concerning nature protection.) 1964:822. (In Swedish.)

Nielsen, I. 1987. Bevaringsarbejdet i andre lande. (Preservation in other countries.) In: Nielsen, I. (ed). Bevar din arv. (Preserve your heritage.) (In Danish.)

Nielsen, V. 1987. Det er på høje tid. (It's high time.) In: Nielsen, I. (ed). Bevar din Arv. (Preserve your heritage.) (In Danish.)

Olwig, K.R. 1990. Naturens synliggørelse og natursynets usynliggørelse. (Making nature visible and making nature perception unvisible.) In: Landet og Loven. Skov– og Naturstyrelsen, Hørsholm. p. 95–100. (In Danish.)

Olwig, K.R. 1987. Parker, skove, kosmologier og naturpolitik: en antologi om det åbne land. (Parks, forests, cosmologies and nature policy: an anthology of the open land.) In: Bramsnæs, A. et al. (eds.). Sådan ligger landet. Dansk Byplanlaboratorium. (In Danish.)

Ormio, H. 1988. Naturskyddsområdena som etnologiska museer. Summary: Nature reserve as an ethnological museum. Nord nytt 33/34: 144–149.

Parker, M. 1980. Vern om faste kulturminner: en samfunnsoppgave. (Protection of concrete cultural remains: society's task.) Universitetsforlaget. (In Norwegian.)

Plan– og bygningslov. (Law concerning planning and building.) 14. juni. Nr. 77 1985. (In Norwegian.)

Reunala, A. 1987. Skog: inte bara trä. Skogens immateriella värden. Summary: Immaterial values of forests. Sveriges Skogsvårdsforbunds Tidsskrift 1/1987: 9–13.

Reunala, A. 1987. The Forest and the Finns. In: Engman, M. & Kirby, D. (eds.). Finland: people, nation, state. Hurst, London. p. 38–56.

Skogbruksloven. (Forestry Act.) Mai 1965. (In Norwegian.)

Skovloven. (Forestry Act.) Lov nr. 383 af 7. juni 1989. (In Danish.)

Thorarinsson, T. 1974. Thjó in lif i en skógurinn dó. Ársrit skograktarfélags Íslands. (In Icelandic.)

Thorsen, S. 1984. Levende fortidsminder. Plejebogen: en håndbog i pleje af naturområder og kulturlandskaber. (Living relics of the past: a handbook for the management of natural areas and cultural landscapes.) Fredningsstyrelsen. (In Danish.)

Tvengsberg, P.M. 1988. Finnskogen brukes: bönder, finner og godseieres utnyttelse av granskogsområdene på Östlandet in Norge 1600–1900. (Forestry in Finnskogen: peasants', Finns' and farm owners' utilization of the spruce forest areas in Östlandet in Norway.) Nord nytt 33/34: 59–70. (In Swedish.)

Wikan, S. 1988. Fleretnisk bruk av grenseskogene i Finnmark. Summary: The poly–ethnic use of border forests in the Pasvik Valley of Finnmark. Nord nytt 33/34: 17–24.

Worsøe, E. 1988. Historiske driftsformer som relikter i det danske landskab. Summary: Historical utilization forms as relicts in the Danish landscape. In: Fortidsminder og kulturhistorie. Miljøministeriet/Skov– og Naturstyrelsen, Antikvariske Studier 9.

61. Forordning om Skovenes Udskiftning, Vedligeholdelse og Fredning i Kongeriget Danmark. (Ordinance concerning trading, maintenance and protection of forests in the kingdom of Danmark.) 1805. (In Danish.)

Personal communications

Guðmundur Ólafson, 1991, Head of the Archeological Department, National Museum of Iceland.

13 Economic value of non–timber forest goods and services

Per–Olov Johansson[1]
Bengt Kriström[2]
Leif Mattsson[3]

Abstract

Many forest goods (or services) have no market price to reflect their value. While timber is priced on the timber market, the forest as an environment for recreation is, in most parts of the Nordic countries, a non–priced public good. The latter results primarily from the Right of Public Access ("Everyman's Right"), which also allows berry and mushroom picking, irrespective of who is the owner of the forest. Non–priced goods include, for example, endangered species. Besides "fully" priced and non–priced goods, there are "partially" priced goods, such as game for hunting. The article summarizes a number of Nordic studies which focus on the value of forest goods other than timber. Basically, two different methods have been used in the evaluation of such goods, namely the Travel Cost Method and the Contingent Valuation Method, the latter being used more often than the former.

Keywords: economic valuation, hunting, forest recreation, fragile forests, endangered species, methodology.

---

[1] Stockholm School of Economics
P.O.Box 6501, S–11383 Stockholm, Sweden

[2] Stockholm School of Economics
P.O.Box 6501, S–11383 Stockholm, Sweden

[3] Swedish University of Agricultural Sciences
Department of Forest Economics
S–90183 Umeå, Sweden

13.1 Introduction

In the Nordic countries, timber production for the sawmill and pulpmill industry is often considered to be the most important function of the forest. A very large portion of forest economic research has therefore been focused on timber production problems. However, forests produce many goods and services, and the relative importances of these are changing in course of time and from place to place. Multiple use of forests was normal practice in the old agrarian society. After a trend towards "single use of forests" during the first half of this century, which can be seen, for example, in the development of "timber–orientated" forest legislation, we are again facing a situation where multiple use is a matter of course.

As is well known, different categories of forest users make different demands upon the forest, regarding, for example, its age structure, flora and fauna. This means that meeting the demands from one category will often affect another category's use of the forest. In other words, multiple use of forests involves external effects, and these may be either negative or positive. In some cases, the external effects are so negative that one type of forest use will necessitate the exclusion of another type. For example, this is the background to the creation of woodland national parks – a forest "shaped" for timber production is contradictory to the kind of environment that certain species of flora and fauna require. Generally, however, external effects are not so negative as to exclude the possibility of multiple use. Then the question is: to what extent one forest use must decrease if another forest use is increased; and if the external effects are positive, the question is: to what extent one forest use could be increased as the other is increased? The overall question is: how should different forest uses or forest functions be balanced against each other?

Answering that question is, of course, fraught with considerable problems. These problems are aggravated by, among other things, the fact that many forest goods (or services) have no market price to reflect their value. While timber is priced on the timber market, the forest as an environment for recreation is, in most parts of the Nordic countries, a non–priced public good. The latter results primarily from the Right of Public Access ("Everyman's Right"), which allows, for example, anyone to pick berries and mushrooms. Non–priced goods also include endangered species in the forest. Besides "fully" priced and non–priced goods, there are "partially" priced goods, such as game for hunting.

Accordingly, for many forest goods the economic value is not to be found in a market price. Basically, two groups of methods are used in the valuation of such goods. A direct method utilizes questionnaires or personal interviews in order to make a sample of individuals to reveal their preferences for the good in question. Put another way, in the absence of a "real" market for the good, it is valuated via a hypothetical market presented to the respondents. This is often referred to as the Contingent Valuation Method (CVM). For a detailed description of different methods, see Johansson (1987). An indirect method means that the value of the good is estimated via complementary private goods. The Travel Cost Method (TCM), introduced by Hotelling (1947) and further developed by Clawson (1959), and Clawson and Knetsch (1966), is a well known example of an indirect method often used to estimate the value of specific recreation areas. The basis for the valuation is thus data of visit rates and travel expenditures obtained by questionnaires or interviews.

A number of Nordic studies on the economics of multiple use of forests have been published in recent years. Some are empirical and some have a theoretical approach. In the following, empirical studies will be summarized in order to give an idea of how to measure the economic value of forest goods (uses) other than timber (production) and the magnitudes of such values. These studies include hunting, forest recreation, fragile forests and endangered species living in forests. To facilitate comparisons, *Table 13.1* supplies basic information on and results from each study (researcher(s), country, year, commodity, and value). Value measures reported are most often WTP and/or WTA, in both cases results from the CVM. WTP stands for willingness to pay for an environmental improvement or for preventing an environmental loss, while WTA stands for willingness to accept compensation for an environmental loss (or for an environmental improvement that fails to appear). In some cases two different WTP measures are reported, continuous and discrete. In a continuous WTP question, the respondent is asked to state the absolute amount of his/her willingness to pay, while, in a discrete WTP question, the respondent is "only" able to reject or accept a suggested amount, which is varied for the individuals in the sample.

The final section focuses on methodological problems, where we discuss WTP versus WTA measures, continuous versus discrete WTP questions, uncertainty, and the overall environmental value.

13.2 Hunting

Mattsson has undertaken two CVM surveys on hunting in Sweden. The first one was undertaken in 1986 and focused on moose hunting in the county of Västerbotten in the northern part of the country (see Mattsson and Kriström 1987). 272 hunters living in the county answered the questionnaire. Two thirds of these were asked about their WTP, and one third was asked a WTA question. The hunters were thus asked to state their maximum willingness to pay to be able to continue their hunting, or to state the minimum amount that they would accept as a compensation for stopping hunting. The figure for the WTP –group was 3,150 SEK on an average per year. The corresponding figure for the WTA –group was 7,400 SEK. The basic valuation question referred to the current year's hunting, but the hunters were also questioned concerning their WTP if the current number of moose was doubled, and

Table 13.1 Studies of non–timber values (arranged in the same order as in the main text).

Publication	Country	Year	Commodity	Method	Measure/Mean value
Mattsson & Kriström (1987)	Sweden	1986	Moose	WTP WTA	3150 SEK/hunter and year 7400 SEK/hunter and year
Mattsson (1990)	Sweden	1987	Moose and other game	WTP WTP	3680 SEK/hunter and year 2370 SEK/hunter and year
Sødal (1989)	Norway	1988	Moose	WTP WTA	3300 NOK/hunter and year 6900 NOK/hunter and year
Ovaskainen et al. (1992)	Finland	1988	Grouse	WTP	1500 FIM/hunter and year
Schei (1991)	Norway	1990	Grouse	WTP	4970 NOK/"tourist hunter" and year
Johansson (1990)	Sweden	1987	Moose	WTP	2530 SEK/hunter and year
Bojö (1985)	Sweden	1985	Nature reserve	WTP TMC	730 SEK/visit 760 SEK/visit
Bostedt & Mattsson (1992)	Sweden	1991	Recreation area	WTP	990 SEK/visit
Christensen (1984)	Denmark	1977	Recreation area	TCM	18-36 DKK consumer surplus/visit
Linddal & Jensen (1991)	Denmark	1987-1988	Recration area	WTP	12 DKK/visit
Sievänen et al. (1992)	Finland	1991	Recreation area	WTP	12 FIM/visit
Mattsson & Li (1993)	Sweden	1991	Goods other than timber	WTP cont. WTP disc.	2420 SEK/individual and year 5800 SEK/individual and year
Kriström (1990)	Sweden	1987	Preservation of fragile forests	WTP cont. WTP disc.	50 SEK/household and year 125 SEK/household and year
Johansson (1989)	Sweden	1987	Endangered species	WTP	85 SEK/household and year
Navrud et al. (1990)	Norway	1989	Selective forestry Preservation of virgin forests	WTP WTP	35-72 NOK/visit 46-96 NOK/visit
Hoen (1991)	Norway	1990	Cautious forestry	WTP	274 NOK/household and year
Saastamoinen (1982)	Finland	1978	Recreation Reindeer herding		Gross receipts of tourism enterprises in the area = 8.9 mill. FIM Total value of reindeer production in the area = 0.67 mill. FIM
Johansson & Zavisic (1989)	Sweden	1989	Reduction of environmental damage (excluding recreation activities)	WTP	1600 SEK/household and year

halved, respectively. In the second study (Mattsson 1990), undertaken in 1987, 1,700 hunters throughout Sweden answered a similar questionnaire. The main difference being that these hunters were asked only about their WTP, not only for moose hunting but also for other game species. For the county of Västerbotten, which is represented in both studies, the average WTP for moose hunting turned out to be almost the same in both studies. This indicates that there is stability in the answers for (consecutive) years. For the country as a whole, the average WTP for hunting game other than moose was 2,370 SEK per year.

Similar studies have been undertaken in Norway and Finland. In a study by Sødal (1989), 1,500 moose hunters in the counties of Hedmark and Ostfold in Norway answered a WTP and a WTA question. Both counties lie in the best moose areas, but have different patterns of land ownership and hunting traditions. The questionnaire design was to a large extent the same as in the first of the Swedish studies, which made it possible to directly compare answers and estimate values in the two countries. Not surprisingly, the results and estimates turned out to be rather similar to those in the Swedish study. The average WTP for moose hunting was about 3,300 NOK per year. The corresponding figure for the WTA question was 6,900 NOK. However, while in the Swedish study the recreational part and the meat part of the total moose hunting value was about 40 % and 60 %, respectively, corresponding figures in the Norwegian study were about 16 % and 84 %.

In a Finnish study by Ovaskainen, Savolainen and Sievänen (1992), questionnaires were sent to hunters in two different regions – Lammi and Keski–Pohja. In this case, the questionnaire design was similar to the second of the Swedish studies, but focused solely on the three common grouse species (family Tetraonide). The average WTP was found to be about 1,500 FIM per year in Keski–Pohja (in 1988), which is similar to the result in the Swedish study concerning these grouse species. In Lammi, the WTP value was about 50 % lower. As in the Swedish study, the recreational part of the total grouse hunting value was more than 80 %. Consequently, the meat value is a minor part. Furthermore, as in the other hunting studies, the results suggest that the hunting value is clearly affected by the stock of game.

Schei (1991) interviewed 52 tourist hunters in the Finnmarksvidda in northern Norway about, among other things, their WTP for grouse hunting in the area. He found an average WTP of nearly 5,000 NOK, i.e. far higher than the figures from Sweden and Finland, as well as less variation in WTP. This difference is explained by the fact that tourist hunters in Finnmarksvidda represent a homogeneous category of hunters, very keen on their hunting, who travel long distances to be able to hunt there.

In the studies mentioned above, hunters were asked about their WTP for the hunting, conditional on the outcome of this year's hunting. Johansson (1990), on the other hand, asked a sample of hunters of their WTP for hunting the forthcoming season. This approach comes closer to the working of a market, since a market forces the consumer to accept or reject an offer conditional on his expectations of the properties of the considered commodity. Economic theory suggests that the resulting ex ante WTP should be lower than the one generated by the "ex post" WTP questions asked in other studies (at least for risk–averse hunters). This was also confirmed by the results from the study – the average (ex ante) WTP for moose hunting was 2,530 SEK per year. The reader interested in detailed comparisons of the two approaches is referred to Johansson (1990).

13.3 Forest recreation

In March 1983 it was suggested that a nature reserve protected from forest harvesting be created in the Vålå Valley, which is a tourist area in northern Sweden. In mountainous regions such as the Vålå Valley, cutting may cause irreversible damage to the environment. For this reason, the Swedish Environment Protection Agency initiated a social cost–benefit analysis of the two scenarios, i.e. preservation versus forestry. The study was carried out by Bojö (1985). To estimate the benefits of preserving the area, 282 visitors were interviewed about, among other things, their willingness to contribute to preservation (compensation for economic losses to the forest owners) and their travel costs. The CVM and the Travel Cost Method (TCM) both produced an average value per visit of 730–760 SEK (including travel costs of 460 SEK). This result is rather similar to that of a study by Bostedt and Mattsson (1992). They studied a tourist area in southern Sweden, Risebo, and estimated (using CVM) the average WTP for the forest environment in the area to be 990 SEK per visit (the average visit in the area lasting 4.5 days).

As in the study by Bojö (1985), TCM was also used in a Danish study undertaken by Christensen (1984), in which he focused on the recreation value of two forest areas –Jægersborg Dyrehave og Hegn and Hareskovene og Jonstrup Vang. These recreation areas are (as opposed to the tourist areas Vålå Valley and Risebo mentioned above) located in densely populated regions, and their use is characterized by "short–time visits". The results indicated that the Jægersborg area, on the outskirts of Copenhagen, was more important. In this area the consumer surplus, i.e. the total recreation value minus actual costs for the recreation, was in the range of 18–36 DKK per visit. This means around 10 million DKK for the total number of annual visits in the area (in 1977).

Within the large Danish project "Skov og Folk" (Forest and People), another study of forest recreation value was carried out by Linddal and Jensen (1991). They focused on an area called Vestmager, and, using a CVM approach, 75 visitors were interviewed about their WTP. The average WTP turned out to be slightly more than 12 DKK per visit. Considering about 475,000 visits per year in the area, the total recreation value of it amounts to nearly 6 million DKK per year. A similar study was made in Finland by Sievänen, Pouta and Ovaskainen (1992). Visitors to the Luukkaa area, just outside Helsinki, were asked to fill out a questionnaire handed over to them while they were in the area. As in the Danish study, a WTP question focused on a hypothetical entrance fee. Here, the average WTP for a visit was about 12 FIM.

A Swedish study by Mattsson and Li (1993) focused on non–timber values of the forests in Västerbotten county. A CVM approach was used, where the respondents were asked both a continuous and a discrete WTP question, as well as questions about preferences and habits in relation to forests. The continuous WTP question resulted in 2,420 SEK per individual per year on average. The corresponding figure for the discrete WTP question was as high as 5,800 SEK (the difference between the results from continuous and discrete WTP questions will be discussed in the final section). Of this value, about two thirds were on–site use value, i.e. attributable to, for example, picking berries, hiking or simply taking walks in the forest, while one third was off–site (indirect) value given by the view of the forest. About three quarters of the on–site use value was based on the Right of Public Access. By combining the WTP data with the data on people's preferences regarding forest attributes, it was also shown that the non–timber value of the forests in the county could be increased considerably from the present level. Such an increase would necessitate a decrease in clearcutting with artificial regeneration in favour of natural regeneration using advance growth or seed trees, as well as a reduction of spruce in favour of broadleaved trees.

13.4 Fragile forests and endangered species

Kriström (1990) investigated methods for assessing the value people place on preserving fragile and virgin forests. In a questionnaire, a sample of Swedes were shown a map depicting eleven areas with such forests, and were informed that the areas are important for recreation (for present and future generations) as well as for many endangered species. According to a continuous valuation question, the mean WTP for preserving the forests in question was about 50 SEK per household per year, and the corresponding figure for a discrete WTP question was 125 SEK (we will come back to some methodological aspects of this study in the final section).

Johansson (1989) asked a sample of Swedes (200 people, out of whom 122 replied) about their WTP for measures taken to save endangered species – animals, birds, and plants – living in Swedish forests. The respondents were asked to make contributions to programs that would save some or all of the endangered species. Four different programs were suggested. First, the respondents were asked about their WTP for a program that would save 50 % of the species. The respondents were then asked about programs which would save

75 % and 100 % of the species, respectively. Finally, the respondents were asked about a program designed in such a way that the probability of the program saving all species is 0.5, and of it saving 50 % of the species also 0.5. It turned out that the results were in accordance with the predictions of economic theory. For example, WTP was an increasing function of the number of species preserved. The average WTP for the program saving all 300 endangered species was 85 SEK per household per year (in the final section we will come back to this study).

A CVM study undertaken by Navrud, Simonsen, Solberg and Wind (1990) focused on different forest management practices in mountainous forests in the area of Hirkjølen in southeast Norway. Clearcut forestry, selection forestry (selective thinning) and preservation of virgin forests were described as alternatives for 252 interviewed persons – 104 hikers, 100 car tourists and 48 cottage tourists. These user groups were asked about their WTP – per visit to the area – for having selection forestry or for preserving virgin forests instead of clearcutting in the area. Hikers showed the highest WTP – a mean value of 72 NOK for selection forestry and 96 NOK for preservation of virgin forests, as compared to clearcut forestry. The corresponding figures for car tourists were 35 and 46 NOK, and for cottage tourists 63 and 55 NOK. This amounts to a total WTP from the three user groups of 3.3 and 4.4 million NOK annually for selection forestry and preservation of virgin forests, respectively. These amounts are considerably higher than the net income from clearcut forestry in the area.

Another Norwegian study of people's valuation of a more cautious forestry and preservation of coniferous forests is reported by Hoen (1991). 1,204 persons throughout Norway were interviewed. Different effects of a more cautious forestry and forest preservation were described to the respondents, and then they were asked to rank these effects. Preservation of endangered plants and animals and preservation of virgin forests for descendants were considered to be very important, while, for example, improved possibilities of picking berries and mushrooms was ranked lower. The mean WTP per year for a more cautious forestry was 274 NOK per household.

Analysis of multiple use of fragile forests has also been the object of a Finnish study, undertaken by Saastamoinen (1982). He studied forestry, outdoor recreation and reindeer herding in the Saariselkä area in Finnish Lapland. This study differs from the others in the sense that neither CVM nor TCM was used to estimate non–priced values. The recreationists' (tourists') actual expenditures in the area were considered as to reflect the recreational value of the area, and it was postulated that this measure offers a realistic basis for comparisons with the economic values of the other land uses. Recreation (tourism) turned out to be the most important land use. The value of total output of this land use amounted to nearly 9 million FIM in 1978. This was 1.8 times greater than that of timber production (the area is close to the timber line) and 12.7 times greater than that of reindeer herding.

13.5 Methodological problems and discussion

Willingness to pay versus willingness to accept compensation, and continuous versus discrete willingness to pay questions

The difference between willingness to pay (WTP) and willingness to accept compensation (WTA) is, according to some researchers, a continuing embarrassment for the Contingent Valuation Method (CVM). While Hanemann (1991) provides an interesting explanation for the possible disparity between these measures from the viewpoint of economic theory, researchers may not have reached a consensus yet regarding why many studies show a large difference between WTP and WTA. Mattsson and Kriström (1987) as well as Søndal (1989) report both WTP and WTA measures in their moose hunting studies, finding that WTA is about 2–3 times larger than WTP. Other studies reveal even larger differences.

The past decade has seen a bit of a shift from the use of the continuous WTP question to the use of the discrete WTP question. These questions differ in how much leeway the respondent is given when answering. In the former case the respondent is, as mentioned earlier, able to state exactly the WTP, while in the latter case, he/she is "only" able to reject or accept a suggested amount – a bid (which is varied across the individuals in the sample). The amount of information received is therefore very different in these two set–ups, and requires very different estimation techniques and study designs. In general, the discrete question approach is more demanding in these two respects, since it requires several assumptions not needed in the continuous question approach.

There are two reasons why the discrete WTP question is currently a popular technique, despite the relative drawbacks associated with it. Firstly, it resembles much more closely our ordinary market decision to reject or accept a given price for a good. The second is that this technique is less vulnerable to strategic responses, since the respondent cannot state a "very large" or a "very small" sum.

The discrete question technique needs two sets of assumptions. The first is how to distribute the bids, i.e. the number of different bids, the magnitude of each bid and the number of respondents (ex ante) to each bid. A bad distribution of the bids can easily lead to biased results, as can be seen by imagining an extreme case when all respondents accept the bids given in the survey. In this case or, if all respondents reject the bids presented, the information is not very useful. Several studies may suffer from a bad ("unfortunate") distribution of the bids (see, for example, Bishop & Heberlein 1979, Johansson & Kriström 1988).

When the design rule has been accepted and the survey results collected, the mean WTP still has to be estimated from the set of "yes"/"no" answers to the proposed bids. Several suggestions have been made in literature on how to tackle this problem. However, the estimated mean WTP may be very sensitive to the statistical assumptions that have to be made, which is one of the drawbacks of the discrete question technique.

This brief discussion suggests that it is probably too early to say which of the two techniques, discrete or continuous WTP questions, will be more predominant in the future. As indicated previously, the two techniques gave different results in a practical application. The discrete question technique tends to give a higher mean WTP than the continuous question technique. It is difficult to explain, disregarding estimation problems, why people may interpret these two types of WTP questions differently. Psychologists have suggested that people tend to use any type of information as an anchor when providing a valuation when there is an element of uncertainty. In the study by Kriström (1990) of preserving fragile and virgin forests, both types of WTP questions were asked. The sample was split into two parts, subsample A and subsample B. Subsample A received a discrete as well as a continuous WTP question, and subsample B only a continuous WTP question. The questionnaires were identical in other respects. A test of the anchoring hypothesis, i.e. whether the mean WTP differs between subsample A and subsample B, was rejected. *Figure 13.1* gives a graphical display of the distributions.Thus, we plot the proportion (0.0–1.0) of "yes" –answers to each proposed bid. It is seen that the distribution for the discrete WTP data is constantly outside the one for the continuous WTP data. Consequently, both the mean and the median will be higher when calculated from discrete WTP data. Similar results were found in the study by Mattsson and Li (1993), as mentioned earlier.

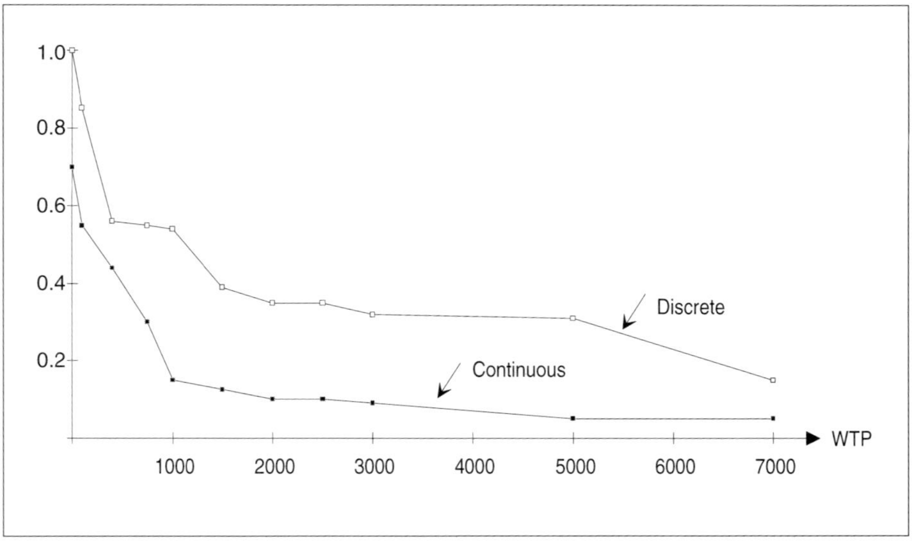

Figure13.1 Willingness to pay distributions – the case of fragile and virgin forests in Sweden.

Uncertainty and the overall willingness to pay for improved environmental quality

Many public sector programs involve various elements of uncertainty. In particular, the outcome of a proposed change in environmental quality is impossible to predict with certainty. Still, most valuation studies "suppress" any uncertainty, or use money measures designed for the certainty case. As mentioned earlier, Johansson (1989) explicitly introduced uncertainty over a program aiming at the preservation of endangered species existing in Swedish forests. Each respondent was asked for his/her WTP for four different programs, out of which one had uncertain (risky) outcomes. The basic idea with the study was to check how respondents consider risk in their valuations. It was found, among other things, that there is a difference in risk attitudes between male and female respondents. For female respondents the results confirm the hypothesis that there is risk aversion. Male respondents, on the other hand, seem to have risk aversion only if many species may become extinct, while they are more inclined to accept risky outcomes if just a few species are threatened.

The study confirms that one can meaningfully ask people about their valuation of public sector programs of which the outcome cannot be known in advance with certainty. Since uncertainty is an inevitable aspect of many environmental projects, further attempts to explicitly introduce risks into the valuation context seems to be an important challenge for environmental economists.

When respondents answer a WTP question focused on a specific environmental commodity, they may fail to recollect all the other environmental commodities they enjoy. In other words, respondents' reports on their WTP may be higher than the specific commodity's "true share" of their "paying capacities" (disposable incomes). Accordingly, to add up the values of different environmental commodities may – if these values are based on separate WTP questions – result in an exaggeration of the overall environmental value. To reduce this problem, the WTP question can be focused on a set of environmental commodities, or on a set of measures taken to reduce environmental damage. The latter approach was used by Johansson and Zavisic (1989), who report an annual average WTP of 1,600 SEK per household in Sweden. This amount of money was interpreted as being "reserved" for measures such as new or improved pollution treatment and purchases of land to preserve endangered species, fragile forests, an open landscape, etc. Thus, the sum does not include for example the value of "pure" recreation activities (hiking, camping, hunting, etc.) or "private" risks (such as radon radiation in houses).

Finally, it should be mentioned that the sample in Johansson and Zavisic's (1989) study (500 adult Swedes of whom 250 returned the questionnaire) was also asked to allocate a 100 million SEK increase of government spending on health care, child care, education, labour market policy, the environment, and other measures. It turned out that the average respondent spent 30 % each on health care and the environment. There are, however, considerable regional differences. Those living in depressed but "clean" parts of the country spent more on labour market policy measures and less on the environment than those living in large cities with more severe environmental problems. This result and other similar results presented by Johansson and Zavisic (1989) lend some support to the hypothesis that people can meaningfully state their WTP for various public sector programs.

References

Bishop, R. & Heberlein, T. 1979. Measuring values of extra market goods: are indirect methods biased? American Journal of Agricultural Economics 61: 926–930.

Bojö, J. 1985. Kostnadsnyttoanalys av fjällnära skogar: fallet Vålådalen. (Cost–benefit analysis of forests close to mountains: case Vålådalen.) Stockholm School of Economics, Stockholm. (In Swedish.)

Bostedt, G. & Mattsson, L. 1992. The value of forests for tourism in Sweden: an empirical study. In: "Forestry and environment: economic perspectives". University of Alberta, Department of Rural Economy, Project Report. Edmonton.

Christensen, J.B. 1984. Recreation economics: some results from a Danish study. In: Saastamoinen, O., Hultman, S.–G., Koch, N.E. & Mattsson, L. (eds.). Multiple–use forestry in the Scandinavian countries. Communicationes Instituti Forestalis Fenniae 120: 52–59.

Clawson, M. 1959. Methods of measuring the demand and value of outdoor recreation. Resources for the Future, Reprint 10. Washington D.C.

Clawson, M. & Knetsch, J. 1966. Economics of outdoor recreation. Johns Hopkins University Press, Baltimore.

Hanemann, W.M. 1991. Willingness to pay and willingness to accept: how much can they differ? American Economic Review 81: 635–647.

Hoen, H.F. 1991. Flersidig skogbruk og barskogvern: resultater fra en nasjonal undersøkelse. (Multiple–use forestry and protection of coniferous forests: results of a national study.) Statens Fagtjeneste for Landbruk, Faginfo 8: 110–119. (In Norwegian.)

Hotelling, H. 1947. Unpublished letter to Director of National Park Service.

Johansson. P.–O. 1987. The economic theory and measurement of environmental benefits. Cambridge University Press, Cambridge.

Johansson, P.–O. 1989. Valuing public goods in a risky world: an experiment. In: Folmer, H. & Lerland, E. (eds.). Valuation methods and policy making in environmental economics. Elsevier Science Publishers, Amsterdam. p. 37–48.

Johansson, P.–O. 1990. Willingness to pay measures and expectations: an experiment. Applied Economics 22: 313–329.

Johansson, P.–O. & Kriström, B. 1988. Measuring values for improved air quality from discrete response data: two experiments. Journal of Agricultural Economics 39: 439–445.

Johansson, P.–O. & Zavisic, S. 1989. Svenska folkets miljöbudget. (The environmental budget of Swedish people.) Ekonomisk debatt 17: 472–474. (In Swedish.)

Kriström, B. 1990. Valuing environmental benefits using the contingent valuation method. Umeå Economic Studies 219.

Linddal, M. & Jensen F.S. 1991. Værdi–undersøgelse af friluftslivet på Vestamager. (Value study of outdoor recreation on Vestamager.) Ugeskrift for Jordbrug 35/36. (In Danish.)

Mattsson, L. 1990. Hunting in Sweden: extent, economic values and structural problems. Scandinavian Journal of Forest Research 5(4): 563–573.

Mattsson, L. & Kriström, B. 1987. The economic value of moose as a hunting object. Scandinavian Forest Economics 29: 27–37.

Mattsson, L. & Li, C. 1993. The non–timber value of northern Swedish forests: an economic analysis. Scandinavian Journal of Forest Research 8(3): 426–434.

Navrud, S., Simonsen, K., Solberg, B. & Wind, M. 1990. Valuing environmental effects of different management practices in mountainous forests in Norway: a survey of recreationists' preferences and willingness to pay. Paper presented at the XIX World Congress of IUFRO, Montreal, Canada, August 1990.

Ovaskainen, V., Savolainen, H. & Sievänen, T. 1992. The benefits of managing forests for grouse habitats: a contingent valuation experiment. Scandinavian Forest Economics 33: 263–274.

Saastamoinen, O. 1982. Economics of multiple–use forestry in the Saariselkä forest and fell area. Communicationes Instituti Forestalis Fenniae 104. 102 pp.

Schei, T. 1991. Environmental quality and benefit: A case study of Finnmark and the grouse resource: the tourist hunter segment. Finnmark Distriktshøgskole, Avdelningen for Økonomi og Administrasjon, Arbeidsnotat 2.

Sievänen, T., Pouta, E. & Ovaskainen, V. 1992. Problems of measuring recreation value given everyman's rights. Scandinavian Forest Economics 33: 231–243.

Sødal, D.P. 1989. Økonomisk verdsetting av elgjakt. (Defining economic value of moose hunting.) Agricultural University of Norway, Department of Forest Economics, Scientific report 1. (In Norwegian.)

14 Multiple–use forestry administration, legislation and interest groups

Katarina Eckerberg[1]

Abstract

The article describes multiple–use forest policy in Denmark, Norway, Finland and Sweden, in particular as it relates to the preservation of nature and recreation values. The outcome of legislative and administrative measures is analyzed in relation to differences and similarities in the forestry sector of these countries. The explanatory factors include relative economic importance of forests, the role of the forest industry and the location of forests in relation to the population, ownership structure and the influence of forest owners. Also, the influence of professional culture and attitudes towards multiple–use forestry among implementing agencies and practitioners, and of environmental groups, play an important role in forest policy decision making. Multiple–use regulations were first introduced in the mid–1970s in Norway and Sweden. General environmental goals are today incorporated in the forestry legislation of all four countries, but with limited enforcement possibilities. Conflicts between environmental protection and commercial forestry developed from the late 1960s onwards. Economic subsidies to forest operations, which counteracted environmental concerns, increased during the 1980s, but have been cut back in the 1990s. Campaigns to educate forest professionals and forest owners in environmental protection were launched in the mid–1980s. Practical results from changes in attitudes still remain, however, to be evaluated.

Keywords: forest policy, timber production, recreation, nature conservation, legislation, administration, non–governmental organizations.

[1] University of Umeå
Department of Political Science
S–90187 Umeå, Sweden

14.1 Introduction [2]

In this chapter, political, administrative and juridical measures for multiple–use forestry are analyzed, in particular as they relate to the preservation of nature and recreation values. We look at in what ways the four Nordic countries[3] of Denmark, Finland, Norway and Sweden have tried to cope with conflicts between commodity and amenity values in forestry. To what extent are their approaches similar or different? By what means do the governments influence the forest owners' management that affects environmental values? To which extent are multiple–use regulations included in the legislation? And if they are included, how does the administration work to enforce such regulations? Which agencies have been designed by the governments to implement multiple–use forest policy? And how do they prioritize environmental concerns in comparison with commercial forestry?

The outcome of legislative and administrative measures related to multiple–use forestry is analyzed for the four countries. Without claiming to present a fully fledged explanatory model, four factors are solicited to cast light on the differences and similarities in multiple–use forest policy. These factors include (1) the forestry sector's relative economic importance, the role of forest industry and location of forests in relation to population, (2) ownership structure and influence from forest owners, (3) professional culture and attitudes towards multiple–use forestry among implementing agencies and practitioners and, finally, (4) influence from environmental groups[4].

The first two factors are described in the introductory analysis of the four Nordic countries (section 14.2), whereas forest–environmental legislation and administration, professional culture and environmental interest groups are examined separately for each country. Following the description of the situation in the four countries according to the four factors outlined above, the final section compares and discusses the findings, including outlining of possible future development.

---

2. Apart from explicit literature citations, this chapter builds on general information from a variety of brochures and internal documentation from forest and environmental authorities, interest groups such as forest owners associations, state forests, and environmental groups from Norway, Finland, Denmark and Sweden. In addition, a total of 40 one-hour interviews have been carried out in August-September 1991 with representatives from the different authorities, interest groups and researchers involved in forest-environmental policy in three countries: 13 interviews in Norway, 15 in Finland and 12 in Denmark. The interviews are tape-recorded and transcribed. The Swedish case draws heavily upon Eckerberg (1990).

3. Iceland is excluded in this analysis because of its very different forest situation by which comparisons of forest-environmental policy are impracticable.

4. The term "environmental groups" or "environmental organizations" is used here to include both ecological, recreational and cultural interest groups, if not explicitly described otherwise.

14.2 The four Nordic countries

Multiple–use forest policy in all the Nordic countries can be described as a multidimensional policy field because of its relationships to many sectoral policy areas. These include such diverse fields as economic, industrial, cultural, forestry, agricultural, environmental and international policy. Examining all these different policies' influences on and relationships to multiple–use forest policy is too big a task for the limited scope of this chapter. However, it is important to bear in mind that a multitude of public and private institutions are involved in the formulation, implementation and monitoring of such policy, and that various economic, administrative and legislative measures from these sectors influence what can be done in forestry practice.

Hence, in each of the countries, land–use policy regulatory instruments include building and planning legislation which is implemented through municipal governments, and, to a various extent, coordinated by regional and national planning units. Moreover, cultural remnants are protected through legislation on ancient monuments. Nature conservation legislation regulates nature reserves and national parks, and outlines major goals for preservation of the environment, along with environmental protection legislation, which contains regulatory measures to protect the environment in industrial operations. Special legislation restricts the use of chemicals and other potential negative effects of forestry operations, such as the impact on water quality and quantity. Hunting regulations restrict permission to pursue wild animals. Special regulations apply to forest grazing by reindeer husbandry. Agricultural legislation directs the management of borderline areas between forest and arable land. Forestry legislation, finally, is the main juridical instrument for controlling forest management. In the following country descriptions, for practical reasons emphasis is placed on forestry, nature protection and land–use planning legislation although other legislation to some extent also affects multiple–use forestry.

Similarly, a collection of government agencies and private institutions is involved in implementing multiple–use concerns in forestry. Only the major ones are described and analyzed in this chapter, that is, the public forestry and environmental agencies, as well as various kinds of other organizations concerned with forest management, both semi–private and private.

The forest situation

The four countries show both similarities and differences in respect to their forest situation. Finland and Sweden are the most dominated by forests, covering 65 % and 57 % respectively of the total land area, whereas forests in Norway cover 37 %, and in Denmark only 12 %. In both Sweden and Finland, forestry accounts for a large part of the country's economy.

Historically, forests in Denmark were almost totally vanished. Cattle and pigs were common in the forest, and trees were frequently cut from the top. Destruction of forests was due to the need for more agriculture and pasture lands. Efforts to stop forest destruction were intensified during the last decades of the 18th century. The 1805 Forestry Act was introduced to preserve remaining high forests. Forest cover increased by 50 % between the years 1907 and 1976 largely due to economic subsidies to afforestation, and a further increase is expected also in the future.

The Right of Public Access traditionally regulated household use of resources from the forest. After 1880, when the sawmilling industry was introduced, forestry became a commercial activity, and small–scale home enterprises gradually disappeared as the wood industry took over their role. Nowadays, total production value from the forest sector in Denmark is 7 % of total industrial production value. In 1980, around 35,000 people were employed within the forest sector (En fremtidig... 1986).

The most productive forests in Norway are located in the southeast, close to the most populated areas, whereas a large part in the central and northern regions is classified as montane forest. These low productive forests are located in the more sparsely populated parts of the country and comprise almost one fourth of the forest land. Like in Sweden and Finland, the montane forest is of great importance for recreation and nature protection, since it is one of the last wilderness areas in Europe (*Figure 14.1*).

Figure 14.1 Montane forests are common in Sweden, Finland and Norway. They have been subject to conflicts between commercial forestry and environmental protection. Photo: K. Eckerberg.

Figure 14.2 Afforestation is now taking place on marginal agricultural land in Sweden, Finland and Denmark. This has become a controversial issue since many of these former pasture lands are highly valued for nature protection and recreation. Photo: K. Eckerberg.

From having played a major role in Norway's economy in the beginning of the century, forest products nowadays account for approximately 10 % of the export income. Some 70,000 people are active within the forest sector, including employment in forest industries. During the last ten years, total employment within the forest sector has drastically decreased. At the same time, the crisis in agriculture has led to the closing down of many farms. This has affected the status of the forestry sector within Norwegian policy in that its role is diminishing.

Finnish forests are the largest per capita in Europe. Starting in the 1950s, large areas of agriculture and pasture land have been forested. Also, more than 5.5 million hectares of wetlands have been drained, leading to an increase in the forested area of 11 % (Report of the Finnish... 1989). The age structure has also changed: clearcuttings and the proportion of young stands have generally increased, and old–growth forests have diminished, especially in the northern and eastern parts of the country. The most productive forests are located in the southern parts of the country, dominated by non–industrial private ownership. State forests are in the northern parts and less productive, with large parts being montane forests and vast marshlands.

In Finland, forestry has always been very important to the country's economy. The forest industries account for almost 40 % of the total value of exports. Because of the strength of the Finnish forest industry, it has had a key role in Finnish politics. The forest sector is the driving power in the Finnish economy, but the "internal" problems and contradictions of the sector (such as nature protection vs. economic exploitation, work vs. recreation) become societal problems and cleavages because of the significance of the sector (Koskinen 1985).

The most productive forests in Sweden are located in the southern and middle parts of the country, whereas large areas in the north are covered with less productive forests, including montane forests. The growing stock is constantly increasing, both because of improved management and afforestation of former farming land (*Figure 14.2*).

The Swedish forest industry still accounts for the largest proportion of export income, despite international competition. The forest sector's share of total export value was 19 % in 1988. Some 150,000 people are employed within the forest sector. Due to the augmenting mechanization in forest operations, the number of forest workers is, however, rapidly declining.

Ownership structure

The proportion of state–owned forests in the Nordic countries is not as great as many foreigners may believe. Denmark has the largest proportion of publicly owned forests. Private ownership dominates in all four countries, but the distribution between private forest companies and individual holdings varies. Private companies prevail particularly in Sweden, whereas individual ownership is most common in Norway and Finland. The size of management units also differs between the countries. The largest clearcuttings, with their direct effect on the environment, are found in Swedish and Finnish forestry. However, forests adjacent to more densely populated areas are often environmentally important, particularly in relation to social and recreational values. Although Danish and southern forestry in the other three countries appears less destructive in terms of the magnitude of forest operations, it can still be provocative if environmental concerns are not sufficiently taken into account.

In Denmark, 34 % of the forests are publicly owned leaving 66 % privately owned. Of the public forests, 88 % are owned by the state and the rest by the counties, municipalities and church. Private forest companies have grown among the private forest owners, but individuals dominate, with 2/3 of the total privately owned area, and a total of 26,000 forest owners. The average size of forest holding for the individual forest owner in Denmark is 9 ha (Helles et al. 1984).

In Norway, only 17 % of the forests are publicly owned. The state owns most of this, with an emphasis on the northern and montane forests. There are also communal forests, where the local farmers have certain rights to utilize them. About 66 % of the forests are owned by private individuals, some 120,000 in total. A majority of them are farmers who combine forestry with agriculture. As much as 83 % of the private forests are units less than 50 hectares. Each management unit is thus only 1.5 hectares on average for the small–forest owners and 5–7 hectares for the large–forest owners (private and public).

The current share between public and private forests in Finland is 64 % owned by individuals, 9 % by private and partly state–owned companies, 4 % common lands owned by various associations (including the church) and the rest is state owned. The historic background to the present distribution of forest land among forest owners deserves special mention. Private ownership of forests adjacent to villages and farms is of ancient origin in Finland. Forests outside of adjacent village or farm land were common lands for centuries. When population increased, many of these were shifted to private or state ownership. The more productive forests became privately owned, while the remote forest lands in the east and north became state forests (Reunala 1987).

Through the Peasant Act in 1918, some 150,000 new independent small farms were created. Also, after the Second World War, 10 % of the total Finnish land area was handed over to small farmers as land for colonization. The result is that Finnish forest lands are largely private: 110,000 farms in 1901 have become 310,000 today (larger than 2 hectares). The state's share of forest lands has diminished from 35 % in the 1920s to 24 % today (Reunala 1987). Private forest owners are increasingly urbanized, and half of them are today living in the cities.

The average holding in private Finnish forestry is 35 hectares, often divided between different locations. In the southern parts, clearcuttings are usually small in size (average some 1.5 hectares), whereas larger cuttings mostly take place in the northern parts on state lands.

Up to 26 % of Swedish forests are publicly owned, and most of these are state forests in the northern parts of the country. Private individuals own 49 %. The remaining 25 % are owned by large private companies and mainly located in central Sweden. The Swedish private forest companies, thus represent a large part of the forestry sector compared to the other Nordic countries, where individual private ownership is more predominant. Forest land is distributed among some 240,000 holdings. One third of individual forest owners do not live on their holdings.

The size of management units varies greatly between the different forest owners. The Swedish state forests' average–size clearcutting is 14 hectares, private forest companies' 15 hectares, while private individuals' is only 2.5 hectares. The difference reflects the magnitude of multiple–use conflicts, since large–scale forestry is much more subject to environmental critique.

14.3 Denmark

Forest–environmental legislation and administration

The Ministry of the Environment was established in 1974. At the same time, the National Board of Forestry was moved from the Ministry of Agriculture, where it used to belong, to the new Ministry of the Environment. A parallel agency was created to deal with nature protection issues, whereas the Board of Forestry continued to be responsible for the management of state forests. The move to the Ministry of the Environment was a political sign which accelerated the acknowledgement of multiple–use goals on public forest lands.

Before 1987, the forestry and environmental protection agencies used to be separate bodies. They were then merged into a new agency, the National Forest and Nature Agency (Skov– og Naturstyrelsen), which took overall responsibility for national forest and environmental policy. To increase interest group influence in forest–environmental policy, the minister appointed a Forestry Council (Skovråd) composed of representatives from forestry, research institutions and nature conservation and outdoor recreation organizations. This Council is to advise the Minister in forestry matters and suggest policy changes.

The Ministry of Agriculture assumes the main responsibility for private forests, although they are also monitored through the National Forest and Nature Agency districts that are under the Ministry of the Environment. The Forestry Act of 1989 is administered by the local Forestry Agency officers, with the assistance of their colleagues in private forestry associations.

The Danish Forestry Society (Dansk Skovforening) was established to assist forest owners in better economic forest management. The Society took the initiative in 1895 to hire consultants who became advisors to small–forest owners, and this was supported by state subsidies in 1919. Another society of forest owners (Hedeselskabet), established in 1866, has initiated large–scale forest plantations as well as smaller plantations and advisory services for the upkeep of these plantations. The work of Hedeselskabet has been supported by state subsidies almost from the beginning, and permanently from the 1880s. The Danish Woodland Owners' Association (De Samvirkende Danske Skovforeninger), established in 1940, today organizes some 7,000 small–forest owners in 16 local associations. The first of these was created in 1904. It is also represented among the 900 members in the Danish Forestry Society.

The three above mentioned forest owners' associations are all consulted in forest policy matters, and have close relations to the National Board of Forestry. In collaboration with the Ministry of Agriculture and the National Forest and Nature Agency, the forest owners' associations produce popular material for the forest owners, informing them of, for example, important legislation and economic subsidies to forest operations.

Danish forest policy since 1989 has two main objectives according to the new Forestry Act: production of wood and wood products and production of environmental and recreational values. The previous forest legislation from 1935 had mainly wood production as its goal, although the preparation of the Act reveals that environmental values had also to be taken into account to some extent. The new Forestry Act thus explicitly introduced multiple–use values, and also emphasized cooperation, advisory services and information as important means to achieve the new goals. Ways to achieve multiple–use management are not specified. Rather, it is a frame law. It is administered by the National Forest and Nature Agency and its 26 state forestry districts.

Through the Ministry of Agriculture, subsidies were initiated in 1978 to private forest owners, for clearing and thinning operations, converting low–productive forests, establishing broadleaved forests for protection (fire or wind), building of forest roads, draining, and forestry planning (the latter introduced in 1989). Of the 200 million DKK (33 million US dollars) that the state subsidizes per year, some 40 % is for converting agricultural lands into forests under various laws. Afforestation subsidies for state lands were introduced in 1989 and for private lands in 1991. Inventories of which areas are suitable for such conversion are carried out by the county (amt) planning units, which are also in charge of regional planning. In this process, zoning of afforestation and agricultural 'minus' areas takes place in consultation with affected interests, including the forest and environmental agencies, land owners and environmental organizations.

The goal has been set of doubling the forest area within the next hundred years. This is partially motivated by the decreased profitability of Danish agriculture, and the government has launched subsidies to reforest marginal agricultural lands. Between 10 % and 20 % of Danish agricultural land is expected to be excessive. This is accelerated by the 1992 Common Agricultural Policy (CAP) of the European Union, by which agriculture production is to be brought down to levels more in line with market demand. Also, according to the new Forestry Act, most of the current forest area is classified as protected which means that it must be kept forested (so called "fredskov").

The first Nature Conservation Act from 1917 gave special authorities power to decide on preservation issues. These authorities were established as independent juridical bodies at county level. In the Nature Protection Act of 1969 (Naturfredningsloven), it is stated that Danish nature and landscape values must be protected as much as possible, and that the Danish population must be given access to these values. There are two major ways to reach the goals. First, the law specifies public access to private forest lands. Private forests are legally accessible only by foot or bicycle, during the daytime, and on roads or paths. Small forests (less than 5 hectares) can, however, be closed according to a general Act on Access (Mark– og Vegfredningsloven). Public access to state forest is more extensive than to private forests, and includes access outside roads and paths (by foot) except for, for example, fenced cultures. Second, since 1980 the Nature Protection Act protects different biotopes that are important to wildlife and plants, requiring special permission from the county administration for any activity that may affect them. This applies to open grasslands ("heder"), water courses, and lakes and marshlands, including forested border zones. The law has been bolstered over time, so that it applies to smaller and smaller areas.

The Danish people do not have the same right of access to private land as the people in the other Nordic countries. This issue, therefore, is subject to political controversy. As long as the population was agrarian, they had personal connections with the private land owners and trespassing was not a problem. But when the population became urbanized, many private land owners started to defend their property rights against "outsiders". In order to develop a common understanding on how the right of public access should be interpreted, a committee was formed in the late 1980s, consisting of governmental as well as non–governmental organizations, including agricultural land owners' organizations. This committee has formulated general recommendations and advice on how to establish special agreements between private owners and interest groups in cases where recreation pressure is particularly strong. They also agreed on enlarging the right of public access to private forest lands to include bicycling, in 1990, and this was introduced by law in 1992. The committee can thus function as a complement to regulate conflicts between land owners and recreational interests.

Environmental protection and monitoring at national level is located within the National Forest and Nature Agency. The Nature Conservation Act is administered by 25 nature conservation districts at county and local levels. Part of the total around 150 million DKK that have been allocated annually since 1984 to protect nature areas has recently been decentralized from national to district level. Currently, 160,000 hectares of land are covered by protection regulations, corresponding to approximately 4 % of the country.

Professional culture

There has been an ongoing change of attitudes within the state forestry administration over the last twenty years, which has recently been speeded up. The merging of forestry and environmental agencies in 1987 meant that professional cultures within the two agencies came closer together, and collaboration increased substantially. New procedures that enhance the participation of public interest groups in state forestry have been introduced, and the environmental movement is now largely satisfied with the current management of state forest lands.

Although recreational use of Danish forests has been appreciated for a long period of time, there has been a scarcity of materials providing information to practitioners, to strengthen this aspect of forest management. A special information committee consisting of representatives from the Ministries of Agriculture and of the Environment, together with the three forest owners' associations, has been formed to develop information materials to support government policy. The stress on soft steering methods is well articulated in the Ministries (Kristiansen 1988). Several popular brochures aimed at forest owners, and a more comprehensive handbook for foresters on how to deal with multiple use, have been produced in 1991 and 1992. Informative leaflets on hiking in state forests have also been widely distributed by the National Forest and Nature Agency.

According to environmentalists, it was only in the late 1980s that multiple–use thinking has penetrated the forestry sector in Denmark. This may be because the environmental movement was not as committed to influencing forestry methods in the 1970s as was its counterpart in other Nordic countries. There was some public discussion on the need to preserve beech forests at that time, but the debate never became politicized to general forest policy matters.

Environmental groups

The most important group is the Danish Society for the Conservation of Nature (Danmarks Naturfredningsforening), which has increased from 50,000 members in 1975 to 270,000 in 1990. It operates through 210 municipal level units, regional bodies and at national level. The organization has close connections with the scientific community. Special committees consisting of researchers on a number of environmental policy areas prepare the organization's viewpoint. Danish legislation since 1937 provides this Society with a special legal standing on issues concerning nature and the environment in that it gives the right to initiate and appeal decisions by local or regional authorities. As a result, most proposals for preservation have originated from the Society for the Conservation of Nature. The Society reviews about 4,000 environmental permits yearly, of which some 2 % are appealed. In addition, it has a leading role in influencing national decision–making in the Ministry of the Environment and in the Ministry of Agriculture, through frequent hearings and membership in consultative bodies.

The Danish ornithological society[5] has also been actively involved in forest policy matters since the mid–1980s. This development coincides with the merging of the two central agencies for forestry and environmental protection. It can be partially explained as a process of increased understanding on the part of the ornithologists that forests, and not only open areas, are also vital bird habitats. From being mainly a scientific society carrying out bird inventories, in the last 20 years the ornithological society has made efforts to influence land use and environmental policy in order to protect bird habitats. Its inventories are now also being used by the environmental protection authorities. For a few years it has had a fairly established relationship with state forestry, but not, so far, with the private sector.

The Council for Recreation (Friluftsrådet) is an umbrella organization for some 90 different outdoor recreation organizations, including sports, scouting, tourism and nature studies. Through the Council, these organizations jointly formulate their policies at national and county level. There is a full–time staff

---

5. Its current membership is around 9,000. From 1960 to 1980 the amount of members increased from some 3,000 to 8,000, but stabilized during the 80s.

at national level that is formally consulted in legislative and other policy matters. At county level, the member organizations cooperate through newly formed "green consulting bodies" to influence land–use planning. These consultations are still not legalized, but take place informally.

14.4 Norway

Forest–environmental legislation and administration

The majority of small–forest owners are represented by the Norwegian Forest Owners' Federation (Norges Skogeierforbund), which started up in 1913. Apart from serving its 56,000 members with technical advice, workers and machines, this association also has close contacts with the Ministry of Agriculture in policy matters. Another forest owners' association, the Norwegian Forestry Association (Norsk Skogsbruksforening) represents only the 250 largest private forest owners, and is even more engaged in lobbying. For example, it has advocated a liberal view on pricing policy for hunting and fishing permits and is also opposed to the present legislative hindrances to land acquisition.

In Norway, there was a ban in 1916 on any new establishment of private forest companies. Therefore, the largest private forest owners in Norway are only a fraction of those in, for example, Sweden. Agricultural policy in the 1960s, which aimed at preserving a flourishing countryside, implied that private ownership of land is reserved for those who actually live on this piece of land. In this issue, the two forest owners' associations are completely divided: the small–forest owners want to keep the rule, while the large–forest owners want to abolish it in favour of bringing market forces to regulate land prices.

Crosscutting public and private forest owners is the Norwegian Forestry Society (Det Norske Skogselskap). It was established to promote planting of new forests in the beginning of the century, when large areas were devastated. After the war, it was this Forestry Association that initiated the government forestry agency. The Forestry Society receives public funding for its information activities. More recently, its focus has changed towards informing the general public about the importance and role of forests and forestry.

The National Board of Forestry (Skogavdelningen) in charge of government forest policy is located within the Ministry of Agriculture (Landbruksministeriet). It includes a central forestry department within the Ministry, county level agencies as well as local level forest officers throughout the country. Each county (fylke) is divided into districts covering one to two municipalities for which a forest officer works as an advisor to local forest owners.

This forestry agency was historically established as the government's advisory service to private forest owners. It has also had a major task in planning and supervising the building of forest roads. In the last decade, its emphasis has changed towards administering government subsidies to various forest operations, including building of roads, forest planning, harvesting in difficult sites, regeneration, and use of herbicides. The total state subsidies currently amount to approximately 130 million NOK (20 million US dollars) per year. The largest increase was in the year 1985, after which it has stabilized. Because of the heavy work load for the forestry agency in distributing these subsidies, part of the advisory services has been taken over by the two forest owners' associations' staff. They collaborate closely with the state forestry agency at local level, and they also receive government support for this work.

The first Norwegian forestry act that required regeneration efforts in order to secure future forest production dates from 1932. It was not until the end of the 60s that increased mechanization and expansion of economic forestry into formerly almost untouched areas required some response in the form of legislative amendments to include environmental protection aspects. In the Forestry Act of 1976, the first paragraph was changed stipulating that due consideration must be given to recreation, landscape protection, animal and plant habitats and areas for hunting and fishing. The Forestry Act is designed as a frame law, with almost no detailed rules. There is very little risk of legal punishment if a forest owner does not follow its intentions, and there is no requirement that he/she announces what forest operations will be undertaken. The tradition to leave the forest owners with "liberty under responsibility" continues to be the principle of Norwegian forest policy. Since 1989, state forestry (Direktoratet for Statens Skoger DSS) in Norway is also regulated by the Forestry Act.

Paragraph 17b, which was added to the Forestry Act in 1976, makes it possible to regulate special areas of importance to recreation and nature conservation interests. Until now, this opportunity has only been used in the Oslomarka area. Another paragraph (19) requires forest owners to give advance notification of certain forest activities when it is assessed by the forestry agency to be necessary for monitoring. It can be applied to selected forest owners or for all forest owners within a district. It has, however, not yet been used to monitor environmental protection.

Many new multiple–use regulations have been issued during the last two years by the Ministry of Agriculture. These are special recommendations to reduce negative environmental impacts attached to the provision of state subsidies for drainage, building of forest roads, and forest planning. For example, in order to receive government subsidies for forest planning[6], plans must now also mention specific measures on how to include environmental interests, al-

6. The forest planning in Norway is not compulsory, but there is a strong pressure also from the forest owners associations to carry this out. In practice, 80-90 % of private forest owners join. The planning is then used as a basis for giving advice to the forest owners. Although subsidized by the state, the plans are not public, but can only be used by the forestry agency if the owner allows.

beit not in much detail, and without consultation with different interest groups. The methodology that is used was initially developed by the Norwegian Forest Owners' Federation (Norges Skogeierforbund) in 1988, and has been adopted as a planning package that is now used in all forest planning, both on public and private lands. This is an example of how the forest owners sometimes even precede government regulations, instead of the other way around.

Although most of the forestry agency officers' work is in terms of advising and discussing with the forest owners what to do, there are some possibilities for stricter surveillance. Most of this monitoring is done through the subsidy system, where operations must be checked at some point in order for the forest owner to receive financial support. In some areas, up to 90 % of all forest operations on private lands may receive some form of subsidy, of which 20–25 % of the costs are paid by the state[7]. In addition, permission from the forestry agency is required for use of herbicides in forestry[8], in which case the county environmental protection agency is also consulted. In practice, use of herbicides has decreased drastically due to the government's decision to abolish economic subsidies to spraying, and also because of a general negative public opinion towards spreading chemicals in nature.

The Ministry of Environmental Protection (Miljöverndepartementet) was established in 1972. At this time, it became impossible for the voluntary organizations to cover the increasing demands for nature protection (Hafsten 1977). Nature protection legislation is, however, of much older origin, starting with protection of single species and particular sites in 1910, protection of national parks in 1954, and of nature conservation areas (other than national parks) in 1970. The Recreation Act regulates protection of recreational areas by the state or municipalities.

The environmental agency does not usually claim to master forestry questions, although its branch for Nature Protection and Management (Direktoratet for Naturforvaltning) in Trondheim is increasingly involving itself in forest policy matters. Currently, there is a division of competence between the environmental and forestry agencies in that the former is involved in the planning of protection areas and pollution issues, whereas the latter mainly takes responsibility for policy matters on those lands where the primary management goal is timber production. There is also a distinct difference in professional culture between the two agencies.

So far, environmental policy has been very centralized in Norway, which has led to tension between national and local interests in many cases where the state has promoted protection and the municipalities exploitation. A program to develop environmental protection work in the municipalities was initiated in 1987 (the MIK[9] programme). The municipalities who participated were

7. Interview with the Director of the County Forestry Agency in Elverum.

8. This permission is only required when the spraying is done from a helicopter. The forest owner can still spray by hand or by tractor without special permission.

supported by government funds to employ special environmental officers. Evaluation of this pilot programme was positive, and showed increased environmental consciousness in municipal decision–making (Jansen 1991). If the program is extended, it may imply that local multiple–use forestry issues can also be monitored at municipal level by the environmental officers. However, the municipalities' current economic situation is extremely strained, and economic interests are therefore most likely to be prioritized over environmental. There is a tendency in Norwegian politics to increase physical planning, but nobody has yet dared to promote policies that will be incompatible with the municipalities' current economy.

The Building and Planning Act (1986) regulates the municipalities' right to control land–use policy within their territory. The municipalities can indicate in their master plans what major goals different areas would have, including recreational, nature protection or climate protection areas. They must, however, lean on the Forestry Act unless the area is protected by nature conservation legislation. Until recently, the municipalities have refrained from trying to influence forest management, but the question is currently raised in the Ministry for the Environment (which administrates this Act) whether building of forest roads should not be subject to municipal planning. According to the Planning and Building Act, municipalities can create their own nature reserves without going through national priority–making, but this possibility has not yet been taken up.

Professional culture

All of the forest organizations have produced information materials and policy statements on multiple–use forestry, mostly from 1987 and onwards[10]. Education of personnel in these issues has taken place to some extent. For example, the forest owners' associations, with support from the forestry agency, have translated and adapted Swedish study material aimed at forest owners[11]. Also, interest is growing among forestry agency staff to follow the direction of their Swedish counterparts in improving their biological and ecological knowledge[12].

---

9. MIK stands for Municipal Environment Protection (Miljövern I Kommunerna).
10. Already in 1978, a special issue of the Forestry Journal "Norsk Skogbruk" was published (volume 24:2) on landscape and nature protection in forestry. Most information, however, has been produced in the last few years.
11. This material is entitled "Richer Forest" (Rikare Skog). The Swedish study material has, however, also borrowed ideas from Denmark.
12. In Sweden, the Forestry Agency personnel have undergone intensive education in forest ecology during 1989-91 based on a specially designed study material produced by the Swedish National Board of Forestry. This has now spread to the border districts in Norway, particularly from Värmland county in Sweden to Hedmark county in Norway.

Although the goal for state forestry changed to include multiple use in 1981, it is only recently that this change is apparent. State forestry has improved its multiple–use management as a result of the pressure from environmental groups, especially as a consequence of the debate on montane forests. They started to discuss with environmentalists the values, use and management of the forests, both at central and local levels. A change of attitudes within the state forest administration has taken place, particularly in the last few years. This is a result of both internal education of the existing staff, and recruitment of non–foresters (including planners, zoologists, and landscape architects) into new positions.

Environmental groups

The largest non–profit organizations that are active in influencing multiple–use forestry are the Norwegian Tourist Association (Den Norske Turistforening), with its 150,000 members and 47 local groups, and the Society for Nature Conservation (Norges Naturvernforbund) with 55,000 members and 144 locally active groups[13]. The Council for Outdoor Recreation (Friluftsrådet) is an umbrella for these and many other related organizations, and represents environmental groups in policy matters, particularly relating to public use of lands. Additional environmental organizations which are active at the central forest policy level are Future in Our Hands and World Wildlife Fund. Underway is a strategy for forest policy that will be a joint product between these organizations, with the Society for Nature Conservation taking a leading role.

A range of multiple–use issues has been debated by environmentalists over the last decades. It started with the size of clearcuttings in urban forests, and continued to include various forest operations, such as use of herbicides and building of forest roads. Another issue is the choice of species in forestry, which particularly affects the Western region. The question is to what extent the former, low–productive, broadleaved forests should be replaced by conifers. The forestry side, which is promoting fast–growing conifers, has however began to reconsider the reforestation strategy not only as a result of pressure from environmentalists, but also because of economic reasons, since the quality of the new species has not been up to expectations.

In the mid 1970s, forest workers were largely in opposition to the environmental movement, and it was the forest workers who stopped the protesters by force, for example in the Oslomarka area. Since then, the attitude among forest workers towards multiple–use forest management has changed, and local protesting groups may now be composed of forest workers and environmentalists in collaboration.

---

13. In 1990. In addition to these numbers, there are 9,000 young people organized in the youth section (Natur og Ungdom).

14.5 Finland

Forest–environmental legislation and administration

Two Ministries are responsible for multiple–use forestry: the Ministry of Agriculture and Forestry, under which the Finnish Forest and Park Service as well as the two National Forest Extension Services are located; and the Ministry of the Environment, which is responsible for more general environmental protection policy. The Finnish Forest and Park Service manages the state forests, but also nature conservation areas. The National Forest Extension Services, one for the Finnish speaking parts of the country (Metsäkeskus Tapio), and one for the Swedish speaking parts (Skogscentralen Skogskultur), supervise and promote private forestry[14]. In addition, private forestry is administered by 19 District Forestry Boards. The National Forest Extension Services and District Forestry Boards are self–governing bodies, consisting of representatives from forest owners, forest workers, government agencies, and the forest industry. The system with District Forestry Boards was created in 1928, and builds on close collaboration with the forest owners. The private forest owners are organized in some 300 Forest Management Associations at local level.

The actual field work of giving advice to the private forest owners is a joint undertaking between these District Forestry Boards and the local Forest Management Associations. The Forestry Boards also administer government subsidies to forest owners, give assistance to planning and carrying out projects concerning, for example, forest road building, drainage, pruning, and regeneration.

Finnish forestry legislation is characterized by a long list of special regulations for different purposes, instead of being collected into a single Forestry Act. For the state forests, there are three recent laws that are of particular interest here: (1) the Nature Livelihood Act (Naturnäringslagen), which was launched in 1984 to enhance the possibilities of the northern peoples to retain their traditional rights to make a living from natural resources[15], (2) the Wilderness Act (Ödemarkslagen) from 1991 designates special wilderness areas, and (3) the law on the Finnish Forest and Park Service from 1991 (this law was up–dated in 1993 when the FFPS was turned into a public funding supported

14. The administration of private forestry is presently being reorganized. The amount of organizations is likely to be reduced by 1996.

15. The Act regulates how state owned (particularly forest) land can be sold or leased to people (maximum size 30 hectares) for use in agriculture, fishing, hunting, reindeer husbandry, berry- and mushroom-picking, handicraft making etc. It is administered through the county agricultural agency. Those who have aquired a piece of land according to this Act also have the right to free grazing and gathering of various plants on state forest lands under the supervision of the forest officer.

commercial enterprise) which sets as its goals the promotion of nature protection and recreation.

The Law Concerning Private Forestry (1967) lacks specific goals both for economic forestry and for multiple use. However, the leading rule is that forests must not be destroyed. The law obliges the owner to regenerate after clear-felling and to follow prudent management principles. A forest owner who does not comply with regulations can be prohibited from carrying out any further forest operations for a period of ten to twenty years. The forest owner is also obliged to make an announcement to the District Forestry Board (seven days in advance) before he/she can proceed with a commercial cutting. When regeneration cuttings are concerned, the forest owner must present a forest plan that is accepted by the District Forestry Board. Such forest planning does not, however, require environmental protection interests to be included.

The Forest Improvement Act of 1987 states that subsidized measures are not allowed to "considerably" affect the environment in a negative way. This came as a result of pressure from environmental interest groups. Government subsidies since 1987 can no longer be used in areas that are designated to be potential protected areas (e.g. for nature protection and recreation) in the future. Drainage operations were then also made subject to special permission from the environmental protection authority.

Since the 1960s, annual development programs have been drawn up, specifying the extent of silvicultural and basic improvement measures. Public funding of forest improvement measures was introduced, however, in the 1920s. In the 1960s, the MERA[16] programmes intensified subsidies to, for example, forest regeneration, drainage, road building and fertilization. The funding level has during recent years been 350–400 million FIM per year (90–100 million US dollars). There are no rules to help the districts prioritize the subsidies towards environmental concerns, but it is left to the discretion of District Forestry Boards and their officers.

In 1985, the Forest 2000 Programme was adopted by the government. This was the first program to include some multiple–use thinking through a specially designed multiple–use working group. However, multiple use was still regarded as a constraint to timber production (Saastamoinen 1987). The program was revised in 1991–92, since it was built on the fear of shortage in wood supply to the industry, and there is now an excess supply of wood because of the economic recession and rapidly increasing growing stock. The revised Forest 2000 Programme pays attention to the environmental consequences of wood production.

The Ministry of the Environment was created in 1983. It includes units for environmental protection and nature conservation, physical planning, and housing (the structure of the Ministry will be changed in 1995). Their connection to the field goes through the regional planning associations and through

---

16. MERA is the Finnish abbreviation for FOrest FInancing.

the environmental units of the county administrations. Usually they do not involve themselves in forestry matters, unless the area is designated for protection or recreation. The county administration has the delegated power from the Ministry to settle nature reserves on private lands, if the private owner initiates the case him/herself. The Ministry of the Environment has initiated special protection programs for peatlands, lake shores, herb–rich broadleaved forests, wetlands and small streams and watercourses. Future programs include old–growth forests, cultural landscapes and cliffs. There is no formal procedure to check whether forest operations affect such areas. However, the District Forestry Board often consults with the County Administration. The environmental protection units at the County Administration are responsible for nature reserve areas through the Nature Protection Act.

The Building Act, 1981, regulates public planning procedures (the Act is presently being rewritten). The 20 regional planning associations carry out planning for general land use. In practice, they do not differentiate between various priority goals in forestry, but only between "protected" forest areas (through nature protection legislation) and commercial forests. Municipal planning in relation to forestry could become a way to regulate urban forestry towards multiple–use management. The municipalities already have the mandate to control cutting of trees within their territory. This has been applied particularly in Helsinki and its surrounding municipalities, and may also spread elsewhere. Depending on the specific rules of each municipality, forest operations within designated forest areas (private[17] or public) can be prohibited unless special permission is given by the municipality beforehand. In the Helsinki area, this has increased public participation in the planning of forest operations. The Building Act is thus formally much more effective than the Forestry Act in regulating multiple use, but can only be used in forests that are particularly important to the municipality residents.

Professional culture

Very few non–forestry professionals are employed within the forest organizations. In the Finnish Forest and Park Service, a change is currently underway due to reorganization that will decrease the staff working on traditional economic forestry, and increase staff working on multiple use, including recreation and nature protection. This change is also enhanced by economic reasons, since net income from commercial forestry is steadily decreasing.

---

17. In private forests, every single forest operation must be dealt with separately for each individual owner. In practice, the Forest Management Association combines several cases and discusses them with the municipal officers so that the administrative work load does not become overwhelming. There are also informal meetings between forest owners' representatives and municipal officers to achieve common guidelines.

Until now, the District Forestry Boards have not intensively campaigned to increase knowledge of multiple–use forestry among private forest owners. However, during the last few years more written material has been produced than ever before. Education of forestry professionals and forest owners has been strengthened. Recently more comprehensive study material has been produced which may increase the level of education within the forest organizations.

Among private forest owners in Finland, there has been a change of attitudes since the beginning of the 70s. At that time, 72 % approved of current forest plantation practices, 64 % were convinced that heavy scarification does not spoil nature, and 33 % thought that forest officers do not consider forest owner interests. In all these areas, there has been a clear change towards favouring environmental concerns (Ulfvens 1985).

By tradition, the state as well as the private forestry professionals have learned to defend themselves against the environmental critique. Their attitudes have been highly in favour of economic prosperity in the forest sector, in particular the pulp industry. Facilitated by their common professional culture, an informal forestry sector quasiboard has been formed that regulates the mutual objectives of the contradictory interests of the basic organizations, including the forest industry, the state forest service, and the labour unions (Korhonen 1991). Such a quasiboard has emerged in Finland, where the role of forest industry is particularly strong, but similar tendencies are also indicated in the other countries.

Environmental groups

The Finnish Association for Nature Conservation was founded in 1938. It became distinctly critical of forest practices in the 1960s. Around that time, nature conservation expanded to include environmental protection. Its current 32,000 members and 204 local associations have remained steady through the 1980s, while membership greatly expanded in the 1970s. The Swedish speaking sister organization is concentrated in the southern and western districts, with 4,000 members in 22 local groups. The local groups often engage in forestry matters. Public interest in Finland over multiple–use concerns can be partially explained by extensive use of forests for berry– and mushroom–picking. In particularly the eastern and northern regions up to 70–90 % of all households are involved in such activities.

Environmentalists started by criticizing clearcuttings, use of chemicals, and drainage of peatlands in the 1970s. The debate has intensified in the late 1980s and beginning of the 1990s. Forest action groups have been formed in different parts of Finland, and the debate has actualized, at a more general level, the question of democratic influence in forest policy. A challenge to forest programming based on ecological, regional and social differentiation within society has been expressed (Lehtinen 1991). Pressure is coming also from international environmental organizations. For example, Finnish forest companies' activities abroad, as well as the environmental effects of their methods at home, are scrutinized by international environmentalists.

14.6 Sweden

Forest–environmental legislation and administration

The National Board of Forestry (Skogsstyrelsen) is the agency responsible for monitoring the Forestry Act, regardless of forest ownership. This is a central government agency with 22 county Forestry Boards (hereafter called Forestry Agency) that carry out daily supervision in 141 districts, with district forest officers. The organization employs about 1,500 people. This organization has recently been revised to respond to the new forest policy. The National Board of Forestry is completely separate from the Forest Service (Domänverket) which manages state–owned forest lands.

The new forest policy was approved by Parliament in May 1993 and entered into force in January 1994. It involved major changes in two respects: first, deregulation of previous management rules and prescriptions, and second, increased emphasis on environmental values. The change was justified by the growing recognition that Swedish forestry was contributing largely to the decline of biological diversity (Skogspolitiken inför... 1992). Timber production and maintenance of environmental values as policy goals were given equal priority in the new Forestry Act. The revised forest policy also implied fewer precise regulations, with the hope of thereby enhancing variation in forest management (En ny skogspolitik 1993).

Multiple–use aspects were first introduced into the Forestry Act in 1975, and in 1979 these were made more specific and accompanied by a penalty clause[18]. Specific requirements and recommendations are spelled out in the Forestry Act for protection of environmental interests in all forest operations. These can be monitored by a notification system for clearcuttings (one month in advance, and two months in sensitive areas). Other types of forest operations cannot be monitored except forest road building in environmentally sensitive areas, and ploughing and drainage of forest land. The latter requires consultation with the environmental section of county administration.

Beginning in 1994, all clearcutting notifications must be accompanied by a brief EIA (environmental impact assessment), stating what special environ-

---

18. The legal penalty is only valid if the Forestry Agency has issued specific prescriptions for each forest operation. The forest owner cannot be punished if such prescriptions were not made beforehand. Only the Forestry Agency is allowed to sue the forest owner, which means that the public can only informally try to convince the Forestry Agency to prescribe what measures are necessary in order to fulfil multiple–use interests.

Figure 14.3 Protection of broadleaved trees at forest clearcuttings is one of the environmental measures contained in the Swedish Forestry Act of 1979. Evaluations have shown, however, that single trees are more often saved for aesthetic rather than pure floristic and faunistic reasons. Here, an aspen tree has been felled. Photo: K. Eckerberg.

mental measures will be taken within the planned clearcut area. This procedure increases the possibility for forest agency officers to give more specific advise on environmental issues to forest owners.

Despite the ambiguous formulation of the environmental protection requirements, they have functioned to increase the awareness of such issues, particularly among forestry professionals. They also play a role in legitimizing the work by both the environmental and forest authorities towards increased environmental protection measures. Several studies of the extent to which these regulations are implemented in forestry practice show that average compliance is, however, not satisfactory. Evaluations of their implementation reveal figures of around half (Eckerberg 1990) or two thirds (Tagen hänsyn... 1991) of what is desired[19]. Aesthetic values are protected to a larger extent than purely faunistic/floristic, and more environmental protection measures are achieved in areas where recreational pressure is high (*Figure 14.3*). The use of highly mechanized equipment proved to be the most significant reason for the low degree of environmental protection measures (Eckerberg 1990).

19. The first follow-up of the Forestry Act regulations for multiple use from the years 1981-84 (Eckerberg 1990) shows approximately the same result as the next follow-up 1989-90 (Tagen hänsyn... 1991). The differences between the two evaluations can largely be ascribed to different methodologies in measuring successful outcome.

Eight regional Forest Owners' Associations (Skogsägarföreningar) assist private forest owners in forest management and coordinate timber trade. Their 87,000 members represent almost half of the total land area owned by private individuals. There is also the Forestry Society (Skogssällskapet), a private organization offering forest management services to municipalities, foundations and private forest owners. The Swedish Forestry Association (Sveriges Skogsvårdsförbund) is a non–profit organization for information and education, geared towards sustainable use of forests. It collaborates with private forestry as well as with environmentalists in multiple–use related issues.

Since 1983, all forest owners were required by the Forestry Act to prepare a management plan but obligatory planning was abolished in the new forest policy of 1994. The state subsidized this planning, since it combined it with forest inventory carried out by the Forestry Agency. The plans are not legally binding, but often used as a basis for advice and follow–up by the Forestry Agency, and they are in principle public documents[20]. Since the mid–1980s, these plans have to some extent included environmental information, but the quality of this information varies at the discretion of the planner concerned. If the Forestry Agency, the environmental agency, or the municipality considers the forest area to be of particular interest, the plan may be extended by public funding to put more emphasis on inventories and management strategies to protect environmental aspects.

State subsidies to forestry have been justified to increase wood production and improve regional policy, with the aim of creating jobs in regions with unemployment. The subsidies amounted to around 240 million SEK (40 million US dollars) annually during the 1980s, but have decreased to 90 million in 1992/93, and to 45 million in 1993/94. This includes regeneration measures, building of forest roads, drainage of wetlands, forest planning, and forest improvement measures for low–productive forests. More recently, subsidies have also been launched by the agriculture agency for afforestation of marginal agricultural lands. Many of these subsidies have been intensely criticized by environmentalists, who claim that they counteract environmental protection. This critique, in addition to economists' evaluations showing little or sometimes even negative effects of the subsidies (Bångman 1990), has influenced the government to cut down the most controversial grants.

Special inventories and conservation programs have been launched by the Swedish Environmental Protection Agency (Naturvårdsverket) over the years, including wetlands, virgin forests, montane forests, and marginal agricultural lands. Some of these inventories have been conducted in collaboration with the National Board of Forestry, such as for virgin and mountain forests. The purpose has been to increase knowledge about environmental values for planning purposes. The National Board of Forestry has recently begun to inventory

20. Although the plans are formally open to the public, in practice they are not designed in a way that non-professionals could make much use of them, especially in relation to multiple use.

wet forests and key habitats for threatened species. New regulations to protect small–size, threatened biotopes has been introduced in the Nature Conservancy Act in 1993, which may affect the way nature conservation measures will further be handled in forestry.

Legislation to protect broadleaved deciduous forest stands was introduced in 1974 for beech forests (Bokskogslagen), and for other deciduous trees including oak and ash in 1984 (Ädellövskogslagen), since these southern forests had drastically diminished. Felling and regeneration of such stands comprising at least 50 % of the regulated species must be preceded by special permission.

The Building and Planning Act, 1986, regulates the municipalities' right to plan their own territory, which may also include forest lands that are of particular local importance. In such areas, they have the right to be consulted by the environmental and forest authorities at county level. The municipalities are to prioritize land–use management goals according to their political wishes, but are dependent on the county–level authorities to enforce such goals. Since 1987, the municipalities have had the delegated right from the county administration to set up nature reserves, but this opportunity has not yet been notably used because of the current economic situation.

Professional culture

The County Forestry Boards were initially created at the regional level in 1905, and became state agencies only in 1980. This explains partially why the Forestry Agency officers tend to advocate the interests of the collective forest sector. Furthermore, the officers have a split role in relation to the forest owners; they base their services to a large extent on commercial activities[21], and are thus dependent on the good–will of the forest owners. At the same time it is their obligation to enforce the Forestry Act.

In practice, legal sanctions are rarely used due to two major reasons: the Forestry Act regulations on multiple use are formulated as a frame–law, which leaves considerable room for different interpretations, and the Forestry Agency is therefore reluctant to resort to court action (Tagen hänsyn... 1991). Furthermore, the Forestry Agency traditionally has a friendly, personal relationship with the forest owners.

The first multiple–use ideas that penetrated into forest administration in the mid–1970s were inspired by landscape protection and aesthetic values geared to increase recreational experience from forests. In the beginning of the 1980s, this changed into more ecological thinking, motivated by up–to–date scientific data, in the form of long lists of threatened species in Swedish forests. Consequently, the Forestry Agency has campaigned along the same lines: first in landscape protection, then faunistic, floristic and ecological values. On

---

21. They administer some 20 major forest nurseries distributed throughout the country and carry out different services to the forest owner on a commercial basis.

its way is a campaign to increase cultural awareness, i.e. protection of artifacts and cultural remnants in the forests. The campaigns have been quite impressive in terms of the number of people within the forest sector attending courses and excursions. In addition, comprehensive educational materials have been produced and widely distributed. During 1991–92, more than 60,000 forest owners, forest workers and entrepreneurs have been educated with the aid of a specially designed study material "A Richer Forest".

The Forest Service as well as the private forest companies have throughout the 1980s educated their personnel in ecology and environmental protection. A large number of colourful brochures and materials has been produced, and new management procedures are being launched in order to increase the amount of multiple–use considerations. This is a long–time effort, since previous clashes with environmentalists, particularly in the 1960s and 1970s, have left deep scars on their public image. Starting at the end of the 1980s, the formerly extremely homogeneous professional staff of foresters has gradually widened out to also include specialists in environmental protection, such as ecologists. Such a development began within the Forestry Agency and spread to the state and private forest companies. Still, however, the professional culture of both the forest owners and their controlling agency can be characterized in terms of a common understanding and hesitation to accept public and "outside" demands for environmental measures that are not within their professional code of ethics.

Environmental groups

Forestry is a major political issue for many of the environmental organizations in Sweden. Their intensive campaign against large clearcuttings and the use of herbicides in the 1970s, and its coverage in mass media was probably the main reason for amendments in the legislation. The Swedish Society for Nature Protection (Svenska Naturskyddsföreningen) is the largest and the most established organization, and has grown rapidly[22] to over 200,000 members by 1991. Many of its 241 local groups are carrying out inventories, and are involved in protecting environmental interests in forestry. The Swedish Ornithological Society actively protects bird habitats in the forests. The hunting associations (with 170,000 members) are active in wildlife protection, and a number of "outdoor life" and homestead societies often engage in forest management issues when forestry activities affect recreational and social interests in their neighbouring forests.

In the 1980s, several additional environmental action groups were formed to defend multiple–use aspects in forestry. These include the movement for the protection of montane forests (FURA), which was most active in the second

22. In 1970 there were 40,000 members, which grew to 68,000 until 1980. From 1989 the growth has stabilized.

half of the 1980s during the political discussions on the extent to which the state should protect montane forests. Local groups have also been formed to carry out inventories of threatened species. The results of these inventories are used in negotiations with forest owners and government authories. A group called "One Step Ahead" within the Society for Nature Protection is active in the very north. The group has, during the last two years, inventoried large forest areas and discovered considerable numbers of rare and threatened forest dependent species that were previously unknown, both within and outside of nature reserves. International cooperation to protect boreal forests was initiated in 1992 at a large conference in Jokkmokk in northern Sweden by founding a new organization, Taiga Rescue Network, which involves non–governmental organizations and researchers.

A comparison of the environmental debate from the 1960–70s with that of the 1980s and onwards shows a major change in attitudes and strategy of both the environmental groups and the forest sector (Löf 1993). Both sides are now better informed of the actual environmental situation, and discussions are becoming increasingly constructive. For example, environmentalist standpoints nowadays frequently appear in forest periodicals, and collaborative excursions and courses have increased.

14.7 Comparison and conclusions

The availability of forests varies greatly for the four countries, and can be compared by forest area per capita: Finland 4.16 hectares, Sweden 2.86, Norway 1.67, and Denmark 0.1 (Börset 1986). Naturally, there is more pressure from different public interests when the accessible forest area is small. But it is also a matter of how the forests have been managed, and to what extent they can offer a range of values and opportunities for recreation, diversified ecosystems, utilization of resources, and symbolic features. The combination of access and the environmental status of existing forests in these countries make up the preconditions for multiple–use conflicts.

In Finland, Sweden and Norway the montane forests are of particular importance to multiple use, and have historically more or less been left outside of commercial forestry. Particularly in the 1970 and 1980s, pressure on these forests increased for wood production, thus creating conflicts between exploitation and protection. Old–growth forests in southern areas were also affected by this conflict. Since most of these forests are owned by the state, environmentalists have felt that they should be managed for the public good to a larger extent than forests which are not state–owned. It has taken the state forestry a long time to react to these demands. In Denmark, as a contrast, the state forests are located close to the more inhabited areas. State forestry in Denmark has, hence, been forced to incorporate various public interests into their management at a much earlier stage.

Starting in the early 1960s, forestry in all of the Nordic countries became more and more mechanized[23], and at the same time the size of the forest industry increased, particularly in Finland and Sweden (*Figure 14.4*). The governments initiated a series of programs to meet the increasing demand for timber for the industry, culminating in the 1970s. The programs promoted forestry practices that were seriously criticized by the growing environmental movement: for example, large–scale clearcuttings, the use of herbicides, drainage of peatlands to grow new forests, and fertilization. In addition, economic subsidies were introduced to stimulate faster growth and more intensive management of forests. Many of these measures counteracted environmental concerns. Both legislative and economic instruments were used to force forest owners towards more commercial management. The new interventionist state policy aimed primarily at private forest owners was supported by the governments in collaboration with the labour unions and industry.

The reaction from the general public on how the new forest practices affected the environment – to a large extent channeled through the new environmental movement – grew at the end of the 1960s, and continues to appear frequently in mass media reports. All four Nordic governments have responded by launching forest–environmental legislation. These include legislative amendments as well as adjustments to the above mentioned government subsidies. Starting in Norway, in 1973, the management goals for the Oslomarka area were altered to include multiple–use aspects. At almost the same point of time (in 1975 and 1976), the Swedish and the Norwegian Forestry Acts were amended, and multiple use was mentioned as worthy of consideration, although wood production continued to be the main goal. General multiple–use goals are today incorporated into forestry legislation to some extent in all the Nordic countries, but there are no or few possibilities of enforcing them. Special regulations on how state subsidies to forestry must not negatively affect environmental protection have been issued in Norway and Finland, and all the controversial forestry subsidies have been abolished in Sweden and Norway. The magnitude of subsidies to measures that are doubtful from a multiple–use perspective has thus decreased drastically. The difference in the 1990s is that

---

23. Norway and Denmark are the least mechanized - although the degree of mechanization in Norway has increased from 23 % in 1985 to 44 % in 1989. Also in Denmark, in the last two years the mechanization in conifer forests and in broadleaved thinnings has grown from almost zero to 40 % in 1991. The goal for the state forests is set to 80 % mechanized in 1995, excluding old beech and oak forest. The private forestry is undergoing the same change of technology. In Finland, the figure is over 60 % and is expected to increase further to some 80 %. In Sweden, in 1990 more than 80 % of clearcuttings are fully mechanized and over 50 % of thinnings. According to Eckerberg (1990), the higher the degree of mechanization in forest clear-cuttings the less environmental features are protected. The above tendencies may therefore lead to less multiple–use concerns in forestry practice in all of the Nordic countries. However, there is also an evolution of "softer" technology such as small and flexible machines with less ground pressure that may counteract this development.

Figure 14.4 The size of clearcuttings increased drastically in the 1960s. This was due to mechanization of forest operations and was one of the first environmental issues that led to clashes between environmentalists and foresters, especially in Sweden and Finland. Photo: K. Eckerberg.

a larger proportion of government subsidies are paid for afforestation of marginal agricultural lands. This may affect biological diversity in a negative way, as well as aesthetic and recreational values. Through inventories and environmental planning, priorities for multiple–use goals could be made by the relevant environmental and agriculture authorities. However, recent changes in Nordic countries' forest policy towards sustainable management, which includes genuine environmental concerns for biodiversity, social, recreational and historic values, still remain to be evaluated in practical terms.

Private forests are predominant in all the Nordic countries, and state intervention to influence the private forest owners' forest management – if it goes against the public good – are therefore crucial. Traditionally, the private forest owners have created and maintained their own organizations for advisory and operational services in forestry. These organizations are still closely linked with the government administrations responsible for carrying out forest policy. Through personal and commercial ties, controlling agencies are in the hands of private forest owners. Respect for private property rights is firmly rooted among forestry administrators, and their steering strategy is based on mutual understanding and soft methods such as information and persuasion. Although the formal set–up of the controlling agency differs in the Nordic countries – pure state agency in Sweden and Norway[24], and based primarily on private, state–funded organizations in the other countries – in practice this difference is not important. The Swedish forestry agency has so far also had a much more rigid and detailed legislation to implement, but in practice they have seldom made use of the possibilities for punishment by the law.

Environmental legislation and environmental authorities have a different approach. Their role is to protect, and in doing so they must compensate the forest owners for the loss of land. Forest land acquisition is expensive, and therefore most nature reserves and national parks have been created on low–productive and remote areas, and primarily on state land. In all four countries, there is now increasing discussion of how the more valuable private forest lands can be protected, including remaining old–growth forests and broad-leaved forests in the middle and southern parts of Finland, Norway and Sweden, and the small patches that are left of such forests in Denmark. Current state budgets for preservation of nature are not sufficient to cover such protection. The environmental authorities are reluctant to interfere through applying nature conservation legislation, since there is little chance that they will be able to purchase them in the foreseeable future.

There is a fundamental difference in professional values between foresters (forest workers, forest administrators, supervisors, directors, etc.) and environmentalists. Foresters regard multiple use in forestry mostly from an economic perspective, whereas environmentalists take an ecological standpoint. Hence, perception of the "problems" in forestry is different depending on the professional values, although they tend to coincide more today than in the beginning of the multiple–use forestry debate in the early 1970s. Both, however, demand that society should interfere when markets fail.

14.8 Perspectives for the future

The political changes and liberal ideology which are sweeping over all Nordic countries in the 1990s cannot fail to also affect the forestry sector. In all these countries, there is a recent tendency towards increased discretion in the policy process, both in terms of decentralization of formal policy decisions, deregulation, and allowing market forces into the multiple–use forestry related arena. There are political discussions in all these countries on privatization of state forests, but so far selling off of state land has only been marginal.

The decline of agriculture, and the relative decline of timber value in relation to public interests in recreation and nature conservation is common to all these countries. This favours development of multiple–use management within the forest sector for purely economic reasons. Afforestation of marginal ag-

---

24. Since 1980, the Swedish forest agency was included into state administration, but its heritage comes from the forest owners (see earlier description). In Norway, the state forestry agency works in close collaboration with the private forest associations.

ricultural lands has already become an environmental issue. It relates both to cultural values and to the preservation of threatened species, since many of these species are dependent on such marginal areas.

The actual incorporation of environmental values into the administration depends heavily on the value perceptions and priorities of the society and its leadership. Unfortunately, some of these alterations in human attitude come about only through environmental crises, or through strong pressures from an environmentally sensitive public. There is a built–in resistance to actually make such changes in the administration that could address the complex nature of environmental problems (Henning & Mangun 1989). This includes a reluctance by forest professionals, who dominate this sector, to hand over responsibility for surveillance of forests to non–foresters, such as biologists and public planners. As earlier mentioned, the set–up of the Danish administration as a joint agency for forestry and environmental protection has helped the Danish forestry sector towards more integrated thinking within this policy arena.

Participation by the public in forest–environmental policy–making at all levels is still at its infancy in these countries. There are few or no formal structures to permit such open policies in the forest planning system. Recent pilot efforts by some forest companies may change this picture, if the regional planning units and the municipalities dare to take up the challenge. This has mostly happened in the most densely populated towns. Judging from the current strained economic realities of most Nordic municipalities, their possibilities of paying economic compensation to conserve natural areas will be very limited in the near future.

Nevertheless, the forest sector, with its specific professional culture and close internal ties, is gradually responding to multiple–use thinking in forest management. This development can be ascribed both to an increased acknowledgement by forestry professionals of the actual economic profitability of managing forests in an ecologically sustainable manner[25], and to information and education campaigns to increase the forest professionals' and forest owners' awareness in multiple use of forests, including preservational and recreational interests. There is still, however, little substantial evidence that the forest professionals' frequent rhetoric about the importance of such values has actually changed their behaviour. Short–term economic realities of the forest owner continue to be the driving force behind their choice of forest management systems. Nordic foresters' professional culture is, however, not unique. Foresters in other countries are also known to suppress conflicts and reject so-

---

25. For example, the forest sector itself has taken up the criticism against large-scale stereotyped management, and recently introduced the concept of "forest site adaption" in all kinds of forest operations, including choice of species and methods of harvesting and scarification according to particular biological conditions at micro level.

cial values which are not easily quantifiable in economic terms (Koch & Kennedy 1991).

Scandinavian foresters were forced to entertain multiple use of forests in their thinking, due to the critique from the environmental movement in the late 1960s and early 1970s. From being dominated by people who largely enjoyed nature in a passive way, nature conservation societies in the Nordic countries began in the 1970s to more actively demand societal measures to preserve natural ecosystems and consider ecological principles in economic activities. The domination of foresters in the nature conservation movement waned, and a new generation of young activists and ecologically educated university students took up the initiative (Reunala & Heikinheimo 1987).

In short, the major pressures for more multiple–use concerns came from the local traditional natural resource management culture (especially the Sámi population in the northern Scandinavia), from ecologists (including scientists), and from various recreational interests. On the other side, the forest owners may have different management goals, depending on whether and to which extent they are economically dependent on their resource, and whether they are public or private.

An analysis of multiple–use forest policy in the Nordic countries suggests that the development of multiple–use legislation has been spurred largely by pressure from the environmental groups, which have been particularly strong in Sweden and Denmark. Their role is changing now when multiple–use concerns have been introduced into forestry and land use planning legislation. Scrutinizing the forest and environmental agencies' implementation and monitoring of new legislation is one of the more important roles of environmental groups. In Denmark, the Society for Conservation of Nature has played an important role in appealing permits given by county councils for development within areas of concern for environmental protection, thus putting pressure on public agencies to implement environmental policy. In all the Nordic countries, environmental organizations are becoming more and more specialized, and are to a larger extent than before backed up by the scientific community in their criticism of current forest practices. This has led to increased mutual understanding between the forest sector and the environmentalists. Still, there are several forest policy conflicts that remain unsolved because they are based on different ideological approaches to commercial use and protection of natural resources. The question of who pays for multiple–use forestry will also continue to be on the political agenda.

Denmark is the only one of these countries which is a member of the European Union (EU). Sweden and Finland are joining in 1995, while Norway will remain outside. There may be several ways that the membership could influence multiple–use forest policy. First, it can be expected that more Europeans will discover the natural beauty of the Nordic countries, and be attracted by the Right of Public Access to land, in particular to forest land. This will imply that more people will be wandering around on private lands, and there will be much greater pressure for recreational adaptations of forest management. In Denmark, this is already happening, and private forest owners are starting to complain about commonly used trails through their property. A similar situation can be discerned in Sweden, with increasing numbers of berry pickers coming from other parts of Europe. There will be a great need for multilingual information on the rights and duties connected to the Right of Public Access.

Another effect of the membership of the EU is that its environmental protection policy will have to be followed by all member states. For example, the directive on protection of wild birds states that 111 different bird habitats in Denmark, of which around 13 are within forests, must be protected from major changes (E.C. directive, 2 April, 1979). This is a clear directive, that can be appealed in court if not followed. It is different in character from voluntary international agreements, such as the Wetland Convention.

Both international agreements and non–governmental organizations will most likely put increased pressure on forestry in the Nordic countries to become more multiple use oriented. Already, the European environmental movement has managed to influence Scandinavian forest companies to take environment seriously. The threat is that otherwise there will be no market for Scandinavian forest products. Eco–labelling has been introduced by some of the forest companies as a response to the consumers' demands for "green" products, and it is likely that this will be further developed and followed by the forest industry as a whole. Influence from the international market and environmental movement thus appears to be more decisive for the implementation of multiple–use goals than national policy has been. Despite the forest sectors' predominance in current forest policy, a major change not only in rhetoric and stated policy goals but also in management practices is probably inevitable. In all of the countries, there is a fear on the part of the forest authorities that if they do not live up to the expectations of the environmentalists, and if the forestry sector is not capable of acting now, with its current freedom, new and more rigid regulations will follow. The other future alternative, which is perhaps inevitable, is that non–foresters will take over more decision–making power within this policy field.

References

Andersson, B. & Hultman, S.–G. 1980. Skogens värden: skogsbrukets roll. (The values of forests: the role of forestry.) Sveriges lantbruksuniversitet och LTS Förlag, Kristianstad. 233 pp. (In Swedish.)

Bångman, G. 1990. Skogspolitiska styrmedel och deras effekter: en litteraturstudie. (The administrative tools of forest policy: a literature survey.) Sveriges lantbruksuniversitet, Institutionen för skogsekonomi, Arbetsrapport 122. (In Swedish.)

Bevarande av faunans och florans mångfald vid skogsbruk. (Preserving the diversity of fauna and flora in forestry.) 1991. Arbetsrapport. Skogsstyrelsen, Jönköping. (In Swedish.)

Biologisk mångfald. (Biological diversity.) 1990. Naturskyddsföreningen, Årsbok 1990, årgång 81, Stockholm. (In Swedish.)

Børset, O. 1986. Skogskjøtsel II. Skogskjøtselens teknikk. (Silviculture II. Techniques of silviculture.) Landbruksforlaget, Oslo. 455 pp. (In Norwegian.)

Darpö, J. 1991. Liv i skogen: reformbehov och förslag till rättsliga förändringar för naturvårdshänsyn i skogen. (Life in forest: the need of reforms and a proposition for juridical changes to improve nature protection in forests.) Sveriges lantbruksuniversitet, Institutionen för skogsekonomi, Arbetsrapport 132. (In Swedish.)

Eckerberg, K. 1990. Environmental protection in Swedish forestry. Avebury Studies in Green Research, Aldershot. 179 pp.

En fremtidig skovpolitik. (Future forest policy.) 1986. Landbrugsministeriet, Betænkning nr. 1090. København. 125 pp. (In Danish.)

En ny skogspolitik. (New forest policy.) 1993. Parliamentary Standing Committee on Agriculture, 1992/93: JOU15, Stockholm. (In Swedish.)

Flersidig skogbruk: skogbrukets forhold til naturmiljø og friluftsliv. (Multiple–use forestry: the relationship of forestry with natural environment and outdoor recreation.) 1989. Norges offentlige utredninger, NOU 1989:10. 139 pp. (In Norwegian.)

Hafsten, U. 1977. Naturvernets århundrede: fra pionerinnsats til regjeringsposisjon, fra punktfredning til nasjonalpark. (A decade of nature protection: from pioneering to the government, from saving small pieces of land to national parks.) Oslo. (In Norwegian.)

Helles, F., Jensen, S.F. & Risvand, J. 1984. Den danske skovsektors samfundsmæssige betydning. (The socioeconomic importance of Danish forestry sector.) DSR forlag, København. (In Danish.)

Henning, D.H. & Mangun, W.R. 1989. Managing the environmental crisis: incorporating competing values in natural resource administration. Duke University Press, Durham and London.

Hultcrantz, L. & Wibe, S. 1991. Skogspolitik för ett nytt sekel. (Forest Policy for the new century.) Finansdepartementet, Ds 1991:31. (In Swedish.)

Jansen, A.–I. 1991. Reform og resultater: evaluering av forsøksprogrammet miljøvern i kommunene. (Reform and results: the evaluation of an exprimental program of environmental protection in communities.) NORAS, Oslo. (In Norwegian.)

Koch, N.E. & Kennedy, J.J. 1991. Multiple–use forestry for social values. Ambio 20(7): 330–333.

Korhonen, M. 1991. Emergence and transformation of a sectoral quasiboard: a case of the Finnish forestry sector. Second draft. Prepared for the 10th European group for organizational studies colloquium "Societal change between market and organisation", Vienna, July 15–17, 1991. Helsinki School of Business Economics, Helsinki.

Koskinen, T. 1985. Finland: a forest sector society? Sociological approaches, conclusions and challenges. In: Lilja, K., Räsänen, K. & Tainio, R. (eds). Problems in the redescription of business enterprises. Helsinki School of Economics Studies, B–73: 45–52.

Kristiansen, L. 1988. Skov–info. Om informationspolitik og formidling til den private skovejer. 3 dels projekt. (Forest –info. Information policy and mediation for the private forest owner. 3 parts project.) Skovbrugsinstituttet, Kgl. Veterinær– og Landbohøjskole, København. (In Danish.)

Lehtinen, A. 1991. Northern natures: a study of the forest question emerging within the timber–line conflict in Finland. Fennia 169(1): 57–169.

Löf, M. 1993. Relationen mellan miljörörelsen och skogsbolagen 1970–1993. (The relations between the environmental movement and forest companies 1970–1993.) Sveriges lantbruksuniversitet, Institutionen för skogsekonomi, Arbetsrapport 176. 60 pp. (In Swedish.)

Report of the Finnish Commission on Environment and Development. 1989. Ministry of the Environment, Commission Report 1989:9.

Reunala, A. 1987. Skogen och finländarna. (Forests and Finns.) Historisk tidskrift för Finland 3: 374–396. (In Swedish.) Also available in English: Reunala, A. 1987. The forest and the Finns. In: Engman, M. & Kirby, D. (eds.). Finland: people, nation, state. Hurst, London. p. 38–56.

Reunala, A. & Heikinheimo, M. 1987. Taistelu metsistä: voimaperäinen metsätalous Suomessa ja muissa maissa. (The fight for the forests: intensive forestry in Finland and other countries.) Kirjayhtymä, Helsinki. 188 pp. (In Finnish.)

Saastamoinen, O., Hultman, S.–G., Koch, N.E. & Mattsson, L. (eds.). 1984. Multiple–use forestry in the Scandinavian countries. Communicationes Instituti Forestalis Fenniae 120. 142 pp.

Saastamoinen, O. 1987. Multiple use and the Forest 2000 Programme. Scandinavian Forest Economics 29: 39–47.

Skogspolitiken inför 2000–talet: huvudbetänkande av 1990 års skogspolitiska kommitté. (Forest policy for the 21st century: main report of the 1990 forest policy committee.) Jordbruksdepartementet, Statens offentliga utredningar, SOU 1992: 76. 343 pp. (In Swedish.)

Tagen hänsyn vid slutavverkningar 1989–91. (Considerations in final fellings 1989–91.) 1991. Skogsstyrelsen, Meddelande nr 4. (In Swedish.)

Ulfvens, J. 1985. Kris i skogen. (Crisis in forests.) Miljöförlaget, Helsingfors. 127 pp. (In Swedish.)

15 Planning of multiple–use forestry

Jyrki Kangas[1]
Lene Kristiansen[2]

Abstract

Multiple–use planning of forest resources is needed in order to facilitate the best possible choice, or at least an approvable mix of several inputs and outputs in an extremely complex production process. Multiple–use planning is always multi–objective. In any comparison of decision alternatives, the goals and objectives placed on the utilization of the forestry production process have to be known. Decision alternatives should be evaluated with respect to each objective, and, furthermore, with respect to overall priority. Recently, multiple–use planning methods have vigorously been developed in the Nordic countries – especially numerical optimization procedures and choice models for both strategic and tactic multiple–use planning. However, these methods have not yet gained general approval in forestry practice; practical multiple–use planning is still mainly based on more or less descriptive approaches. Nowadays, the most crucial problem in applying analytical multiple–use planning approaches is the lack of production functions, or other models which could be used to measure and evaluate different products and benefits – quantitative as well as qualitative ones – not only at national or regional but also at forest holding and tree stand level.

Keywords: forest planning, public participation, integrated forestry, decision analysis, landscape planning, multi–objective optimization.

---

[1] Finnish Forest Research Institute
Joensuu Research Station
P.O.Box 68, FIN–80101 Joensuu, Finland

[2] Danish Forest and Landscape Research Institute
Skovbrynet 16, DK–2800 Lyngby, Denmark

15.1 Introduction

Forestry is called multiple use of forest resources when several goods and services are produced and utilized. Planning of multiple–use forestry is, thus, multi–objective forest planning.

In the Nordic countries, during the last few decades, planning of managed forests has mainly been planning for wood production only. Timber management planning is, however, only a part of forest planning. Considerable parts of forest planning include, for example, design of forest landscape, wildlife management planning, recreation management planning, and consideration of biological diversity. In multiple–use planning of forest resources, all parts are integrated into one planning process. That is why multiple–use forest planning has also been called integrated forest management planning (e.g. Kreutzwiser & Wright 1990).

According to both empirical and theoretical investigations, objectives other than those based solely on wood production are carrying more and more weight in forestry decision–making. Nowadays, not only the public but also private forest landowners, both industrial and non–industrial ones, value multiple–use aspects of forests (e.g. Hyberg & Holthausen 1989, Lönnstedt 1989, Karppinen & Hänninen 1990, Kreutzwiser & Wright 1990). Recently, much attention has been paid to biological diversity in particular and how it can be taken into account in forest planning (e.g. Hunter 1990, Millar et al. 1990, Probst & Crow 1991, Kangas & Kuusipalo 1993).

Public pressure on multiple use of forests has increased around the world, and not least in the Nordic countries. Due to the general consciousness of the use and management of forests, public participation in forest planning is often called for. In Nordic countries, this holds true in particular for forests owned by the state.

In this chapter, multiple–use planning of forest resources is discussed, and some examples of attempts to develop multiple–use planning methods in the Nordic countries are briefly presented. Consideration of planning processes is limited to planning within the forestry sector only. For example, other land–use planning systems in the respective countries, such as municipal and regional planning, are not examined.

15.2 Planning facilitates decision–making

The aim of forest planning, as of any planning process, is to provide support to decision–making so that, in the decision making process, the best possible production program leading to an efficient mix of outputs can be chosen. The product of forest planning is usually a forest plan: a recommendation of treatment schedules for forest stands located in the area in question, and some alternative plans with information on impacts of implementing the plans. In this chapter, planning is considered from the point of view of strategic and tactic planning, mostly at a forest area or a forest holding level.

A rational decision maker chooses the decision alternative which maximizes his or her expected utility, determined on the basis of the information available to him or her, related to the decision alternatives. Utility is influenced by all attributes which have value to the decision maker (for example, net income from timber production, nature conservation considerations, scenic beauty, etc.); it is a measure of subjective desirability. The best plan is the one which produces the greatest utility when realized.

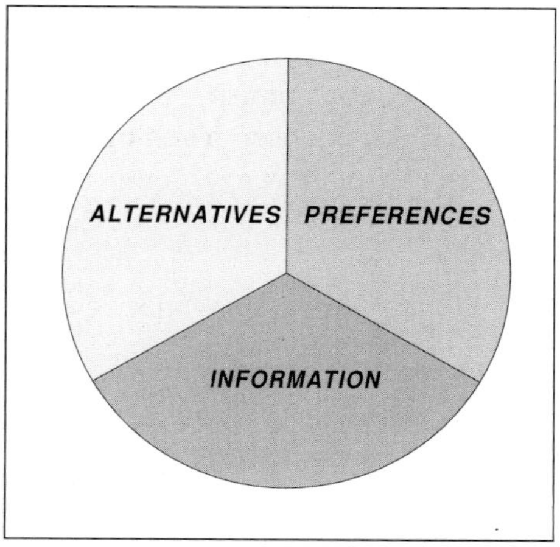

Figure 15.1 The decision basis (Bradshaw & Boose 1990). Each part of the decision basis should be sound in order to facilitate the best choice.

For choosing the best plan, perfect knowledge is required of both the decision alternatives available, and information on the consequences associated with these alternatives, as well as of the objectives and preferences of the decision maker (*Figure 15.1*). Furthermore, decision alternatives have to be evaluated with respect to each objective, and also with respect to overall utility. Because preferences of decision makers vary from one person to another, and from one decision situation to another, components of utility, and their weightings, have to be determined case by case.

A planning process includes, at least, the following phases:

1) Structuring the decision problem.
2) Describing the decision alternatives.
3) Determining objectives and preferences of decision makers.
4) Assessing possible impacts of each decision alternative.
5) Evaluating decision alternatives with respect to each objective.
6) Evaluating and comparing decision alternatives with respect to overall utility.
7) Compiling a plan.

The part of the planning process which precedes the compilation of a plan is often called decision analysis (e.g. Keeney 1982). Its aim is to lend support to decision making in decision problems which are too complex for the intuitive use of common sense. Kangas et al. (1992) broke down forestry decision analysis into five main phases from the point of view of practical applications:

1) Identification and structuring the decision problem.
2) Analysis of interests, including the preferences of the decision maker as well as those of the interest groups or citizens concerned.
3) Detailed description of the forest (e.g. forest inventory).
4) Examination of the production possibilities, and generation of the decision alternatives by applying, for example, simulation calculations.
5) Evaluation and comparison of the decision alternatives on the basis of phases (2), (3), and (4).

If a reliable and versatile enough decision support has been obtained in the decision analysis, a forest plan can be compiled.

In multiple–use forestry, different outputs – goods, services, and other benefits – are consciously produced in the same production process. Different goods and values are usually not measurable or otherwise expressable, accurately enough, in the same units. However, all relevant factors which have an effect on utility have somehow to be made commensurable. If this is not done, comparison of decision alternatives with respect to overall priority is difficult, if not impossible.

The importance of different outputs vary from one planning situation to another, and the choice of a forest plan in a multi–objective case is sensitive to changes in the importance ratios of outputs and objectives. Objectives and their importance, and, thus, the optimal choice, depend on the standpoint from which the planning problem is viewed: weighting schema may be different from a global than from a national point of view – to say nothing about a private landowner's or citizen's opinions. The best production program can only occasionally be found without clarifying the objectives and their weights, and taking them carefully into account in planning.

Decision analysis in planning of multiple–use forestry is more complicated than in planning of single–use forestry. Besides the capacity of the human brain, numerical analysis is often needed in decision analysis of complex planning processes. Numerical calculations and computerized planning systems have been found necessary in timber management planning. It is obvious that numerical approaches are also needed in solving multiple–use planning problems having timber production as one output among others; they are usually far more complex tasks than timber management planning processes alone.

When quantitative objectives and units are considered, a multiple–use planning problem can be solved by applying the approaches of mathematical programming, such as goal programming or other multi–objective linear programming approaches (e.g. Kilkki 1985, Mendoza et al. 1987, Gong 1992, Kangas & Pukkala 1992). Often, however, multiple–use values are qualitative by their very nature, and their quantification is difficult – sometimes even impossible – in such a way that each decision alternative could be evaluated reliably enough with respect to them. Some modern planning techniques, for example the Analytic Hierarchy Process (AHP), can deal with qualitative objectives without converting them into absolute measures (Saaty 1980, Kangas 1992a).

Recently, much attention has also been paid to quantification of qualitative benefits of forests in monetary terms (e.g. Mattsson 1990, Naskali 1992, Ovaskainen et al. 1992). This information would be useful in multiple–use planning. Any quantification method used in forest planning should, however, simply and cheaply, produce measures by which each combination of standwise production programs could be evaluated. So far, most results of research on economic evaluation of non–timber benefits are not applicable in standwise planning, but merely at national, or regional, forest policy level. However, monetary values of non–timber forest benefits can, perhaps, also be applied in the determination of the "socially justified" weights of objectives in planning at stand level. These weights could be applied, in particular, for state forests.

15.3 Some background to present approaches applied to forest planning in the Nordic countries

Traditions and present approaches of multiple–use planning of forest resources are not similar in all the Nordic countries; different forestry concepts create different planning traditions, and different planning traditions easily lead to different planning approaches and methods applied.

Within each country, planning approaches and practices applied differ with regard to the very nature of the area to be planned. There is a greater willingness to pay for the planning of, for example, nature conservation areas and recreation areas than of forests which are mainly reserved for timber production. Deepening and diversifying planning processes should not cost more than the increased value in utility, attained via deepening and diversifying the process. For example, public participation and ground vegetation mapping can be carried out in recreation areas, but they are nowadays regarded as too expensive to be generally applied to the planning of privately owned commercial forests.

In the Nordic countries, there is a relatively long tradition in research of timber management planning, starting with calculations based on the earliest national forest inventories in Finland and Sweden at the beginning of the 20th century. Also, in the last few decades, Scandinavian researchers have been in the forefront of methods for timber management planning, especially in the use of numerical optimization techniques (e.g. v. Malmborg 1967, Kilkki 1968 & 1985, Eriksson 1983, Jacobsson 1986, Kilkki et al. 1986, Lappi & Siitonen 1985, Hoen 1990 & 1992, Kangas & Pukkala 1992, Lappi 1992, Pukkala & Kangas 1993, Valsta 1993). In planning calculations both at national and forest holding level, computerized cutting budget methods and packages have been applied (e.g. Siitonen 1983, Bengtsson & Lundström 1987, Pukkala 1988 & 1993, Spross & Walan 1990, Jacobsson & Jonsson 1991). Most of the systems developed for use at forest holding level are based on compartmentwise ocular inventory, simulation of tree stands, numerical optimization, and compilation of a plan with standwise treatment schedule recommendations (Päivinen et al. 1992).

Quite recently, researchers in Nordic countries have also analyzed and developed advanced methods and approaches for multiple–use planning (e.g. Pukkala 1988 & 1993, Kangas 1991 & 1992a & 1992b, Gong 1992, Nuutinen & Pukkala 1992). However, analytical multiple–use planning methods have gained only limited application in forestry practice so far. Instead, mathematical programming has recently been approved as an approach to be used in practical timber management planning (Päivinen et al. 1992).

In forestry practice, perhaps the most progressive multiple–use planning methods, in the Nordic countries, are used in Norway. A working group under the Directorate for Nature Management in Norway has made a proposal for an action program with guidelines for multiple–use inventory and further integration of multiple use in the forest plan. The work is described in Lorentsen (1987) and Søgnen (1989). The guidelines are adopted by the Norwegian Forest Owners' Association and the Directorate for National Forests, and a computer program for inventory on stand level is developed.

The method has to fulfil the following requirements (Søgnen 1989):

- Multiple–use management programs must be presented together with other silvicultural treatment plans.
- Multiple–use inventory and the presentation of registration must be done at stand level.
- Multiple–use inventory must be sufficient to protect essential interests,
- The method must be adjustable to variable weighting of multiple–use interests.
- The method must be applicable along with traditional forest inventory.
- Multiple–use considerations in focus must be easy to recognize for the forest owner.

These requirements are valid also in other countries.

In order to describe the various perceptions of planning in the Nordic countries, the following will shortly summarize legislation connected to forest planning, the overall planning policy, the different objectives that – within each country – it has traditionally been hoped to achieve, and other major considerations reflecting the differences of management planning practices in different countries.

Norway

In Norway, a long–term forest plan has been compiled for most forests. However, Norwegian legislation does not require forest plans or any other announcements of management or forest plan operations. On the other hand, the state subsidizes the preparation of forest plans. In order to qualify for such support, the plan must include multiple–use interests. A special planning package with an inventory system adopted by the Forest Owners' Association is generally used. There is no obligation to involve the public or interest groups in the planning processes or in the decision–making process.

According to Eid (1990), the forest planning paradigm in Norway is based on a belief in a rationalistic planning process. But, as Hoen (1990) states, the sum of the decisions made by the forest owners without any forest policy will most likely conflict with the preferences of society. Nevertheless, in practice, interaction between the decision maker and the planner is quite rare (Eid 1990).

In Norway, the Forestry Act states the management goals as timber production as well as multiple–use forestry. However, the main decision attributes when compiling a plan are, usually, timber production considerations, but especially in the mountain areas, recreation and nature protection are of great importance. Forests on altitudes above a specified contour line are defined as protected forests, the management of which is restricted by regulations. The localization of this contour line is currently under revision. In mountainous forests, the objectives for the national preservation plans include, broadly defined, trying to protect typical as well as rare Norwegian landscapes.

Sweden

Until quite recently, all Swedish forest owners were required to prepare a management plan. Planning was done by the Forest Agency and subsidized by the state. The municipalities had the right to plan their own forests, if they were of particular local importance. The goals for the municipality forest management were required to be in accordance with national policy but could be adjusted by the environmental and forest authorities at county level.

Swedish forest policy is just now under critical discussion, and new definitions of forest policy, and its effects on forest planning, can be expected. It is obvious that Swedish forest management plans will, and should, in the future, also include environmental information to some extent. Recently much attention has been paid to biological diversity. However, there is no standardized method for multiple–use inventory, and the quality of registration therefore strongly depends on the planner's knowledge and interest in the subject. In general, so far, the public or the interest groups are not consulted. The main goal for Swedish forest planning will continue to be the maintenance of timber production possibilities of the forests.

Iceland

In Iceland, the situation is considerably different from the other Nordic countries. The forested area is very small, and it is mainly owned by the state. Forest planning at this stage is mainly afforestation planning, and it was initiated by the Iceland Forestry Service in 1986. Multiple use is recognized and integrated in the afforestation plans. The involvement of the public is not dealt with in a systematic way, though nature conservationists are consulted.

Forested area in Iceland has been categorized into four main land–use classes. In protected forests the goals include, for example, protecting soil against erosion and restoring vegetation at sites where it has been destroyed. In production forests, timber, firewood and Christmas trees are produced. Recreation forests are mainly reserved for short–term visits close to urban areas and longer visits for camping. Shelterbelts and woodland include crop cultivation areas, areas reserved for domestic animals, farmsteads, and urban areas.

Forests in Iceland provide various intangible benefits which are difficult to attach a monetary value to, but are undeniably real. The following benefits can be mentioned: new landscapes, diversified environment, shelter from wind, the pride of living in a forested country, possible changes in living habits. The present forest plantations in Iceland are few and small compared to the area of the country. But they are extremely important as strategic bridgeheads, which achieve the above mentioned objectives.

Denmark

Forest planning is not compulsory in Denmark, but planning has a very long tradition. Long–term planning is carried out in all state forests and in private forests to a great extent. Planning is more infrequent only in small forests under 50 hectares. Danish forest plans are quite detailed with inventory and registration of compartments down to 0.2 hectares. However, inventory is only undertaken for wood production.

State forest planning, carried out by the Forest and Nature Agency, takes multiple use into consideration. Nature conservation areas, historical monuments and cultural heritage are mapped, but, so far, the scope of the inventory system for multiple–use interests is only to mark out the specific area with a code denoting special interest. An appendix to the forest plan is made with guidelines for treatments in such areas. The management choice is made by the Agency itself, and the plans are not official documents. During the last few years it has been common to invite representatives of the municipalities concerned and different interest groups to let them propose plans for the forest management.

Also in Denmark, the Forestry Act states the management goals to be wood production as well as multiple–use forestry. In general, the Danish planning tradition leans upon descriptive approaches rather than numerical ones.

Finland

In Finland, all forest owners are required to announce cuttings to the District Forestry Board. Furthermore, forest regeneration plans have to be approved by the Board, or, as an alternative, the Board can accept a forest management plan. In the latter case, when implementing the plan, no separate approval is required. Having a forest management plan is not obligatory, but most forest owners have a long–term plan. Multiple–use inventory or other descriptions of multiple–use interests are not officially required in the plan. In state forests, planning covers the whole forest area.

Until recently, forest plans only included timber management considerations. Nowadays, planning instructions also weight multiple–use aspects, both in private forestry and in state forests. Even industrial forest land owners give due consideration to multiple use, in particular in forests used for recreation. In state forests in Finland, special attention has been paid to game management. In the private sector several experiments of incorporating multiple–use aspects into forest plans have been carried out.

As well as in Sweden and in Norway, timber management planning approaches applied in Finland are rather progressive. For example, simulation and mathematical programming are widely used and tested in practical forestry (Päivinen et al. 1992). Nevertheless, planning tools and methods in forestry practice are rather undeveloped with regard to multiple–use considerations. So far, no analytical multiple–use planning methods are used in practice in Finland. Only some simple rules are added to the general management policy: for example, no clearcuttings by the roadside, no cuttings in certain areas close to rivers, etc. Compilation of a forest plan still depends greatly on the opinions and expertise of planners, and there is no warranty of reaching the ideal goals formulated in planning instructions. Recently, methods for taking varying objectives into account in the compilation of a forest plan have been developed in forest research, also for multiple–use planning (e.g. Kangas & Pukkala 1992, Pukkala & Kangas 1993).

Participative planning is applied only in specific forest areas, such as nature conservation areas and recreation areas. In practice, the process of participative planning most often consists of only two meetings, where interest groups and citizens have an opportunity to state their opinions concerning the goals, and planning and management practices in the area in question. No analytical approaches to participative planning have so far been applied, although some experiments on the use of such methods have been carried out (Kangas & Matero 1993).

15.4 Attempts to develop multiple–use planning methods and systems in the Nordic countries

Even if multiple use has not yet become an integrated part of planning practice, all the Nordic countries have worked to create multiple–use planning systems. These planning systems work on different planning levels, and most of them have been developed to ease multiple–use planning and make it adequate in the given situation, and, therefore, the systems may not apply in other countries or on other planning levels and with other objectives. However, a proper planning system should be flexible with regard to objectives and their weightings.

The Analytic Hierarchy Process as an approach to strategic multiple–use planning

In Finland, applications of a decision analysis method called the Analytic Hierarchy Process (AHP) have been presented (e.g. Kangas 1991 & 1992a, Kangas & Kuusipalo 1993), and also tested in forestry practice. The AHP is based on a general theory of ratio scale estimation (Saaty 1977 & 1980, Harker & Vargas 1987). By using the AHP, a ratio scale can be constructed on the basis of pairwise comparisons for describing the importance of objectives and the priorities of decision alternatives.

Advantages of the use of the AHP in multiple–use planning include the ability to deal with both quantitative and qualitative decision attributes and to make them commensurable, and flexibility with regard to the setting of objectives. The approach is easy to apply, although the calculation technique might seem rather complicated. Furthermore, sensitivity analyses are easy to carry out. The main disadvantage of the standard version of the AHP is the small maximum number of decision alternatives to be compared in a planning calculation. Due to this, the AHP is most suitable to strategic forest planning with a limited number of management strategy alternatives. For example, environmental impact analyses can be carried out using this approach.

As an illustration of the AHP approach to multiple–use planning, a case study of Kangas (1992a) is summarized. In the study, a forest plan was chosen for a forest holding owned by a non–industrial forest landowner. The case study area of 86 hectares was located in western Finland. The land owner's utility – which was to be maximized – was broken down into three objectives. The objectives the landowner laid on the forest area were maximizing the monetary benefits from wood production, maximizing the beauty of the forest landscape, and producing as good habitat as possible for certain game species. Because all the objectives were not reachable at the same time, the optimal compromise solution between these three objectives was sought.

By using the timber management planning package MELA (Siitonen 1983), six alternative production programs, consisting of treatment schedule recommendations at standwise level, were generated for the forest area. The starting point was that the objectives and their weightings could not properly be taken into account in calculations made by a planning package based on linear programming.

In any application of the AHP, the decision situation is first organized as a decision hierarchy (*Figure 15.2*). At the uppermost level, there is the general goal of decision making – in this case study, maximizing the utility of the forest land owner. The second level consists of decision objectives. The lowermost level consists of decision alternatives – in this case, alternative forest management strategies for the next 20 years. If needed in order to evaluate the decision alternatives with respect to the objectives, objectives are broken down into more detailed decision attributes between the levels of objectives and decision alternatives. Objectives and attributes are described in more detail until each decision alternative can be evaluated with respect to each detailed decision attribute.

In the second stage, the importances of objectives are compared in a pairwise manner with respect to the overall utility. In the comparisons a certain verbal scale and its numerical analogue is normally applied. As an alternative, graphical interface with barlengths expressing the importance of objectives can be used (e.g. Pukkala & Kangas 1993). The question is, which of the two objectives compared is more important, and how much is it more important. A

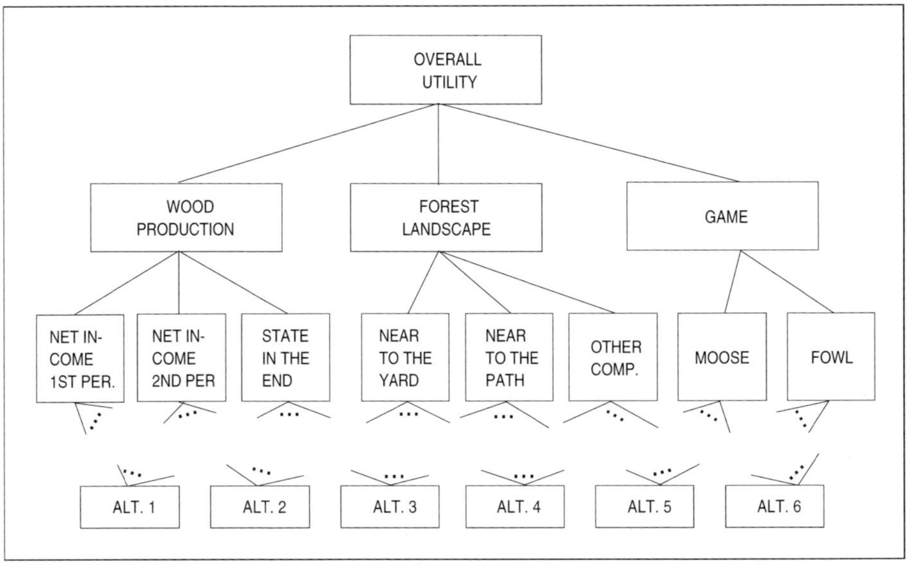

Figure 15.2 Hierarchy for maximizing the utility of a non–industrial forest landowner in a case study by Kangas (1992a).

matrix of pairwise comparisons, consisting of numerical importance ratios, is constructed. Using this matrix as input, the relative importances of objectives are estimated using the eigenvalue technique: the right eigenvector of the largest eigenvalue constitutes the estimation of relative weights of objectives. The importances are scaled to sum up to one. In the case study, the importances, determined by the land owner, were: wood production 0.63, forest landscape 0.28, and game management considerations 0.09.

Using the same technique, the importances of decision attributes with respect to the objective they refer to are estimated. For example, the importance of forest stands near the yard, forest stands near a path going through the forest, and other forest stands in the area, with respect to the forest landscape as a whole, were compared in a pairwise manner, and the relative importance of these decision elements were calculated. The importances, with respect to the forest landscape were, respectively, 0.65, 0.25, and 0.10. In the case study, the decision element of wood production was broken down into three more detailed elements, whose importances with respect to wood production were determined by the land–owner: net income from timber sales in the first 10 year period (relative importance with respect to wood production (0.40), net income from timber sales in the second 10 year period (0.20), and the state of the forest at the end of the second 10 year planning period, measured by the value of the growing stock at that moment (0.40). Correspondingly, the decision element of game was further broken down into two elements: priority of habitats with respect to moose (0.33), and priority of habitats with respect to wildfowl (0.67).

In the next step, decision alternatives are evaluated with respect to those decision elements which are not broken down into more detailed attributes. The same method of pairwise comparisons and the eigenvalue technique is applied. The question is, which of the two management strategies compared is

Table 15.1 Priorities of alternative management strategies with respect to the decision attributes which have not been broken down into more detailed attributes.

Decision attribute	**Priorities**					
	1	2	3	4	5	6
Net income, 1st period	0,02	0,19	0,36	0,05	0,11	0,27
Net income, 2nd period	0,03	0,18	0,10	0,33	0,30	0,06
State at the end of 2nd period	0,26	0,22	0,10	0,10	0,12	0,21
Landscape, near to yard	0,33	0,20	0,10	0,08	0,15	0,15
Landscape, near to path	0,29	0,15	0,19	0,08	0,23	0,06
Landscape, other stands	0,38	0,28	0,18	0,07	0,20	0,07
Habitats, moose	0,08	0,39	0,26	0,15	0,04	0,08
Habitats, wildfowl	0,18	0,09	0,32	0,23	0,14	0,05

more preferred with respect to the decision attribute in question, and how much more is it preferred. Instead of this kind of comparisons, numerical values, describing the priority of decision alternatives, for example, money units, can be applied. The result of matrix operations is the estimation of the relative priorities of the decision alternatives with respect to decision attribute (*Table 15.1*). Comparisons of strategies with respect to habitats of game species were carried out by an expert on game management, because the land owner could not evaluate them himself.

Table 15.2 Global priorities of management strategies determined on the basis of preferences of the landowner.

Strategy	Global priority
1	0,179
2	0,198
3	0,193
4	0,120
5	0,154
6	0,167

After the relative importance of the objectives, the importance of the detailed decision attribute with respect to the objectives they refer to, and relative priorities of management strategies with respect to each decision attribute have been estimated, priorities of strategies with respect to overall utility can be calculated. The management strategy with the greatest global priority is the most recommendable forest plan. In the case study, the recommended management strategy was strategy number 2 (*Table 15.2*).

The AHP has also been applied to participative planning and conflict management in state forests at Ruunaa nature conservation area in eastern Finland (Kangas 1992c, Kangas & Matero 1993) (*Figure 15.3*). Some properties of the AHP have been utilized in determining the weights of objectives in multi–objective mathematical programming (Kangas & Pukkala 1992), and in a heuristic optimization method for tactic forest planning (Pukkala & Kangas 1993). The eigenvalue technique and pairwise comparisons have been applied in the estimation of parameters of multi–attribute utility functions in the analyses of forest regeneration alternatives (Kangas 1992b), in calculating a biodiversity index to be used in forest planning (Kangas & Kuusipalo 1993), and in estimation of landscape preferences (Kangas et al. 1993).

Incorporation of amenity values into forest management planning

Pukkala (1988) has developed a method by which amenity values can be incorporated into forest management planning carried out using mathematical programming. The management plan is based on treatment schedules simulated for each forest stand over a 20 year planning period. The amenity is divided into two sub–criteria: within–stand amenity, and amenity of landscape when viewed from afar, i.e. as a distant scene.

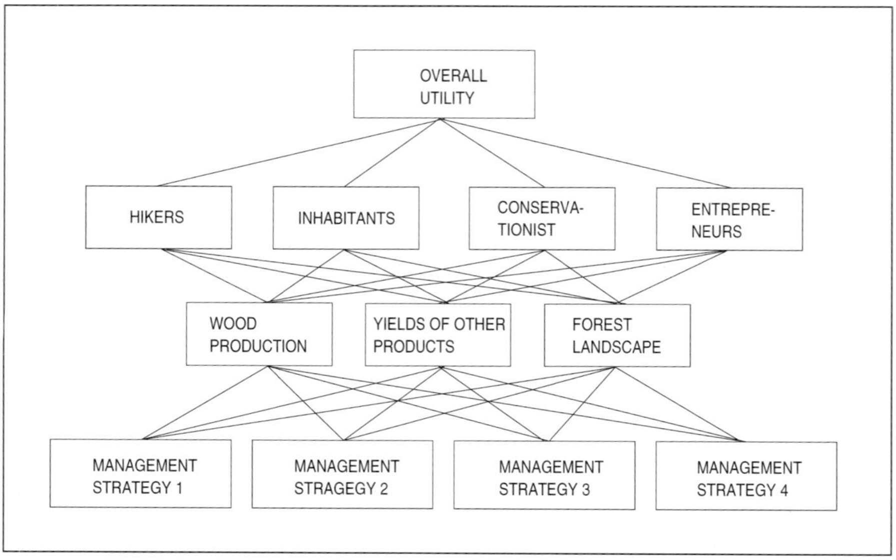

Figure 15.3 A simplified example of hierarchies of participative planning problems when a representative democracy approach is applied (Kangas 1992c). A level of players is added to the second level of the hierarchy. The importances of the criteria are assessed separately by each concerned interest group.

The within–stand amenity is calculated as an adjective sum, which is estimated for each forest stand, and for each treatment schedule considered, at three time points. For expressing the within–stand amenity no supplementary measurements in the forest are needed, as the amenity values are calculated using known stand data as predictors, i.e. such characteristics as total volume of trees, mean height of trees, proportions of different tree species, etc. The mean adjective sum of the whole area in a selected year can be applied as an objective or constraining variable of optimization. The far–view amenity is assessed on the basis of computer illustrations. The illustrations show the predicted temporal change of landscape when implementing a particular management plan.

Management alternatives are developed at stand level through simulating a number of possible treatments for each stand. On the basis of the simulated alternatives, mathematical optimization is carried out in order to evaluate the consequences of each individual alternative, and to find the optimum combination of treatment alternatives. In the case study, the management period of 20 years was divided into two 10 year sub–periods.

The following parameters were chosen as optimization parameters in the case study: net income during the first 10 year period, net income during the second 10 year period, total stemwood volume at the end of the second 10 year period, and mean amenity at the end of the second 10 year period. Amenity was described as an adjective sum according to Savolainen & Kellomäki (1981) and was estimated for the individual stand to be:

HV = 49.04 + 0.3344H + 1.398nT + 0.03370Vl, where
HV = amenity value (adjective sum)
H = mean height (m)
nT = number of tree species
Vl = volume of hardwood (m^3/ha)

Generally, in the computations of Pukkala (1988), the adjective sum varied between 50 (low amenity) and 67 (high amenity). In order to make the range of variation more understandable, the adjective sums were scaled between O (very poor) and 50 (very good). The mean amenity for the forest as a whole was stipulated to be:

$$gHV = \frac{\sum HV \times A}{\sum A}$$

where

gHV = the mean forest amenity based on stand level computation
A = stand area

Far–view amenity for a forest area does not compare with gHV, because, among other things, gHV does not describe the variation in stand characteristics. Therefore, the far–view amenity was assessed subjectively on the basis of computer drawn landscape illustrations which were shown to a number of persons who scaled the illustrations from 0 (very poor) to 50 (very good) (see also Pukkala & Kellomäki 1988) (*Figure 15.4*).

This method, which includes simulation and creation of alternatives at stand level, leads to high flexibility in forest management. However, the optimization results are very dependent on the acceptability of the amenity computations applied. In addition, the scale applied in amenity evaluations does not, in principle, fulfil the requirements of mathematical programming techniques used in the study.

Plan for municipality forest management

In Sweden, Lidestav (1990) has developed a planning technique called PIKS for forests owned by municipalities. PIKS is a combination of two existing planning techniques. One is a position analysis technique (Söderbaum 1984, 1986) which is mainly used in constituting a general framework coverage for the planning process, while the other is a forestry microcomputer–based timber management planning package called PLAN–20 (e.g. Ekvall 1986).

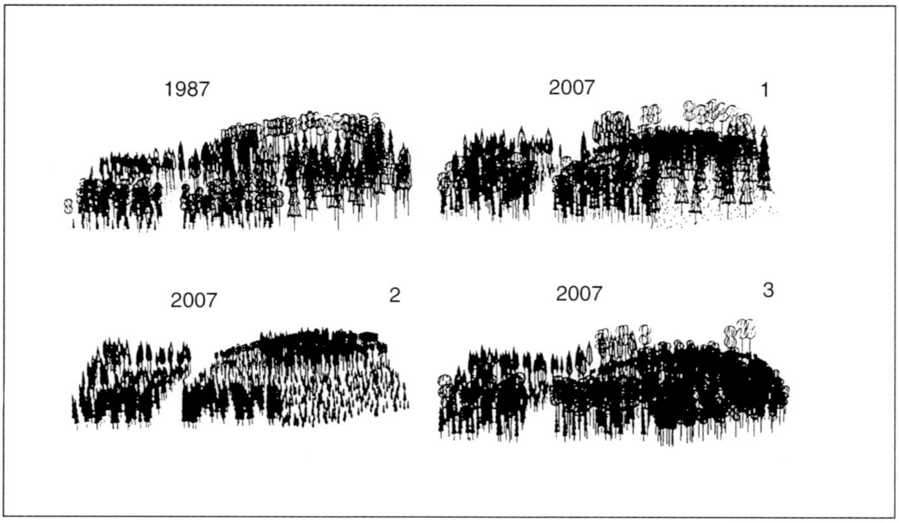

Figure 15.4 An example of the visualization of the effects of cuttings on forest landscape using computer graphics. The actual landscape and the predicted landscape after 10 years in three different management plans (Pukkala 1988).

The fundamental idea behind PIKS is that planning should be based on democratic principles. The following must, therefore, be noted when preparing the planning technique:

1) The municipality has a large multi–sector role.
2) Local conditions must be considered.
3) Consequences of alternate management and operational forms must be shown.
4) The plan must reveal conflicts.
5) The planning process must facilitate direct interaction between different interest groups.
6) Possibilities for non–experts to join in the decision–making process must be emphasized.
7) Investigations and modifications/follow–ups must be carried out at low costs.

The planning technique tries to combine two existing but very different planning methods, a traditional forest planning system and a holistic planning method. PIKS concentrates on identifying relevant systems and comparable parameters and contains examples of how to carry out the following analyses:

1) Calculation and comparison of alternative management programs in regard to the area covered by restrictions, logged volume, amount of logging carried out by municipality, average wood price, income from forestry, employment, and mechanization.

2) Structuring and identifying systems influenced by the choice of management program at different levels: local – within community, local – outside community, entire community and region level. The influenced features include, among others:

- Land, as a physical system (timber production, production of berries, mushrooms, huntable wildlife, protection of valuable areas, recreation).
- Forestry system (community forestry, industrial forestry).
- System for timber supply and conversion.
- Monetary system (direct, indirect).
- Energy supply system (leisure time, fuel wood, cutters, emergency service).
- Labour market system (open, protected).
- Social system (work satisfaction, outdoor environment).
- Institutional system (change in former practice and thereby change in political statement).

3) Analysis of the effects of the management programs on:

- Biological dimensions (forest ecosystem, conditions of landscape and nature protection).
- Aesthetical dimensions.
- Monetary dimensions (forest management, community management, private economics, industrial economics).
- Social dimensions (work opportunities, work environment, recreation, environment for housing).
- Know–how dimensions (e.g. alert in case of war).
- Historical dimensions (e.g. including traditional features difficult to change).

4) Analysis of interests shows what interests a management program will favour. This means also an identification of the direction and the goals which the specific interest attempts to promote (*Table 15.5*).

The PIKS system was illustrated and tested by a case study. Four different plans were prepared for a forest area in a municipality in central Sweden having about 20,000 inhabitants and 4,500 hectares forest land. The planning period was 10 years. After preparing the alternative management programs, systems which might be influenced by the choice of plan were identified.With an interest analysis, interests which were favoured or disfavoured by implementing one or the other plan were clarified. The order of preference was given for each interest. This ordinal information was arranged as a matrix. On the basis of the matrix, possible conflicts and consensuses were examined. By applying an effect analysis, the plans were afterwards compared with regard to the expected effects. As a conclusion of the case study, none of the four management programs proved to be clearly the best or worst.

PIKS can be seen as an approach to organizing the planning situation with its most evident influences, interests, effects and impacts. The final choice

Table 15.5 Examples of interests and connected goals when the PIKS system is applied (Lidestav 1990).

Employer	Administration/ supervision	Highest possible degree of control
	Office work	Simplest routines
	Work planning	Highest flexibility
	Output/product	Highest quality
	Net income from logging	Highest possible income
Employees	Work load	Minimize heavy and risky work elements
	Work contentedness	Varied, qualifying work
	Salary	Highest possible payment
Society	Landscape care	Smallest possible amount of clearcuttings
	Flora & fauna	The greatest species diversity + protection of certain species/ biotopes
	Taxation	Lowest tax rate
Special interests	Residences	Best outdoor environment
	Wood industry	Best supply of suitable products
	Wood industry	Lowest price per m^3

should always be preceded by evaluating and weighting of all effects and interests considered. Using PIKS, effects and interests can be clarified, but, however, PIKS does not state who should be in charge of weighting the effects and interests and determining the final operational goals.

PIKS develops a base for decision making, but it does not tell us how to maximize overall utility. PIKS puts more trust in the planner's brain than in computerized methods of making different dimensions of decision–making commensurable and in evaluating alternative forest plans with respect to overall utility. Implementation of PIKS as a participative planning approach is quite planner oriented, and there is no warranty of the public having an effect on the choice of management plan.

After all, PIKS is a sophisticated effort to connect forest planning with the needs of the surrounding society. Common use of PIKS and the AHP approach to participative forest planning would be worth developing and testing. Use of the AHP might eliminate some disadvantages of PIKS, and vice versa.

Multiple–use planning of the Hou Skov forest area

In Denmark, an approach to the planning of a privately owned forest area with remarkable public interests has been developed (Kristiansen 1993). The approach is still being elaborated. Up to date, only some preliminary tests have been carried out. The study tries, for its part, to make the concept of multiple use, as mentioned in Danish forest legislation, and the discussion of multiple–use planning, specific and operational.

The approach combines the traditional descriptive planning approach with more modern planning methods. An important question is where and how descriptive forest planning processes can successively be applied and/or developed so that multiple–use aspects can properly be taken into consideration. The study deals with registration methods, including the relevance of individual interests for registration, and an comprehensible presentation of registrations. Modern methods considered include linear programming and goal programming as optimization techniques.

A case study is carried out at a forest area called Hou Skov. In the planning process, registrations and suggestions for operational purposes take place in two stages. The first stage is an excursion with local representatives of different interest groups. The purpose is to map out interests in the forest, and to discuss future operation possibilities. In the second stage, a more detailed registration is carried out in areas where a deeper analysis of interests is relevant. The assessment of relevance should not be taken as an argument for or against certain interests, but as a recognition that not all multiple–use elements are relevant in all forest stands. The final objectives are determined by the forest owner together with the planner. In Denmark, as in the other Nordic countries, it is the owner's right to choose among and give priorities to various wishes and interests of the public in relation to his or her own objectives, as long as his or her overall priorities still conform to the requirements of the law.

At the Hou Skov area, the priorities of the interest groups can be listed after an excursion. The forest owner regards the forest, on the one hand, as an economic investment, and, on the other hand, as a property having a high amenity value. The owner has bought the area not only for wood production purposes, but also because of its beautiful scenery and its beech stands. For conservation reasons, two areas are given high priority by the owner: old beech forests, where the intention is to preserve the gene pool intact, and some areas which will be left "relatively untouched" for protecting, for example, wildlife.

Interests other than those of the forest owner at Hou Skov are as follows:

1) <u>Landscape aesthetics</u>. The landscape values of the forest are strongly connected with the visitors' interests, but also with the owner's objective to increase the amenity value of the forest.
2) <u>Outdoor recreation</u>. The unorganized as well as the organized use of the area for outdoor recreation is limited. A large majority of the visitors are

local people. The main activity is walking, often with a dog. In addition, a local orienteering club is interested in using the forest area.

3) Natural values. Generally, more hardwood is wanted, especially beech and oak. It is desirable to preserve or re–create meadows and the present grazing areas at Hou Skov. It is also desirable to preserve the old beech stands and have slow natural reproduction. Two streams in the area have a characteristic stream flora, which are worth protecting, as well as a few rare botanical and ornithological species in the area.
4) Environmental protection. Interests mentioned in 3) included also some interests concerning environmental protection. For environmental reasons it is desirable to minimize the use of fertilizers and chemicals.
5) Cultural history. The cultural–historical values of Hou Skov include old methods of cultivation as well as barrows, arterial roads, dikes, property lines, ponds, etc. Actual maintenance of the cultural monuments is not considered necessary, except grazing of the meadows.

In addition to the above mentioned interests, future conditions and interests also appear in a region plan of the county authorities, and local plans for the municipality. These include, for example, the building of a camping site with a capacity of 500 guests, and the establishing of bicycle routes and public green spaces in the planning area.

For the computational phase of the approach, a detailed level for the registration is laid down, and measurable parameters for the individual interests, as well as interaction between the interests, are defined. At the registration, the following three principles can be applied (listed in order of priority):

1) Counting or other quantitative measures are applied when possible and suitable. If the quantitative value can be priced, this is exploited. For example, wood production is registered in the traditional manner as money units determined on the basis of volume, assortment, and unit prices of wood.
2) When quantitative measures are not practicable, not even at sample plot level, one or a few resources which can indicate a value for the rest of the resources are quantified. For example, biotope quality can be used as a measure of natural value. Biotope quality can be determined, for example, on the basis of birds which breed in tree hollows, dead and dying trees, and some special indicator species.
3) If it is not possible, practicable, or suitable to start a more definite registration work, a subjective area related value of the resource is fixed on a scale from 0 (low) to 3 (high). A similar subjective registration can be made by means of descriptive and graduating words, and worked out on an area basis, line basis, or as individual points.

In planning calculations, the objectives for the forest area are determined in measurable units. Requirements concerning profitability of wood production can be determined as in any forest planning system. On the basis of the analysis of multiple–use interests, additional objectives can be assessed as fulfilment amounts, for example, for grazing area, beech forest area and protected area.

Mathematical programming methods can be applied in supporting the decision–making. However, the planning process can also be carried out without numerical optimization. When analysing the consequences of the plan, an estimated technical and financial status for the forest at the end of the planning period is reached. The new status is compared with the management objective, so that it can be checked whether the plan harmonizes with the objectives or not. Using mathematical programming techniques, alternative efficient solutions can be quickly produced, if the status does not fit. In addition, sensitivity analysis can be carried out using optimization calculations; for example, if the objective is changed, or if the emphasis between the individual interests or the subjective registration of the value of individual interests is changed.

The final plan is a negotiation solution, a compromise between the owner and the interest groups. Results of optimization calculations can be used as useful information in the negotiations.

15.5 The perspectives of multiple–use planning

In the Nordic countries, the first steps in the way of multiple–use planning have been taken. So far, multiple–use planning has, however, mostly served the developing and testing of alternative planning approaches rather than used as a realized forestry practice. Multiple–use has usually only been integrated into forestry by simple rules concerning forest management practices. In some special cases, such as nature conservation areas and research projects, a panel of experts has participated in the planning.

In the future, multiple use will be an integrated part of nearly all forest planning. Besides the public and forest land owners' weighting of multiple benefits of forests, the international community also needs to consider biological diversity and other non–timber benefits and uses of forests to be included in planning. In order to carry out multiple–use planning calculations simply, cheaply and rapidly, but at the same time accurately enough, it would be recommendable to apply computerized multiple–use planning methods and systems. The incorporation of several experts into the planning process, as required by properly performed descriptive approaches to multi–objective planning problems, is usually too expensive and time consuming.

Up to date, because of the lack of analytical methods which would be generally applicable, simple and cheap to use, and flexible enough with regard to the setting of objectives, descriptive approaches have frequently been applied in practical multiple–use planning. However, the shortcomings of a pure descriptive approach, compared to numerical methods, are obvious. For example, descriptive approaches are very planner oriented, and, when they are applied, usually only the planner can control the planning process. Numerical methods improve communication, speed up the planning process, and make it repeatable. Without analytical means, evaluation of decision alternatives with respect to the objectives and preferences of the decision maker is difficult. Furthermore, descriptive methods can usually consider only a few alternative production programs, and, because numerical simulation is not possible, future states of forest are difficult to assess.

The greatest advantage of a descriptive approach is its flexibility. It is not limited to rough numbers and quantities in presenting the planning problem. Nevertheless, proper planning is not possible without managing – either manually or by applying numerical tools and methods – the data and the information describing the forest as well as the objectives and preferences. The capacity of the human brain is often too limited to solve alone complex planning problems of multiple–use forestry.

In each planning process, the final choice is made by a human being, by applying a more or less descriptive approach. The aim of any plan, and system, method, tool, etc., applied in a planning process, should be to further as good decisions as possible in that final choice. Nowadays, and most probably also in the near future, the best solutions to problems in multiple–use planning of forest resources can be attained by the common use of both descriptive and numerical methods, i.e. by applying so–called hybrid approaches.

Crucial problems in multiple–use planning of forest resources include: 1) determining the objectives and their importances, 2) evaluating the decision alternatives with respect to each objective, and 3) making the objectives and the evaluations of decision alternatives commensurable (Kangas 1992a). Nowadays, as mentioned above, some methods for clarifying the objectives and preferences, and taking them into consideration in the evaluation of alternative forest plans are already available – not only in strategic planning but also in calculations at forest stand level. However, these methods are products of forest research, and most often require some refinement before they can be applied to forestry practice.

At the moment, the most crucial problem of multiple–use planning is the evaluation of decision alternatives with respect to different goods and services of forests in a way which is applicable in planning calculations at standwise level. In computerized approaches, effects and influences of implementation of alternative forest plans should be modelled. For example, production functions of different picked products or game species are not yet available, which could be used to estimate priorities of different production programs on the basis of stand and area characteristics. In principle, expertise could be utilized in a case where there is a lack of objective information to be used in the evaluation. There is also a shortage of methods rapid and simple enough for estimating subjective preferences, although some prototypes of such methods have been presented (Kangas et al. 1993). Recently, in Finland, attempts to develop methods for modelling expertise as well as subjective preferences for forest planning purposes have been carried out. Prototypes of such models can be expected in the near future.

The compilation of a proper multiple–use forest plan requires a more versatile description of the forest area to be planned compared to the planning of timber production. For example, the presence of dead or rottening wood is important with regard to many rare and vulnerable species, and the characteristics of ground vegetation are of value when assessing priorities with respect to wildlife habitats. In forest inventory made for timber production planning, attention is paid only to living trees. If, for example, biological diversity is to be taken into consideration, quantities of different kind of dead and rottening wood should be measured. In addition, because the future state of a forest should also be assessed, development of these quantities should be predicted.

Spatial data is often needed in multiple–use planning. For example, management of mating and nest sites is crucial for some wildlife species, and a variety of forest stands with regard to stand characteristics is an important characteristic when the amenity values of a forest area are assessed. Geographic information systems (GIS) have been applied in forest planning for managing spatial data (e.g. Nalli 1992, Nuutinen & Pukkala 1992). Many commercial forest planning packages with spatial data management options are nowadays available.

Besides being more difficult, multiple–use planning is more expensive than single–use planning. The rule of maximizing utility is also valid when it is a question of deepening and diversifying forest planning processes. The marginal cost of deepening and diversifying should not be higher than the marginal utility obtained. Research on multiple–use planning has produced a number of planning methods and approaches. It remains for practical foresters to decide whether to invest or not in applying these methods to real planning tasks of forestry.

References

Bengtsson, G. & Lundström, A. 1987. Beskrivning av HUGIN–systemet. (Description of the HUGIN –system.) Sveriges lantbruksuniversitet, Institutionen för Skogstaxering, Utdrag ur kommande slutrapport. 12 pp. (In Swedish.)

Bradshaw, J.M. & Boose, J.H. 1990. Decision analysis technique for knowledge acquisition: combining information and preferences using Aquinas and Axotl. International Journal of Man–Machine Studies 32: 121–186.

Eid, T. 1990. Long term forest planning: economical and biological production possibilities of a forest. Agricultural University of Norway, Department of Forestry, Doctor scientiarum theses 9/1990. 143 pp.

Ekvall, H. 1986. PLAN–20: ett skogligt värderings– och planeringssystem. (PLAN – 20: a valuation and planning system for forestry.) Sveriges lantbruksuniversitet, Institutionen för skogsekonomi, Rapport 43. (In Swedish.)

Eriksson, L.O. 1983. Column generation applied to long range forestry planning models. The Swedish University of Agricultural Sciences, Department of Operational Efficiency, Report 155. 38 pp.

Gong, P. 1992. Multiobjective dynamic programming for forest resource management. Forest Ecology and Management 48: 43–54.

Harker, P.T. & Vargas, L.G. 1987. The theory of ratio scale estimation: Saaty's Analytic Hierarchy Process. Management Science 33: 1385–1403.

Hoen, H.F. 1990. Theoretical and empirical studies of long range forest management planning. Agricultural University of Norway, Department of Forestry, Doctor scientiarum theses 23/1990. 144 pp.

Hoen, H.F. 1992. GAYA–LP: A PC–based long range forest management model. Paper presented at EURO XII/TIMS XXXI Joint International Conference, Operational Research/Management Science, Helsinki, Finland, June 28th–July 1st.

Hunter, M.L. 1990. Wildlife, forests, and forestry: principles of managing forests for biological diversity. Prentice Hall, Englewood Cliffs, N.J. 370 pp.

Hyberg, B.T. & Holthausen, D.M. 1989. The behavior of nonindustrial private forest landowners. Canadian Journal of Forest Research 19: 1014–1023.

Jacobsson, J. 1986. Optimization and data requirements: a forest management planning problem. Sveriges lantbruksuniversitet, Avdelning för Skogsuppskattning och Skogsindelning. 143 pp.

Jacobsson, J. & Jonsson, B. 1991. The forest management planning package: experience from applications. Swedish University of Agricultural Sciences, Department of Biometry and Forest Planning, Report 21. 38 pp.

Kangas, J. 1991. Menetelmä metsäojitusvaihtoehtojen hyötyvertailuun. Summary: A method for utility comparison of forest drainage alternatives. Suo 42(3–4): 49–59.

Kangas, J. 1992a. Multiple–use planning of forest resources by using the Analytic Hierarchy Process. Scandinavian Journal of Forest Research 7: 259–268.

Kangas, J. 1992b. Choosing the regeneration chain in a forest stand: a decision analysis model based on multi–attribute utility theory. University of Joensuu, Publications in Sciences 24. 230 pp.

Kangas, J. 1992c. Public participation in forest management. An application of the Analytic Hierarchy Process. Paper presented at EURO XII/TIMS XXXI Joint International Conference, Operational Research/Management Science, Helsinki, Finland, June 28th–July 1st.

Kangas, J. & Kuusipalo, J. 1993. Estimation and use of biodiversity in forest management planning. In: Linddal, M. & Naskali, A. (eds.). Valuing biodiversity. Scandinavian Forest Economics 34: 124–135.

Kangas, J., Laasonen, L. & Pukkala, T. 1993. A method for estimating forest landowner's landscape preferences. Scandinavian Journal of Forest Research 8: 408–417.

Kangas, J. & Matero, J. 1993. Ruunaan luonnonsuojelualueen jako aarni– ja puisto–osiin. Kokemuksia AHP –menetelmän käytöstä osallistuvassa metsäsuunnittelussa. (Dividing the Ruunaa nature conservation area into wilderness and park areas. Experiences of the use of the AHP in participative forest planning.) Metsäntutkimuslaitoksen tiedonantoja 449. 44 pp. (In Finnish.)

Kangas, J. & Pukkala, T. 1992. A decision theoretic approach applied to goal programming of forest management. Silva Fennica 26(3): 169–176.

Kangas, J., Päivinen, R. & Varjo, J. 1992. Integroitu metsäsuunnittelu. (Integrated forest planning.) University of Joensuu, Faculty of Forestry, Research Notes 2: 1–34. (In Finnish.)

Karppinen, H. & Hänninen, H. 1990. Yksityistilojen hakkuumahdollisuuksien käyttö Etelä–Suomessa. Summary: Actual and allowable cut in nonindustrial private woodlots in southern Finland. Folia Forestalia 747. 84 pp.

Keeney, R.L. 1988. Decision analysis: an overview. Operations Research 30: 803–838.

Kilkki, P. 1968. Income–oriented cutting budget. Acta Forestalia Fennica 91. 54 pp.

Kilkki, P. 1985. Timber management planning. University of Joensuu, Faculty of Forestry, Silva Carelica 5. 160 pp.

Kilkki, P., Lappi, J. & Siitonen, M. 1986. Long–term timber production planning via utility maximization. TIMS studies in the Management Sciences 21: 285–295.

Kreutzwiser, R.D. & Wright, C.S. 1990. Factors influencing integrated forest management on private industrial forest land. Journal of Environmental Management 30: 31–46.

Kristiansen, L. 1993. Planlægning af flersidig skovdrift. (Planning of multiple–use forestry.) A research project plan. (In Danish.)

Lappi, J. 1992. JLP – A linear programming package for management planning. The Finnish Forest Research Institute, Research Papers 414. 134 pp.

Lappi, J. & Siitonen, M. 1985. A utility model for timber production based on different interest rates for loans and savings. Silva Fennica 19(3): 271–280.

Lidestav, G. 1990. PIKS: Planeringsinstrument för kommunägd skog. (PIKS: Planning instrument for community owned forest.) Sveriges lantbruksuniversitet, Institutionen för Skogsteknik, Research Notes 191. 128 pp. (In Swedish.)

Lönnstedt, L. 1989. Goals and cutting decisions of private small forest owners. Scandinavian Journal of Forest Research 4: 563–574.

Lorentsen, Ö. 1987. Flerbrukshensyn i skogsbruksplanen. (Multiple–use aspects in forestry plans.) Direktoratet for naturforvaltning, DN–rapport 8a: 1–41. (In Norwegian.)

von Malmborg, G. 1967. Ekonomisk planering av det kombinerade skogs– och jordbruksföretaget. (Economic planning in the combined forestry and agriculture enterprise.) Jordbrukets utredningsinstitut, Meddelande 6: 1–129. (In Swedish.)

Mattsson, L. 1990. Hunting in Sweden: extent, economic values and structural problems. Scandinavian Journal of Forest Research 5: 563–574.

Mendoza, G.A., Bare, B.B. & Campbell, G.E. 1987. Multiobjective programming for generating alternatives: a multiple–use planning example. Forest Science 33: 458–468.

Millar, C.I., Ledig, F.T. & Riggs, L.A. 1990. Conservation of diversity in forest ecosystems. Forest Ecology and Management 35: 1–4.

Nalli, A. 1992. Monikäytön suunnittelumenetelmä. (A method for multiple–use planning.) University of Joensuu, Faculty of Forestry. Mimeograph. 48 pp. (In Finnish.)

Naskali, A. 1992. The concept of existence value and wilderness preservation. Scandinavian Forest Economics 33: 207–230.

Nuutinen, T. & Pukkala, T. 1992. Long–term forestry management with landscape illustrations. The Joint FAO/ECE/ILO Committee on Forest Technology, Management and Training. Seminar on the use of information systems in forest management. Garpenberg, Sweden 14–18 September 1992. (In press).

Ovaskainen, V., Savolainen, H. & Sievänen, T. 1992. The benefits of managing forests for grouse habitats: a contingent valuation experiment. Scandinavian Forest Economics 33: 263–274.

Päivinen, R., Kangas, J. & Varjo, J. 1992. Katsaus metsätalouden suunnitteluun Suomessa ja Ruotsissa. (An overview of forest planning in Finland and in Sweden.) Metsäntutkimuslaitoksen tiedonantoja 406. 52 pp. (In Finnish.)

Probst, J.R. & Crow, T.R. 1991. Integrating biological diversity and resource management. Journal of Forestry 89(2): 12–17.

Pukkala, T. 1988. Methods to incorporate the amenity of landscape into forest management planning. Silva Fennica 22(2): 135–146.

Pukkala, T. 1993. Metsäsuunnitteluohjelmisto MONSU. (A forest planning program MONSU.) Mimeograph. 42 pp. (In Finnish.)

Pukkala, T. & Kangas, J. 1993. A heuristic optimization method for forest planning. Scandinavian Journal of Forest Research. (In press.)

Pukkala, T. & Kellomäki, S. 1988. Simulation as a Tool in Designing Forest Landscape. Landscape and Urban Planning 16: 253–260.

Saaty, T.L. 1977. A scaling method for priorities in hierarchical structures. Journal of Mathematical Psychology 15: 234–281.

Saaty, T.L. 1980. The Analytic Hierarchy Process. McGraw–Hill, New York. 287 pp.

Savolainen, R. & Kellomäki, S. 1981. Metsän maisemallinen arvostus. Summary: Scenic value of forest landscape. Acta Forestalia Fennica 170. 74 pp.

Siitonen, M. 1983. A long term forestry planning system based on data from Finnish national forest inventory. In: Forest inventory for improved management. University of Helsinki, Department of Forest Mensuration and Management, Research Notes 17: 195–207.

Söderbaum, P. 1984. Objectives, values and ideologies in economics: towards a normative theory of institutional economics. The Swedish University of Agricultural Sciences, Department of Economics and Statistics, Report 231: 1–55.

Söderbaum, P. 1986. Beslutsunderlag. Ensidiga eller allsidiga utredningar? (Basis for decisions. One–sided or all–sided investigations?) DOXA, Serie för kommunal ekonomi och organisation 9: 1–196. (In Swedish.)

Søgnen, S.M. 1989. Skogsbruksplan med flerbrukshensyn. (Forestry plans with multiple–use approach.) Aktuelt fra Statens fagtjeneste for landbruket 1989(4): 105–111. (In Norwegian.)

Spross, R. & Walan, B. 1990. Dataprogram för skogsbruksplanering: en jämförelse av modeller utvecklade för privatskogsbruket. (Data program for forestry planning: a comparison of models for private forestry.) Examensarbete i skogsindelning. Sveriges Lantbruksuniversitet, Umeå. 40 pp. (In Swedish.)

Valsta, L. 1993. Stand management optimization based on growth simulators. The Finnish Forest Research Institute, Research Papers 453. 51 pp. + 4 appendices.

16 Multiple–use research

Olli Saastamoinen[1]
Marjatta Hytönen[2]

Abstract

Important and useful multiple–use forestry research is carried out in numerous sciences and even inside forestry research it is difficult to delineate multiple–use research exactly. However, it generally refers to research which integrates two or more forest uses and functions. The main part of the article describes the history, organizational development and major multiple–use research efforts in the Nordic countries from the late 1960s. Recent and on–going changes in forest and environmental legislation in the Nordic and other countries reflect the changing values and demands of people. They also create new challenges for forestry research. Important research areas in the future will include the consequences of European integration. There is also a need for increased research efforts focusing on social processes, changing values and value conflicts over forests and forestry in the Nordic countries.

Keywords: integrated forest management, history, research organizations, research projects, internationalization, social values.

---

[1] University of Joensuu
Faculty of Forestry
Box 111, FIN–80101 Joensuu, Finland

[2] Finnish Forest Research Institute
Unioninkatu 40 A, FIN–00170 Helsinki, Finland

16.1 Introduction

The contents of multiple–use forestry have been expressed in different ways in different times and have depended on the cultural context. The contents of multiple–use research also change along with the demands of societies. The functions and uses of forests are numerous and dynamic. There has always been, and increasingly will be, a lot of scientific research done on the different components of forest ecosystems and on the varying values and uses of forests within many branches of basic and applied sciences.

Where should we draw the line between multiple–use research and other forestry research? Twenty years ago most of us regarding ourselves as multiple–use researchers were inclined to think that basic biological research, for example the identification of plant and animal species of forests, was to be left outside the concept. But what about now, when biodiversity conservation is emerging as one of the major objectives of forestry? It seems that biological species research is becoming one of the basic components of multiple–use research.

It must be clearly recognized that there is a lot of extremely important and useful research done for multipurpose forestry in numerous sciences as in basic and applied ecology, botany and zoology, soil science, geography, landscape architecture, sociology, history, political sciences and economics, just to name the most obvious. It is impossible to do justice to all the relevant research in a short article like this. It is also clear that for the purposes of the present survey a line of some kind must be drawn. An organizational borderline will be adopted; the focus will be on multiple–use research carried out in forestry research organizations and the faculties and university departments of forestry in the Nordic countries. However, in some cases relevant research done in some other organizations will be mentioned.

No doubt even the field inside forestry research is too wide to be covered properly. Wood production is one of the many functions of forests. The bulk of forestry research has been and still is concentrated on it. Many studies dealing with it are clearly relevant for the integration of forest uses. Research on the rapidly growing impacts of industrial emissions on forests and the increasingly interesting functions of forests in the carbon cycle also provide useful basic information for the purpose of integration.

This survey will be further limited to that part of forestry research which can be identified in the Nordic context as being most relevant when integrating two or more forest uses or forest functions:

1) in joint production; two or more uses occur in the same area at the same time,
2) by segregating uses spatially as alternative processes; the total area is allocated for specific uses or combinations of uses, or
3) in sequential production; different uses are organized according to time scale using either one of the two above mentioned approaches.

The multiple–use concept presented above is wider than most other commonly presented definitions. It is also useful to emphasize that the concept in this wide sense still has an objective content in the Nordic countries without having been negatively labelled as the specific policy of a specific interest group.

It is necessary to limit the period to be surveyed. The multiple–use concept entered the Nordic forestry discussion gradually during the sixties (Hytönen 1995). Consequently, the related research was initiated in the late sixties. Research from that time will be considered here. However, it should be recognized that many earlier works have been and still are relevant in many aspects of studying forest uses and value relationships.

In the following, multiple–use research will be surveyed in each Nordic country. A short overview of the history and organizational development as well as of major research efforts until now will be presented. Existing organizations and research activities will also be described and, where possible, some future prospects will be outlined.

16.2 Denmark

Multiple–use research in Denmark had its formal beginning in 1975 with the establishment of the "Forest and Folk" (Skov og Volk) –project at the Danish Forest Experiment Station. The project was initiated and financed during 1974–1980 by the Danish Agricultural and Veterinary Research Council. It included national, regional and local estimations of recreational use of forests. Research was also done on people's preferences, recreation modelling and recreation economics (Koch 1978, 1980, 1984a, Koch & Jensen 1988). The aim of the project was to develop a better basis for political and administrative decisions in the field of forest recreation in Denmark.

Even now this detailed study represents a rare if not unique case in the world because recreational use of most forest areas of the whole country (3/4 of all forest areas larger than 50 hectares) were inventoried. According to the results of the study, the annual number of forest visits has been estimated to be approximately 50 million (Koch 1978, 1980). This allows to a general conclusion that the recreational value of many forest areas in Denmark is far above their wood production value. The value assigned by the public to forest areas was calculated using consumer surplus estimates reached by Clawson's travel cost method.

The other important non–timber forest use in Denmark is hunting. As much as 3,6 % of the population are hunters and nearly all forest areas in the country can be rented for hunting purposes. In addition, the losses caused by deer to wood production are significant (Koch 1984b). A special area of multiple–use forestry research is the stress caused by outdoor recreation activities, especially orienteering, on game and other animals (Jeppesen 1984).

There are two non–timber uses in Denmark that are clearly of less importance in other Nordic countries. One is the production of Christmas trees and decoration greenery which has developed because of the good markets in nearby central European countries. The other is the protection function of forests. As Denmark is a flat peninsula and a group of islands between the North Sea and the Baltic Sea, nearly all forests have protection functions as wind shelters. The establishment of dune plantations for protection against sand drift on the long west coast of Jutland has traditionally been given much emphasis in Danish forestry research. Christmas trees and decoration greenery production clearly reduce timber production and thus represent a real trade–off case in forest production. In many cases, forest greenery is privately produced and is a preferred choice because of good profitability.

Due to the geography, land ownership history and the dominance of agriculture in land use, Denmark has many features in common with central European forestry. One of those is the traditional concern given to the management of cultural landscapes and nature conservation. Consequently, forest is a frequent element in the prevailing agricultural landscape of the country. As in other Nordic countries, the research related to landscape has mostly been done outside forestry research organizations.

Besides the above mentioned main fields of research, also other studies in multiple use have been carried out from the seventies onwards. These have dealt with the theory behind multiple–use forestry (Helles 1977), the values of forestry for society (Helles et al. 1984), landscape aesthetics (Borup 1991), nature conservation (Naturen i skoven 1989, Naturpleje i skov 1989), cultural values (Thomsen 1988), advice on management for multiple values (Koch & Kristiansen 1991), and planning of multiple–use forestry (Kristiansen 1994). Recently, attention has been paid to economics of forest environment (e.g. Helles & Linddal 1994).

Research organizations

Due to the small economic importance of forestry in the national economy, forestry research is a relatively small research sector, which – as the Working Group for Forest Research (1989) stated – is closely related to other considerably larger research fields (Forskning og formidling 1989). Therefore, the Working Group felt it necessary to ask whether there are sufficient grounds to maintain a separate organization for forestry research. However, it was concluded that forestry in Denmark has many special features, which make it rational to have an independent Danish forestry research organization. Among those specialities, high emphasis was given to non–timber and non–material aspects such as biological, environmental, cultural, historical, aesthetic and recreational values.

The Working Group suggested that the existing relatively small research and development units of forest and landscape research should be merged into one institution. Consequently, three research institutes, the Danish Forest Experiment Station, the Danish Institute of Forest Technology, the Danish Institute of Park Technology, and some functions from two other organizations were rearranged in 1991 into a new research unit, the Danish Forest and Landscape Research Institute (Forskningscentret for Skov & Landskab). The institute is supervised by the Ministry of Agriculture.

The aim of the Danish Forest and Landscape Research Institute is to carry out research, development and dissemination of knowledge for the benefit of forestry, including production of greenery and Christmas trees as well as park and landscape management. The institute attaches importance to multiple–use of forests and landscape and to environmental conditions (Forskningscentret... Undated).

The organizational environment and the structure of the institute, including subject groups, is presented in *Figure 16.1*. The most relevant research from the multiple–use point of view is done in the Department of Park and Landscape and the Department of Christmas Trees and Greenery. The research interests of the former are the following: re–establishment of natural conditions, park and landscape management, park and landscape maintenance, urban ecology, composting, outdoor recreation and green tourism.

The Royal Veterinary and Agricultural University (Den Kgl. Veterinær– of Landbohøjskole – KVL) has also a significant role in multiple–use forestry research. Forestry education and related research was carried out in the Unit of Forestry until the end of 1993. The Unit has contributed to the development of the multiple–use forestry concept from the point of view of economic theory (e.g. Helles 1977, Helles et al. 1984).

In the beginning of 1994, the Unit of Forestry was merged with the Units of Economy and Landscape to form a new entity called the Department of Economy, Forest and Landscape (Institut for Økonomi, Skov og Landskab). The overall objective of the new Department is to strengthen the socioeconomic and ecological information basis in management, planning, administration and resource utilization in the fields of agriculture, horticulture, forestry and "open lands" (det åbne land). The research projects of the Department are divided into two groups: 1) economics of agriculture and 2) forest and landscape. The latter has seven priority areas with ongoing and planned projects (Strategi for forskning... 1993):

1) Landscape development and regulation in open land.
2) Economic, political and planning aspects of multiple–use forestry.
3) Expressions and basic ideas of landscape architecture.
4) Establishment and management of forests, green areas and nature reserves.
5) Interdependence of forest management, properties of wood, the value of wood and its industrial utilization.
6) System ecological, genetic and ecophysiological aspects of forestry and landscape management.

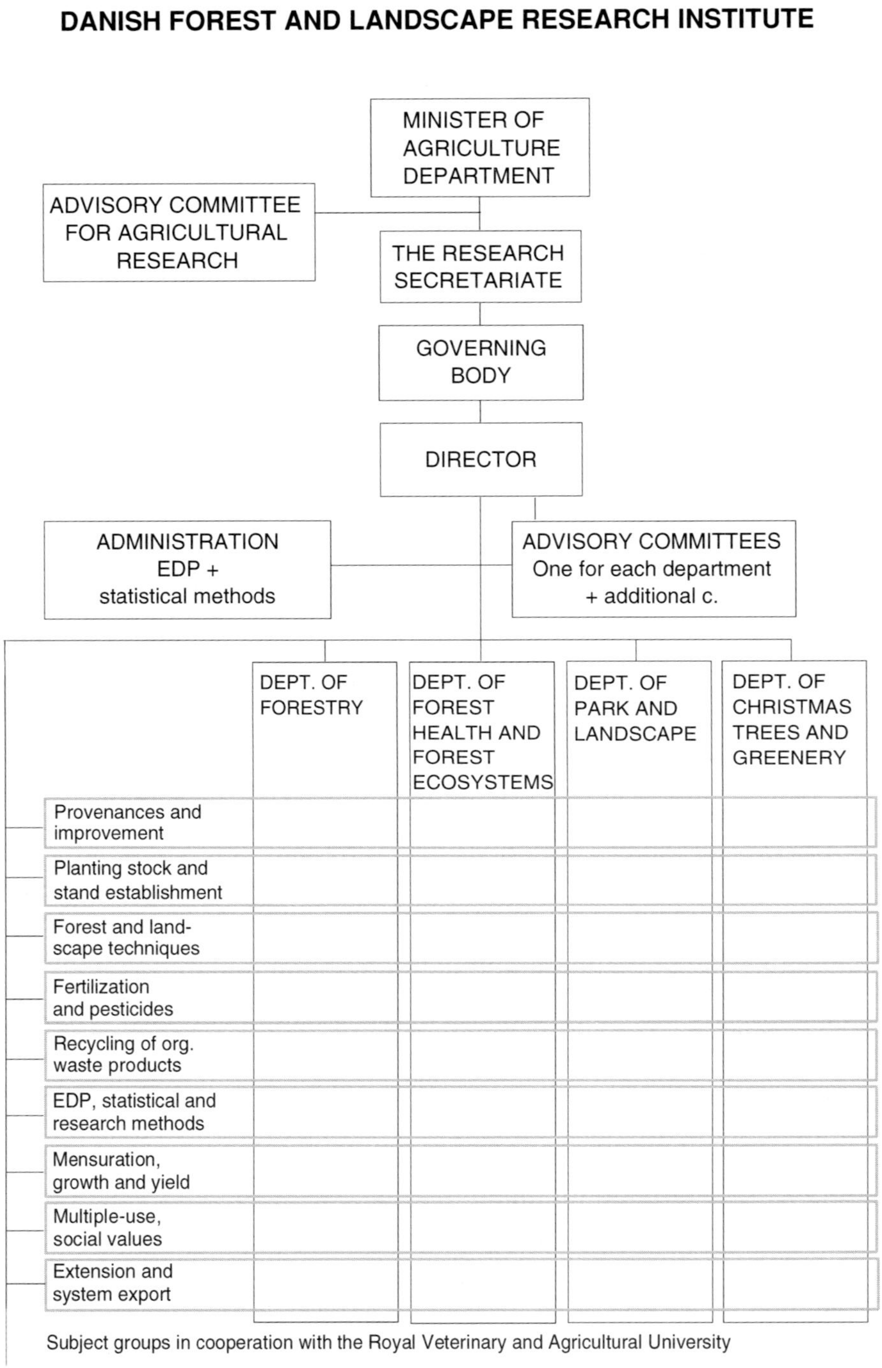

Figure 16.1 Organizational chart of the Danish Forest and Landscape Research Institute.

7) Sustainable forestry and land–use management in developing countries with special emphasis on socioeconomic, political and biological aspects.

The present situation is a kind of transition period during which the new objectives and working methods of the Department will be further developed. Presently, the three different units are still situated in different places. In future, most functions of the Department will be situated in one place. Furthermore, the new Department has a cooperation agreement with the Danish Forest and Landscape Research Institute (DFLRI). This cooperation will be strengthened when the DFLRI will move to a new building in Hørsholm. The new building will also accommodate forest and landscape researchers from the university (Strategi for forskning... 1993).

16.3 Finland

In Finland, multiple–use research was started at the turn of the 1960s and 1970s. The first studies were initiated and carried out in the Faculty of Agriculture and Forestry of the University of Helsinki, in the Departments of Social Economics of Forestry, Silviculture and Forest Mensuration. A little later, research was also begun in the Finnish Forest Research Institute (Metsäntutkimuslaitos – METLA) where the first multiple–use studies were carried out in the Department of Forest Economics and the Department of Peatland Forestry. The first permanent multiple–use researcher position was established in 1972 at the Institute's Rovaniemi research station which is situated in northern Finland.

Although multiple–use research was one of the priorities of the Finnish Forest Research Institute (FFRI) from the early seventies onwards, the resource base grew slowly during the first 15 years. It was not until 1988 that the first leading level multiple–use researcher position was established. This and the start of the Multiple–use Research Program in 1990 can be seen as decisive steps in giving multiple–use research a steady organizational status.

In the following, two issues are of interest: why was multiple–use research initiated just at that time and why did it grow slowly during the first 10–15 years although the demands for such research were more than obvious?

A major reason for beginning multiple–use research just at the turn to seventies was the heated discussion on the scenic, ecological and environmental impacts of intensified silvicultural and forest improvement operations (Reunala & Heikinheimo 1987). The reason for the intensification of wood production was the actual and predicted wood shortages induced by large–scale capacity increases in the forest industry in the 1950s and 1960s. Conflicts between forestry and the environmental movement were at that time unavoidable, and multiple–use research met an evident social demand.

The main reason for the slow and modest resources development can be found in the organizational rigidity of the Finnish Forest Research Institute, where – as perhaps in many other research organizations – new fields of research have difficulties in the establishment process unless organizational changes are made to facilitate reallocation of funds. Those organizational conditions were not created until 1988, although the legislation concerning FFRI was changed to include "research on different forest uses and protection of the environment" as early as in 1986.

In 1983, a working group for multiple–use forestry was established to participate in the preparation of the Forest 2000 Programme, which was the national long–term plan for developing forestry and forest industries in Finland. In its report, the group makes recommendations, for example, concerning the development of multiple–use research and university education (Metsien moninaiskäytön... 1985). However, generally speaking the program did not take an active role in the promotion of multiple use (Saastamoinen 1987). The same holds true also for the monitoring and revision report which was published in 1992 (Metsä 2000... 1992, The representation of the revised... 1992). Although the reports discuss multiple–use and nature conservation, survey the environmental impacts of forestry and forest industries, and propose the introduction of Environmental Impact Assessment for forest improvement work, it can be concluded that, until recently, the official forestry institutions and administrations have been rather conservative and timber–minded in their policy concerning multiple–use forestry.

In addition to the forestry organizations, the Ministry of the Environment, which was established in 1983, has had an active role in supporting multiple use. Its policy in regard to forests has included production of nature conservation programs and promotion of outdoor recreation. In 1985, the Ministry published a research program for outdoor recreation studies which resulted in an increase of related research in the Finnish Forest Research Institute and in the Nature Conservation Research Unit of the National Board of Waters and the Environment.

An important document published by the Ministry of the Environment was the report on environmental policy in Finland (Environmental protection... 1988). The report contained background information for an OECD survey on environmental protection and policy in Finland (Environmental policies... 1988). The final OECD report, among other things, made recommendations concerning multiple use and environmental forestry.

Research organizations

The primarily state–financed Finnish Forest Research Institute (FFRI) is one of the largest forestry research organizations in Europe. It carries out research in all fields of forestry. FFRI has good facilities for multiple use as well as for other forestry research. The two national parks in Lapland (Pyhätunturi and

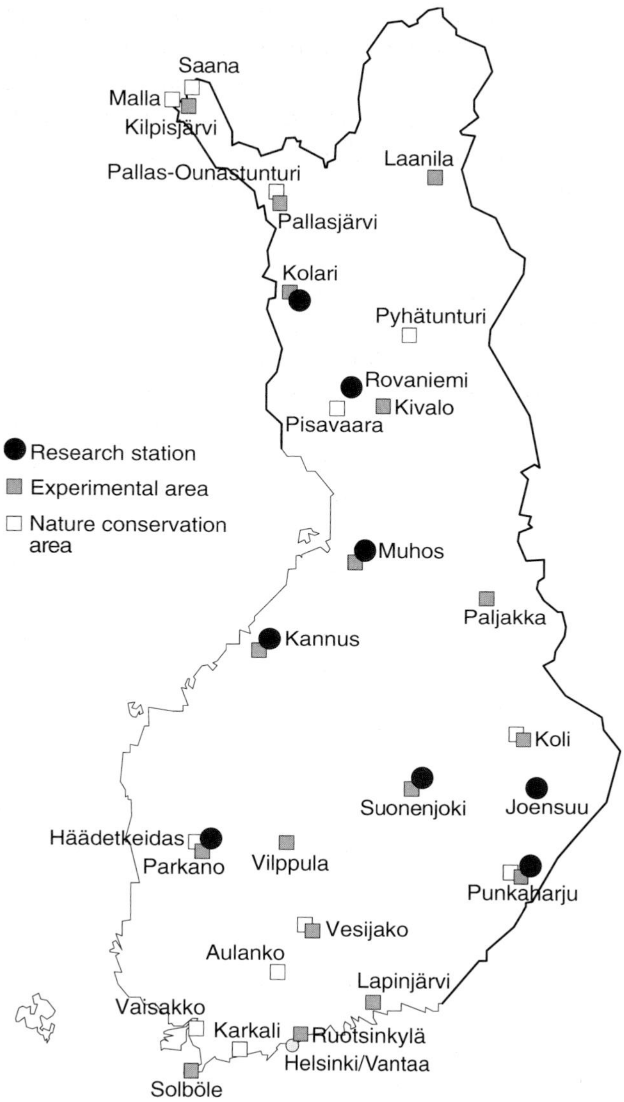

Figure 16.2 Locations of the Finnish Forest Research Institute.

Pallas–Ounastunturi), other nature conservation areas and research forests located throughout the country, and the network of eight regional research stations, provide good possibilities for effective field work and close contacts with practical organizations and other research institutions (*Figure 16.2*). About one third of all researchers in FFRI are located in research stations.

In 1990, a special research program was established for multiple–use research (Kuusipalo et al. 1989). The program gathered studies dealing with multiple use into one administrative unit, and a separate budget was provided for it. Nowadays it comes under the Department of Forest Resources, which

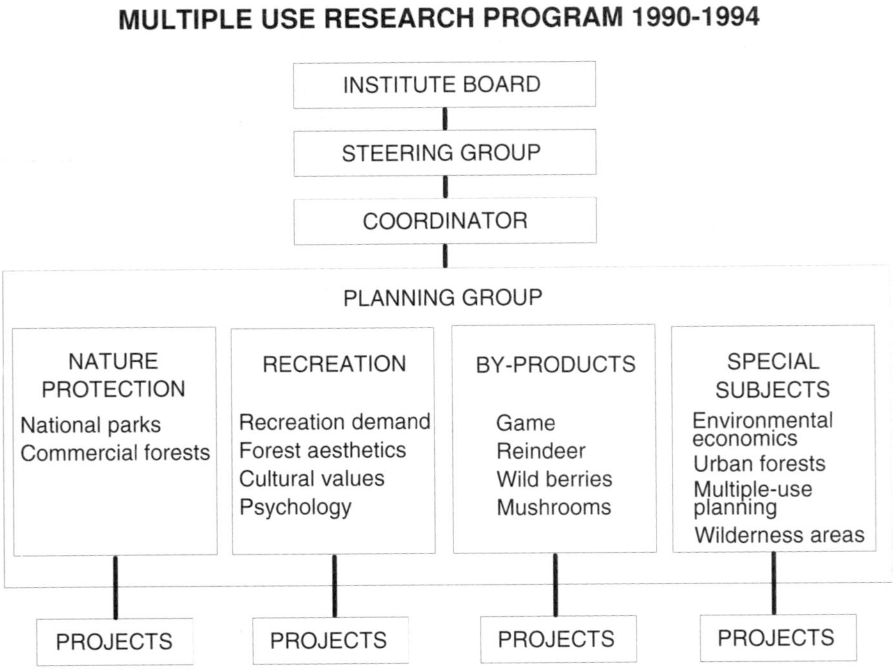

Figure 16.3 Structure of the Multiple–use Research Program 1990–1994.

was created in 1991 as a result of the reorganization of FFRI. It has altogether 12 full–time researchers and some technical staff (*Figure 16.3*).

The first Multiple–use Research Program of FFRI will be finished by the end of 1994. Right now a new research program is being planned. According to present plans, the new program will emphasize the integration aspect of multiple use. The main themes will be planning methods for multifunctional forestry, forest landscape, management of wilderness areas, and economics and sociology of recreation.

The major locations for multiple–use research are the Joensuu research station in eastern Finland, the Rovaniemi research station in Lapland, and the headquarters of FFRI in Helsinki. In Helsinki, urban forestry and outdoor recreation are two major fields of research. The focus of urban forestry is on silviculture. In addition to the two above mentioned topics, studies are being carried out on public attitudes on forests including personal and cultural values. This research is supported by the University of Helsinki, because it has been possible to specialize in multiple use in the Department of Forest Ecology from 1991 onwards.

The coordinator of the Multiple–use Research Program of FFRI is located at the Joensuu research station, where the main areas of research are forest berries and mushrooms and planning of multiple–use forestry. The Joensuu research station also benefits from its location on the university campus just beside the Faculty of Forestry of the University of Joensuu.

The Rovaniemi research station near the Arctic Circle is the largest research station of FFRI. It is located in an environment where non–timber forest uses have traditionally had special importance in regional economy and where the bulk of nature conservation and other protection areas are to be found. Its present multiple–use research topics deal with reindeer husbandry, game management, nature conservation, wilderness areas, recreation, tourism and environmental economics (e.g. Naskali et al. 1993, Sepponen & Lohiniva 1994).

Rovaniemi is also the home town of a few other research institutes concerned with northern nature and people. The most important from the multiple–use point of view is the Arctic Centre of the University of Lapland, which actively participates in international cooperation. The Centre is in charge of the national wilderness research program, which is a co–operative umbrella–type program bringing together many smaller projects in several research institutes and universities (Hallikainen & Jokimäki 1992).

The Nature Conservation Research Unit of the National Board for Waters and the Environment is engaged in environmental research related to water resources, nature conservation and biodiversity. The Unit coordinates the multidisciplinary research program called "The Finnish Biodiversity Research Programme" (Jäppinen & Väisänen 1993). The program involves researchers from various organizations from all over the country. It also cooperates with FFRI.

There are two faculties providing academic education in forestry sciences in Finland: the older one at the University of Helsinki, and the younger one at the University of Joensuu. The Faculty of Agriculture and Forestry at the University of Helsinki has educated professional foresters since 1908 and has had a decisive position in initiating multiple–use research in Finland. The first doctoral dissertation specializing in multiple–use forestry was approved in the Department of Social Economics of Forestry in 1982 (Saastamoinen 1982). Presently it is possible to specialize in integrated forest management or in some subfield of multiple–use forestry (e.g. landscape management, urban forestry) only in the Department of Forest Ecology. Since 1991, students have been allowed to compile their own study plans in which they can include courses from various departments of the Faculty, from other faculties and universities, and also from abroad.

The Faculty of Forestry at the University of Joensuu was established in 1982. The faculty has five major subjects, of which in two subjects – in silviculture and in forest management planning – the students can specialize in multiple–use forestry and environmental management. There is also an increasing body of research in the field, mainly on the silvicultural aspects of multiple use, forest landscape simulation, forest management planning, and on the economics of multiple use, with some extensions also to tropical forestry. The first dissertation written by the faculty's "own nursling" dealing with forest regeneration and multiattribute utility theory had a multiple–use dimension (Kangas 1992). A position of senior lecturer was established in 1988 for multiple use and environmental management. Now there are plans to change this to a professorship.

A new international forest research organization was established in Joensuu in eastern Finland in 1993. The European Forest Institute (EFI) is an independent, non–governmental organization founded by 12 universities and forestry research institutes from ten European countries (Illi 1994). The priority areas for the research and development projects at the EFI are the following: 1) forest sustainability, 2) forestry and possible global climate change, 3) structural changes in markets for forest products (wood and non–wood) and services, 4) forest policy analysis, and 5) forest sector information services and research methodology. The present projects of the Institute include a comparative study of forest biodiversity in Finland and northwest Russia and a study of the use of forest resources and environment related attitudes in five European countries and parts of the USA (EFI News 1994).

Evaluation of the Multiple–use research program of FFRI

FFRI ordered an international evaluation of the new research program in 1991. It is the first evaluation specifically focused on multiple–use research in the Nordic countries (Driver & Peterson 1992). Because the recommendations and conclusions by the evaluators may have some validity in other Nordic countries as well, they deserve to be cited. Recommendations are not arranged in any order of priority and are presented here in a shortened form:

- Multiple–use forestry is a rapidly growing concern throughout the world and is of critical importance for Finland. Therefore its research should be expanded and strengthened.
- The program should include both research of non–timber products and of the problems of joint production and integrated ecosystem management.
- Systematic and effective research planning should include 1) a mission statement 2) problem analyses and 3) detailed study plans; these three stages should be a part of a formalized planning process that includes peer review and executive approval by administrative officers.
- Research teams should be formed; many of the researchers are scattered in different locations in Finland and isolated from like–minded colleagues.
- If funding can not be increased in future years, serious consideration should be given to reducing the scope of the program to avoid spreading resources too thinly.
- Some of the most important research problems involve social, economic and behavioral science. This is seldom recognized by traditional forest sciences. This problem exists worldwide, not just in Finland.
- FFRI should be more responsive to the public than it has been until now.

Future challenges

There is no doubt that multiple–use forestry and research will become more important in the 21st century. There are many reasons for this: continuous urbanization of society, increasing environmental awareness of people, changes in forest ownership structure and the probably decreasing level of stumpage prices which will reduce the opportunity costs of non–timber uses. The future membership of Finland in the European Union is expected to increase tourism and recreational demands on Finnish forests and other nature resources. Also, rationalization of wood harvesting and forest industries, and reduction in the employment of foresters and forestry engineers, seem to strengthen the change of attitudes in forestry organizations in favour of multiple use and environmental forestry. The same kind of pressures are coming from consumers' demands. Environmentally sound forest operations are becoming assets of competitiveness for forest industry products in the hard world of international trade. A further evidence of the ongoing trend is that at the end of 1994, FFRI made a decision to change the vacant professorship of business economics of forestry to that of environmental economics which includes multiple–use forestry.

16.4 Iceland

Early destruction of woodland due to charcoal making and sheep grazing has caused massive soil erosion and loss of other vegetative cover in Iceland. Today only about 1 % of the land area is covered by birch dominated woodland (Blöndal 1991). Because of this, protection and land reclamation have been the major objectives of forestry in the country. From the multiple–use point of view, the Icelandic case is therefore sharply different from other Nordic countries and offers a highly interesting insight on some specific multiple–use problems.

Due to the decreasing number of sheep in the eighties, it has been possible to fence off several areas formerly grazed by sheep to protect the existing birch remnants and for regeneration. These enclosures have considerable value as recreation areas for the urban population. In addition, a pilot project has been initiated in sparsely vegetated and barren areas by the Icelandic Forestry Association to create small birch forests.

Although forestry has been practised since the turn of the century, no formal research work was carried out until the 1950s, when the Forest Service appointed a forestry graduate to initiate silvicultural research. The Forest Service Research Station was officially opened in 1967. Gradually it has expanded and become a very active institute (Loftsson 1991) (*Figure 16.4*). Research on natural regeneration and other establishment techniques for birch is now one of the research priorities of the Iceland Forestry Service (Blöndal 1991). Other topics are the problems of choice and introduction of

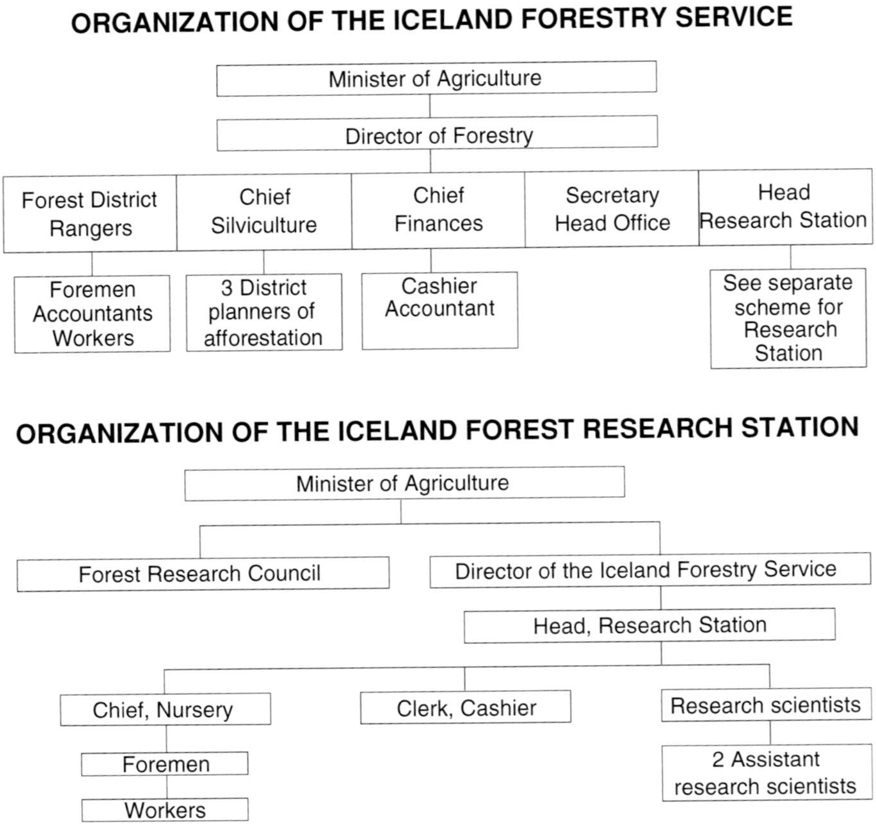

Figure 16.4 Organizational charts of the Iceland Forestry Service and the Iceland Forest Research Station.

exotic species and improvement of the classification of the land according to wood production potential.

While forestry research is concentrated on solving the problems of reforestation and land reclamation, and therefore can hardly be interpreted as multiple use research as such, it is anyway important to notice the multitude of benefits expected from reforestation (Blöndal 1991): soil protection, higher grazing value of vegetation under tall forest, minor products for specific needs of the community and a more desirable environment for recreation, camping sites and summerhouses.

16.5 Norway

Norwegian people are known among the Scandinavians for their enthusiasm for the outdoor life. Various surveys indicate that at least two thirds of the population enjoy some kind of outdoor recreation in forests at some time during the year (Aasetre 1992). The first studies in forestry which can be regarded as belonging to multiple–use research were mainly focused on outdoor recreation (e.g. Strand 1967, Haakenstad 1972, Hofstad 1976a, Hofstad 1976b).

Recreation pressures were also the major cause for the most comprehensive multiple–use study at its time in Scandinavia. The aim of the study was to analyze and plan the combined recreational and forestry utilization of Oslomarka, an area of nearly 1,700 km^2 situated in the capital city of Oslo and surrounding municipalities, and serving the recreational needs of about 15 % of Norwegians (Hofstad 1976a, Svendsrud 1977). The results of the Oslomarka studies and discussions led to a more restrictive forest legislation and to more consultative and participatory approach in planning and decision making concerning the area (Opheim 1984). Hofstad's (1976b) study on the Oslomarka conflicts was the first dissertation on multiple–use economics in Norway.

As a whole, the Oslomarka study can be considered to be the first example of a comprehensive and multifocused area–specific forestry study, the results of which were used in real economic and practical decision making. However, after this intensive and comprehensive research period in Oslomarka, multiple–use issues received less attention in Norwegian forestry research. Multiple–use research was not made part of the institutional framework of forestry research or of university organizations.

In the 1980s, biologists entered the scene and firmly demanded research to save endangered species in forests. Consequently, today research related to multiple–use forestry deals also with biological questions.

Research organizations

There are two major institutions carrying out multiple–use forestry research: the Norwegian Forest Research Institute (Norsk Institutt for Skogforskning – NISK) and the Agricultural University of Norway (Norges landbrukshøgskole – NLH) with its Department of Forestry. Multiple–use research, especially within the Oslomarka study mentioned above, was an important part of forestry research activities at the Agricultural University of Norway in the 1970s. The Norwegian Forest Research Institute was not yet very much engaged in multiple–use research at that time. It concentrated more on forest ecology and various aspects of wood production.

A radical change came about by the mid–1980s. Environmental and multiple–use related studies now play a substantial part in the activities and image of the Norwegian Forest Research Institute. The Institute coordinates a major joint research program "Forest ecology and multiple–use forestry". Furthermore, it has been decided to integrate multiple–use aspects in all relevant research projects of the institute, besides carrying out specially dedicated studies (Strategisk plan 1993, Norwegian Forest... 1994). NISK also has the Hirkjølen experimental area to be used as a demonstration area for ecological relationships and multiple–use of forests. It is situated in montane forest about 280 km north of Oslo (Solbraa 1990, Hirkjølen... 1993).

The Department of Forestry of the Agricultural Universtity of Norway was established in 1898. It offers normally two lines of specialization, namely silviculture and forest resource economics and management. During the last 40–50 years, research has become an integral part of its activity (Department... Undated). The present research of the Department includes studies on alternative silvicultural methods, socioeconomic consequences of multiple use, and political processes and means of carrying out multiple use (Virkomhetsplan 1994).

In 1990, NISK and NLH with its Department of Forestry entered into close administrative and professional cooperation under the name SkogForsk. The aim of the reform was to increase scientific competence in the fields of biological, technical and social forestry research and education. For example, the staff of the Institute will be involved in teaching in NLH (Formål, organisasjon... 1994, Strategisk plan 1993). To streamline future research, SkogForsk has published project planning guidelines (Prosjektkvalitet... 1993).

The third major institute participating in multiple use and related research is the Norwegian Institute for Nature Research (Norsk Institutt for Naturforskning – NINA). It was established in 1988 by the Ministry of the Environment. NINA is divided into five research divisions: mammalian ecology, terrestrial ecology, conservation biology, aquatic ecology, and the East Norway Division. In addition there is a research program for outdoor recreation (Norwegian Institute... 1993). From the multiple–use point of view, the most important research topics are conservation of natural habitats, species preservation, game ecology, and outdoor recreation.

NINA's outdoor recreation program has been designed to provide information on the following: 1) What determines the recreational habits of the general public? 2) How do Norwegians as well as foreigners feel about measures introduced to organize recreational activities? 3) How do outdoor people react to environmental encroachment? 4) What are the effects of official regulations of outdoor recreational activities? (NINA... undated). The program was evaluated in 1993 and it was decided to continue research on the man/nature dichotomy, based on a broader social science approach, and by drawing more direct connections with ecology. Other future activities include research on Arctic issues; NINA opened its own Department of Arctic ecology in Tromsø in March 1994. Also, Norway's cultural heritage research will become part of NINA's organizational structure in 1994 (Annual report 1994).

The "Forest ecology and multiple–use forestry" –research program was established in 1989 and will be completed by the end of 1994. It is a joint effort of NLH, NINA and NISK and the Universities of Bergen, Trondheim and Oslo. The aim of the program is to collect and publish new ecological and economic information for deploying of multiple–use forestry in Norway. It will focus especially on the impacts of the fragmentation of forest structure on flora and fauna, and the economic consequences of multiple–use adaptations. The program includes 15 projects, the results of which will be published in a handbook once the program is completed (Skogsøkologi... 1992, Årsberetning... 1994).

Future plans and challenges

Norwegian research policy has come under intensive scrutiny in the 1990s. Evaluations and action plans have been produced in various fields of research, including forestry (e.g. Forskning... 1992/93, Gornitzka 1992, Økt verdiskapning... 1994). The Research Council of Norway states that environmental aspects and the relationship between timber production and other uses of forests will be central themes in forestry research in the future. The Council's report groups the most relevant research areas into three categories: 1) forest and environment, 2) forestry and industry, and 3) forest and society. It is recommended that research be carried out in research programs, and should consist of theoretical research, applied research and development work (Økt verdiskapning... 1994).

16.6 Sweden

Early multiple–use related research in Sweden was mainly initiated and carried out by professor Dr. Lars Kardell in the Department of Environmental Forestry in Uppsala. Being from a silviculturist background, it was natural for him to orientate into the problems of the complicated relationships between timber production and other forest uses. The early publications of the Department of Environmental Forestry dealt widely with berries and mushrooms, recreation, landscape management, nature management and conservation in forestry.

A nationwide inventory of forest berries in Sweden was carried out in connection with the national forest survey from 1974 to 1977. It was the first of its kind in the sphere of minor forest products in the Nordic countries (Eriksson et al. 1979). Hultman (1983ab) was the first to write a doctoral dissertation on people's perceptions of forest landscapes in 1983. Since then, landscape management and landscape preferences have been given much emphasis in Swedish research. Recent research in the Department of Environmental Forestry has concentrated on forest recreation in different locations, nature perceptions and landscape management in forestry.

There are also other departments in the Swedish University of Agricultural Sciences carrying out multiple–use research and education. The Department of Landscape Planning has been engaged in forest landscape research. The studies have been dealing with stand structures and their development, and with the problems of urban forestry. Axelsson Lindgren (1990) has investigated the perception of forest stands from recreational and landscape planning points of view.

The most traditional way to utilize forests is no doubt by hunting. Even now, hunting enjoys a high status in practical forestry and education. The Department of Wildlife Ecology has been pioneering in flora and fauna studies. The Department of Ecology and Environmental Research is engaged in systems ecology of forests but is also studying nature protection problems. The Department of Silviculture has been mainly focusing on plantation forestry and the effects of silvicultural treatments on stand and site. Recently, increasing research effort has been devoted to alternative (without clearcutting) methods of silviculture, which could be used especially in areas having special importance from a multiple–use point of view (Falck 1991). Also silvicultural problems of urban forests have been covered.

There is a shortage of knowledge and practical experience of alternative regimes of silviculture, mainly because from the early 1950s alternative management practices were restricted by legislation. Now, the official forest policy is changing from strict governmental control aiming at highest possible timber production to a more flexible attitude to the use of forest land (Falck 1991).

The Department of Forest Economics in Umeå has been active in multiple–use research. The first doctoral dissertation related to multiple–use forestry in Sweden was Leif Mattsson's work on forest history and on the relationship between forestry and reindeer husbandry. It was published in 1981. Later on, research has concentrated on policy aspects of multiple use and environmental forestry, environmental economics and economics of hunting. For example, environmental protection in Swedish forestry was the topic of a dissertation by Katarina Eckerberg (1990).

Research organizations

In Sweden, forestry research and the university education of foresters have since 1915 been carried out in the Royal College of Forestry (Skogshögskolan). The College was merged with the National Forestry Research Institute (Statens skogsforskningsinstitut) in 1962. In 1977, the College and research activities were joined together with the Royal College of Agriculture and the Royal College of Veterinary to form the Swedish University of Agricultural Sciences (Sveriges lantbruksuniversitet). Forestry became one of the three faculties of the University (von Segebaden 1978).

The Faculty of Forestry is responsible for most academic forest education and for about 70 % of forestry research. The remaining research is done by the Forestry Research Institute of Sweden, other universities, companies, and other smaller institutions. The faculty is spread over four locations. Education and research is practised in Umeå, Garpenberg and Uppsala. Recently, a research unit in Alnarp has begun forest and landscape research in southern

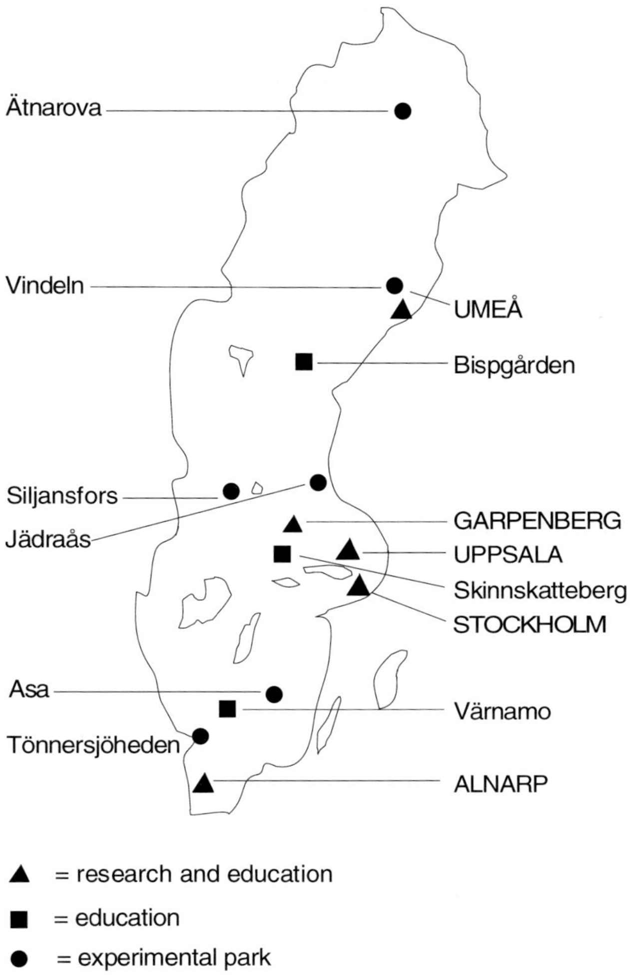

Figure 16.5 Locations of the Faculty of Forestry of the Swedish University of Agricultural Sciences.

Sweden. The faculty has six experimental areas from Gällivare in the north to Halmstadt in the south. In addition, the faculty has three separate units for the training of forest technicians. It is also responsible for information and extension activities (*Figure 16.5*) (Bäckström 1990).

The research of the Faculty is divided into five research programs, each of which includes research projects from several departments. They are 1) economics, 2) inventory and forest resources, 3) forest site yield, 4) forest technology and wood utilization, and 5) nature and the environment. Most of the multiple–use forestry research is done in the nature and environment research program, the aim of which is to "gain knowledge on and to develop methods for production, management, preservation and utilization of those utilities and environmental values which, in addition to timber production, are of general importance for land use". The program includes studies on the following topics (The Faculty... 1991):

- Basic knowledge on ecosystems, species and gene resources and principles for their continuing care and preservation.
- Production of environmental values and utilities such as wild game, berries, mushrooms, recreation areas and cultural landscape.
- The protective role of forests against erosion, wind damage, etc.
- Consequence evaluations, e.g. of new silvicultural methods.

The Forestry Research Institute of Sweden (Skogforsk – Stiftelsen Skogsbrukets Forskningsinstitut) was established by merging two small research institutes, the Forest Operations Institute (Forskningsstiftelsen Skogsarbeten) and the Institute of Forest Improvement (Institut för Skogsförbättring) in 1992. The new institute is situated in Uppsala. It is financed by forest companies, foundations, the state, and the National Federation of Forest Owners. Its aim is to promote economically profitable and ecologically sound forestry (Skogsägarnas Riksförbund... 1994, Skogforsk... 1993).

Future plans and challenges

The Faculty of Forestry of the Swedish University of Agricultural Sciences presented the program "Swedish Forest Research to the Year 2000" in 1989 (Svensk skogsforskning... 1989). It proposes a significant increase in research funds and resources for thirteen specified study areas of strategic importance. Although the concept of multiple use is not explicitly mentioned in that context, the substance of the following two research areas is especially significant from multiple–use point of view:

1. Production and environment. Most forest areas meant for economically important wood production also contain recreational and environmental values. Therefore, research which integrates production and environmental aspects will have a high priority in the coming years.
2. Nature and environment. This area focuses on the environmental impacts of forest operations caused, for example, by the drainage of peatlands and by the planting of exotic species. Natural forests and the conservation of forest flora and fauna will be studied. The faculty has also set up lists of rare animals and plants threatened in forest ecosystems, and this work will be developed.

Other strategic areas are the following: 3) production optimization, nutrient leakage, vitality fertilizations, 4) air pollution, 5) alternative silvicultural methods, 6) biotechnology 7) wood quality, 8) long–term field experiments, 9) forests and energy, 10) engineering and forest technology, 11) forests in national and private economy, 12) forests in a historical perspective, and 13) forests in a global perspective. Some of these other strategic areas also have linkages with multiple–use research (Svensk skogsforskning... 1989, The Faculty... 1991).

Another recent research initiative further illustrates the trends in forestry research. It has been proposed that a research program on natural resources and environmental economics related to area–based industries be carried out as a joint effort of several institutes: the Department of Forest Economics (Umeå) and the Department of Economics (Uppsala) at the Swedish University of Agricultural Sciences and the Department of Economics and the Department of Political Science at the University of Umeå, the Swedish University of Commerce and the Beijer International Institute of Ecological Economics (Stockholm) (Mattsson et al. 1991).

The proposed program consists of two main studies with two subproblems:

1. A choice and balancing problem in forest utilization

1a. Socioeconomic analysis of the multiple benefits of forests

The study includes research on methodologies to quantify non–market forest benefits, the value of non–market benefits in timber production forests, the value of forests for tourism, the existence value of plants and forest wildlife, and the costs caused by adjusting timber production activities in order to produce and maintain also other forest values. Finally, with the help of the attained research results, synthesizing analyses will be done.

1b. Control methods in forest policy

This substudy examines the community level planning and decision making for nature conservation and outdoor recreation in forestry and agricultural lands near cities. Another problem area is the effectiveness of the economic ruling methods (e.g. environmental fees for chlorine emissions) in forest industries and the distribution of impacts of the fees on environment, industry, forest owners and sawmills.

2. Discharge problems in the watershed areas of the Baltic Sea

2a. Willingness to pay, cost effectiveness and socioeconomic consequences

The substudy mainly deals with the problems of water resources management in the industries influencing the Baltic Sea. It also examines, among other things, the role of agriculture and forestry in watersheds.

2b. The economic consequences of the biodiversity changes

The proposed research agenda of this study is concentrated on the ecology and economics of biodiversity conservation. It includes three major issues: the value of biodiversity, the driving forces behind biodiversity loss and the elements of a strategy for biodiversity conservation. These topics also characterize the present trends in other Nordic countries (e.g. Helles & Linddal 1994).

The Faculty of Forestry of the Swedish University of Agricultural Sciences produced a discussion paper for the planning of future research. It is a collection of researchers' ideas concerning the development of research. Multiple–use is present to some extent but most of the proposals are dealing with biological, economic or technical aspects of timber production, and wood–processing and marketing (Elmberg & Sallnäs 1993).

16.7 Research for the future

The increasing importance of non–wood benefits and multiple use has been stated so many times during the last two decades that one should perhaps remain somewhat skeptical about them. Science has never been very good at predicting sudden, structural changes in economy, politics or social values. One should not exclude the repulsive possibilities of chaos, disorder and disintegration with their unaccountable consequences on forest uses and forest benefits.

However, while not forgetting a doomsday scenario, all available evidence is for a continuous trend of increasing weight being given to multiple–use forestry compared to wood production oriented forestry, and for changes between the weights of forest uses composing multiple use (e.g. Koch & Kennedy 1991, Opheim 1992, Krott & Zimmerman 1992, Driver & Peterson 1992). Recent and ongoing changes in forest and environmental legislation of the Nordic and other countries reflect the new values and demands. A recent example is the new forest policy of Sweden, which is based on extensive research and forward–looking analyses (Skogspolitiken inför... 1992). It places the maintaining of biodiversity on a par with the traditional production objective to direct future Swedish forest policy.

New challenges and demands for forestry research are strengthened by legislation. Problems are derived from changing economic structures, changing values of people, and from the changing natural environment. New problems also appear from the logic of the scientific progress itself: a solution for one problem may generate new questions to be answered, or offer resources to be allocated for research in new fields.

In each Nordic country, there are, no doubt, even in the future certain, nationally, regionally or locally specific research needs, which should be anticipated and met in due course of time. These needs can only be identified in each country. It is possible that a major part of future research in each country will be directed to country–specific problems.

In this article, the focus of interest is on those future areas of research which one can assume to be of similar interest in most Nordic countries. All these countries have reached about the same level of economic and social development, and share the problems in forest use which are typical to all industrial and urbanized societies. They have relatively similar traditions in forest use and everyman's rights.

Some of the challenges for the remaining years of this century are by no means new. Societies still need a theoretical basis and methodology to measure and compare the benefits of forests – and sacrifices needed to attain those benefits – in a rational and consistent way.

Similarly, there is a continuing need to improve and enlarge knowledge of how to attain the desired benefits from our forests. Whether this science (or art) is simply called silviculture, ecosystem engineering, ecosystems management or knowledge of multi–commodity production functions does not matter. What matters is the increasing need to manage forest and related ecosystems so that the growing and diversifying needs of people and societies could be sustainably and cost–efficiently met.

It is also likely that multiple–use research will be more integrated than earlier with other and more traditional fields of forestry research: forest ecology, silviculture, entomology, and so on. On the other hand, integration will occur with sciences of other fields as well: such as sociology, economics, political science, landscape aesthetics and wildlife biology.

The challenge is evident: climatic change will affect as much multiple benefits of forests as it does forest ecosystems and wood production. With forest ecosystem changes, everything changes: non–wood products, ecosystem influences and services, landscape values, recreation environment and biodiversity values.

Whatever will be the schedule and actual organizational process, the economy and politics are likely to support further integration in Europe. The Nordic countries, especially Sweden, Finland and Norway, have been traditional suppliers of sawnwood and pulp and paper products to other parts of Europe. What may be expected is that the forests and other natural environments of northern Europe will increasingly be in demand for the recreational needs of the heavily populated urban centres of the rest of Europe. To what extent is recreational demand from outside likely to increase, and what are the interests and hopes of central European travellers for northern forests? What are the policy changes needed to adequately respond to the economic, social and environmental problems which are likely to arise? For example, what will be the fate of the traditional Right of Public Access in Scandinavian forests?

All Scandinavian countries participate actively in development cooperation. In many developing countries, the multiple benefits of forests have traditionally been, and will continue to be, economically and socially comparable to wood, even if the figures in national accounts may give another picture. Future Nordic forestry research should assist developing countries to assess and develop the non–timber products and services of the forests and also support developing ways to integrate wood and non–wood uses of forests for the benefit of people while sustaining and even strengthening the resource base.

The latter issues should remind us that the challenges for research in multiple–use forestry are not only increasingly urgent but also increasingly universal. Forestry research will have growing role not only in saving the forests of our planet but also in promoting wiser management of them.

Finally, one should recognize that wiser management necessarily refers to something that can be called socially optimal. This involves the numerous conflicting, and difficult to treat, but nevertheless real, social values. No doubt there will be a need for research efforts focusing on social processes, changing values and value conflicts over forests and forestry in future multiple–use research in the Nordic societies.

References

Aasetre, J. 1992. Friluftsliv og skogbruk: en litteraturstudie. (Outdoor recreation and forestry: literature review.) Norsk institutt for naturforskning, Utredning 34. 52 pp. (In Norwegian.)

Annual Report 1993. 1994. Norwegian Institute for Nature Research, Trondheim. 29 pp.

Årsberetning 1993. (Annual report 1993.) 1994. Norsk Institutt for Skogforskning, Ås. 151 pp. (In Norwegian.)

Axelsson Lindgren, C. 1990. Upplevda skillnader mellan skogsbestånd: rekreations– och planeringsaspekter. Summary: Perceived differences between forest stands: recreation and planning aspects. Dissertation. Stad & Land 87. 443 pp.

Bäckström, P.–O. 1990. Svensk skogsforskning under 1990 –talet. (Swedish forestry research in 1990s.) Sveriges Skogsvårdsförbunds Tidskrift 3: 12–21. (In Swedish.)

Blöndal, S. 1991. Socioeconomic importance of forests in Iceland. In: Alden, J., Mastrantonio, J.L. & Odum, S. (eds.). Forest development in cold climates. Plenum Press, New York. 13 pp.

Borup, A. 1991. Landskabelige hensyn og fremtidige skovbryn. (Landscape aspects and future forest edges.) Den Kgl. Veterinær– og Landbohøjskole, Frederiksberg. 181 pp. (In Danish.)

Department of Forest Sciences. Undated. Agricultural University of Norway, Ås. 2 pp.

Driver, B.L. & Peterson, G.L. 1992. Evaluation of the Multiple–use Research Program of the Finnish Forest Research Institute. The Finnish Forest Research Institute, Research Papers 438. 71 pp.

Eckerberg, K. 1990. Environmental protection in Swedish forestry. Avebury studies in green research, Aldershot. 179 pp.

EFI News, Vol 2. 1994. European Forest Institute, Joensuu. p. 5, 11.

Elmberg, J. & Sallnäs, O. 1993. Förslag till framtida programområden för forskning vid Sveriges lantbruksuniversitets skogsvetenskapliga fakultet. (Propositions for future research program areas in the Faculty of Forestry of the Swedish University of Agricultural Sciences.) Manuscript. 11 pp. (In Swedish.)

Environmental policies in Finland. 1988. OECD, Paris. 230 pp.

Environmental protection in Finland: national report. 1988. Ministry of the Environment, Environmental Protection Department, Series A 67/1988. 361 pp.

Eriksson, L., Kardell, L. & Ingelög, T. 1979. Blåbär, lingon, hallon: förekomst och bärproduktion i Sverige 1974–1977. Summary: Bilberry, lingonberry, raspberry: occurence and production in Sweden 1974–1977. Sveriges lantbruksuniversitet, Avdelningen för Landskapsvård, Rapport 16. 124 pp.

The Faculty of Forestry: activities and development. 1991. Swedish University of Agricultural Sciences. 24 pp.

Falck, J. 1992. New aspects of multiple use in education and research at the Department of Silviculture in Sweden. Norwegian Journal of Agricultural Sciences, Supplement 8: 5–29.

Formål, organisasjon, ledelse, arbeidsform. (Objectives, organization, leadership, way of working.) 1994. Norges landbrukshøgskole, Institutt for Skogfag, Ås. 5 pp. (In Norwegian.)

Forskning for felleskapet. (Research and common interests.) 1992/93. Kirke–, utdannings– og forskningsdepartementet, St. meld. nr. 36. 167 pp. (In Norwegian.)

Forskning og formidling: skovbrug. (Research and extension: forestry.) 1989. Landbrugsministeriet, Rapport fra en arbejdsgruppe. 28 pp. (In Danish.)

Forskningscentret for Skov & Landskab. (Danish Forest and Landscape Research Institute.) Undated. Folder. 4 pp.

Gornitzka, N. 1992. Organizational structures of science in Norway: a system in transition. Norges allmenvitenskaplige forskningsråd. 9 pp.

Haakenstad, H. 1972. Skogbehandling i et utfartsområde: en opinionsundersøkelse om Oslomarka. Summary: Forest management in an area of outdoor life: an investigation of public opinion about Oslomarka. Meldinger fra Norges landbrukshøgskole 51(16): 1–79.

Hallikainen, V. & Jokimäki, J. 1992. Suomen erämaatutkimusohjelma 1993–1996. Summary: Finland's wilderness area research program 1993–96. Arktisen keskuksen tiedotteita 7. 72 pp.

Helles, F. 1977. Om teorin bag flersidig produktion i skovbruget. (The theory behind multiple production in forestry.) Dansk Skovforenings Tidsskrift 62(3): 179–198. (In Danish.)

Helles, F., Jensen, S.F. & Risvand, J. 1984. Den danske skovsektors samfundsmæssige betydning. (The socioeconomic importance of Danish forestry sector.) DSR's forlag, København. 230 pp. (In Danish.)

Helles, F. & Linddal, M. (eds.). 1994. Proceedings of the biennal meeting of the Scandinavian Society of Forest Economics. Gilleleje, Denmark, November 1993. Scandinavian Forest Economics 35. 426 pp.

Hirkjølen: flersidig bruk av fjellskog. (Hirkjølen: multiple use in a mountain forest.) 1993. Norsk Insitutt for Skogforskning, Ås. 28 pp. (In Norwegian.)

Hofstad, O. 1976a. Economic and sociological analysis as a basis of the multiple use planning for Oslomarka. Proceedings of the XVI IUFRO World Congress, Division IV. p. 143–152.

Hofstad, O. 1976b. Konflikter ved flersidig bruk av skog. (Conflicts in multiple–use of forests.) Dissertation. Agricultural University of Norway, Department of Forest Economics. 173 pp. (In Norwegian.)

Hultman, S.–G. 1983a. Allmänhetens bedomning av skogsmiljöers lämplighet för friluftsliv. 1. Bedömning på plats eller i bild? Summary: Public judgement of forest environments as recreation areas. 1. Judgement on site or from photos. Sveriges lantbruksuniversitet, Avdelningen för landskapsvård, Rapport 27. 97 pp.

Hultman, S.–G. 1983b. Allmänhetens bedomning av skogsmiljöers lämplighet för friluftsliv. 2. En rikstäckande enkät. Summary: Public judgement of forest environments as recreation areas. 2. A national survey. Sveriges lantbruksuniversitet, Avdelningen för landskapsvård, Rapport 28. 91 pp.

Hytönen, M. 1992. Metsien monikäytön tutkimus Suomessa 1970–1990: tiivistelmäbibliografia. Multiple–use forestry research in Finland 1970–1990: an annotated bibliography. Metsäntutkimuslaitoksen tiedonantoja 430. 395 pp. (Partly in English.)

Hytönen, M. 1995. Origins, evolution and significance of the multiple–use concept. In: Hytönen, M. (ed.). Multiple–use forestry in the Nordic countries. The Finnish Forest Research Institute, Helsinki. p. 43–66.

Illi, A. 1994. European Forest Institute studies Europe's forests and forest policy. Paperi ja puu/Paper and Timber 76(1): 1–2.

Jäppinen, J.–P. & Väisänen, R. 1993. Luonnon monimuotoisuuden tutkimusohjelma LUMO. (Biodiversity Research Programme LUMO.) Väliraportti 31.5.1993. Vesi– ja ympäristöhallituksen monistesarja Nro 441. 113 pp. (In Finnish.)

Jeppesen, J.L. 1984. Human disturbance of roe deer and red deer: preliminary results. In: Saastamoinen, O., Hultman, S.–G., Koch N.E. & Mattsson, L. (eds.). 1984. Multiple–use forestry in the Scandinavian countries. Communicationes Instituti Forestalis Fenniae 120: 113–118.

Kangas, J. 1992. Metsikön uudistamisketjun valinta: monitavoitteiseen hyötyteoriaan perustuva päätöksentekomalli. Summary: Choosing the regeneration chain in a forest stand: a decision analysis model based on multi–attribute utility theory. University of Joensuu, Publications in Sciences 24. 230 pp.

Koch, N.E. 1978. Skovenes friluftsfunktion i Danmark. I del. Befolkningens anvendelse af landets skove. Summary: Forest recreation in Denmark. Part I: The use of the country's forests by the population. Det Forstlige Forsøgsvæsen i Danmark, vol. 35: 285–451.

Koch, N.E. 1980. Skovenes friluftsfunktion i Danmark. II del. Anvendelsen af skovene, regionalt betragtet. Summary: Forest recreation in Denmark. Part II: The use of forests considered regionally. Det Forstlige Forsøgsvæsen i Danmark, vol. 37: 73–383.

Koch, N.E. 1984. Skovenes friluftsfunktion i Danmark. III del. Anvendelsen af skovene, lokalt betragtet. Summary: Forest recreation in Denmark. Part III: The use of forests considered locally. Det Forstlige Forsøgsvæsen i Danmark, vol. 39: 121–362.

Koch, N.E. 1984a. Project "Forest and Folk": a summary. In: Saastamoinen, O., Hultman S.–G., Koch, N.E. & Mattsson, L. (eds.). Multiple–use forestry in the Scandinavian countries. Communicationes Instituti Forestalis Fenniae 120: 44–45.

Koch, N.E. 1984b. Multiple–use forestry: Danish statistics. In: Saastamoinen, O., Hultman, S.–G., Koch N.E. & Mattsson, L. (eds.). Multiple–use forestry in the Scandinavian countries. Communicationes Instituti Forestalis Fenniae 120: 33–38.

Koch, N.E. & Jensen, F.S. 1988. Skovenes friluftsfunktion i Danmark. IV del. Befolkningens ønsker til skovenes og det åbne lands udformning. Summary: Forest recreation in Denmark. Part IV: Preferences of the population. Det Forstlige Forsøgsvæsen i Danmark, vol. 41: 243–516.

Koch, N.E. & Kennedy J.J. 1991. Multiple–use forestry for social values. Ambio 20(7): 330–333.

Koch, N.E. & Kristiansen, L. 1991. Flersidigt skovbrug: et idékatalog. (Multiple–use forestry: a handbook of ideas.) Skov– og Naturstyrelsen, Hørsholm. 39 pp. (In Danish.)

Kristiansen, L. 1994. Planlægning af flersidig skovdrift. (Planning of multiple–use forestry.) Forskningscentret for Skov & Landskab, Lyngby. Manuscript. (In Danish.)

Krott, M. & Zimmerman, W. 1992. From deforestation to air pollution: new challenges for sustained yield policy. In: Gundermann, E. (ed.). Sustained yield: historical, economic and political aspects. Proceedings of interdivisional and divisional sessions of Division 6 and 4. IUFRO Centennial Berlin–Eberswalde, Germany 31.8–4.9.1992.

Kuusipalo, J., Reunala, A., Salo, K. & Sepponen, P. 1989. Metsien monikäytön tutkimusohjelma: kokonaissuunnitelma 1990–1994. (Multiple–use research program: plan for 1990–1994.) Metsäntutkimuslaitos, Joensuu. 47 pp. (In Finnish.)

Loftsson, J. 1993. Forest development in Iceland. In: Alden, J., Mastrantonio, J.L. & Odum, S. (eds.). Forest development in cold climates. Plenum Press, New York. p. 453–461.

Mattsson, L., Petrini, F., Johansson, P.–O, Löfgren, K.–G. & Mäler, K.–G. 1991. Naturresurs– och miljöekonomi i samband med areella näringar: ett förslag till forskningsprogram. (Natural resources and environmental economics in connection with livelihoods: a proposal for a research program.) 23 pp. (In Swedish.)

Metsä 2000 –ohjelman tarkistustoimikunnan mietintö. (Report of the revising committee of Forest 2000 Programme.) 1992. Maa– ja metsätalousministeriö, Komiteamietintö 1992:5. 112 pp. (In Finnish.)

Metsien moninaiskäytön työryhmän raportti. (Report of the multiple–use forestry working group.) 1985. Talousneuvosto, Metsä 2000–ohjelmajaosto, Helsinki. 59 pp. (In Finnish.)

Moen, H.O. 1992. Norsk institutt for skogforskning: organisasjon, ledelse og samarbeid. (Norwegian Forest Research Institute: organization, leadership and cooperation.) In: NISK: Norsk institutt for skogforskning: 75 år. (NISK: Norwegian Forest Research Institute: 75 years). NISK, Rapport fra Skogforsk 12. p. 5–13. (In Norwegian.)

Naskali, A., Ovaskainen, V. & Sepponen, P. 1993. Ympäristötaloustiede monikäytön tutkimusohjelmassa. (Environmental economics in the Multiple–use Research Program.) Metsäntutkimuslaitos, Rovaniemi. 30 pp. (In Finnish.)

Naturen i skoven. (Nature in forest.) 1989. Skov– og Naturstyrelsen, Hørsholm. (In Danish.)

Naturpleje i skov. (Nature management in forest.) 1989. Skov– og Naturstyrelsen, Hørsholm. (In Danish.)

NINA: Norwegian Institute for Nature Research. Undated. NINA, Trondheim. 10 pp.

Norwegian Forest Research Institute. 1994. Norwegian Forest Research Institute, Ås. 14 pp. (Folder.)

Norwegian Institute for Nature Research. 1993. Norwegian Institute for Nature Research, Trondheim. 15 pp.

Økt verdiskapning fra skog. Skogbruk og skogindustri: perspektiver, utfordninger og områder for forskning. (Increased value from the forest. Forestry and forest industry: perspectives, challenges and topics for research.) 1994. Norges Forskningsråd, Oslo. 40 pp. (In Norwegian.)

Opheim, T. 1984. Notes on the Oslomarka. In: Saastamoinen, O., Hultman, S.–G., Koch N.E. & Mattsson, L. (eds.). Multiple–use forestry in the Scandinavian countries. Communicationes Instituti Forestalis Fenniae 120: 39–43.

Opheim, T. 1992. The demand on research in forest economics from the policy makers' point of view. Proceedings of the biennal meeting of the Scandinavian Society of Forest Economics, Gausdal, Norway, April 1991. Scandinavian Forest Economics 33: 5–14.

Prosjektkvalitet ved SkogForsk. (Quality of projects in SkogForsk.) 1993. Norsk institutt for skogforskning/Institutt for skogfag, Ås. 32 pp. (In Norwegian.)

The representation of the revised Forest 2000 Program: Finnish forest policy in the 1990's.1992. Finnish Forestry Association, Helsinki. 26 pp.

Reunala, A. & Heikinheimo, M. 1987. Taistelu metsistä: voimaperäinen metsätalous Suomessa ja muissa maissa. (The fight for the forests: intensive forestry in Finland and other countries.) Kirjayhtymä, Helsinki. 188 pp. (In Finnish.)

Saastamoinen, O. 1982. Economics of multiple–use forestry in the Saariselkä forest and fell area. Communicationes Instituti Forestalis Fenniae 104. 102 pp.

Saastamoinen, O. 1987. Multiple use and the Forest 2000 –programme. In: Hänninen, R. & Selby, A. (eds.). Proceedings of the biennal meeting of the Scandinavian Society of Forest Economics, Porvoo, Finland, May 1987. Scandinavian Forest Economics 29: 39–47.

von Segebaden, G. 1978. Förord. (Foreword.) In: Skogshögskolan 150 år: problem och idéer i svensk skogbruk 1828–1978. (College of Forestry 150 years: problems and ideas in Swedish forestry 1828–1978.) Sveriges lantbruksuniversitet, Almänna skrifter nr 2. (In Swedish.)

Sepponen, P. & Lohiniva, S. 1994. Luontomatkailu: uusi tutkimuskohde. (Nature tourism: a new research object.) Metsäntutkimuslaitoksen tiedonantoja 488: 113–122. (In Finnish.)

Skogforsk. 1993. Skogforsk, Redogörelse 8. Back page information.

Skogøkologi og flersidig skogbruk: et tverrfaglig forskningsprogram. (Forest ecology and multiple–use forestry: a multidisciplinary research program.) 1992. Norsk institutt for skogforskning, Ås. 4 pp. (In Norwegian.)

Skogsägarnas Riksförbund: näringspolitisk bokslut. (The National Federation of Forest Owners: business policy summary.) 1994. In: Skogsåret 1993–94. (Forest year 1993–94.) Skogsägarnas Riksförbund, Stockholm. p. 82–85. (In Swedish.)

Skogspolitiken inför 2000 –talet: huvudbetänkande av 1990 års skogspolitiska kommitté. (Forest policy for the 21st century: main report of the 1990 forest policy committee.) 1992. Jordbruksdepartmentet, Statens offentliga utredningar, SOU 1992:76. 343 pp. (In Swedish.)

Solbraa, K. 1990. Hirkjølen Forsøksområde. Summary: Hirkjølen Experimental Area. Norsk institutt for skogforskning, Rapport 7. 27 pp.

Strand, H. 1967. Økonomiske synspunkter på rekreasjon. (Recreation from an economic viewpoint). Meddelelser fra det Norske Skogforsøksvesen 22(84): 161–187. (In Norwegian.)

Strategi for forskning ved Institut for Økonomi, Skov og Landskab. (Strategy for the research in the Institute of Economy, Forest and Landscape.) 1993. Den Kgl. Veterinær– of Landbohøjskole, Frederiksberg. 41 pp. (In Danish.)

Strategisk plan 1993–1996. (Strategic plan 1993–1996.) 1993. Norsk Insitutt for Skogforskning, Ås. 10 pp. (In Norwegian.)

Svendsrud, A. 1977. Utnytting av et skogsområde til konkurrerende formål. (Utilization of a forested area for competing purposes.) Norges landbrukshøgskole, Ås. 24 pp. (In Norwegian.)

Svensk skogsforskning mot år 2000. (Swedish forestry research towards the year 2000.) 1989. Sveriges lantbruksuniversitet, Skogsvetenskapliga fakulteten, Rapport 8. 177 pp. (In Swedish.)

Thomsen, A. 1988. Skovens fortid i fremtiden. (The past of forest in future.) Skov– og Naturstyrelsen, Hørsholm. (In Danish.)

Virkomhetsplan for 1994. (Action plan for 1994.) 1994. Norges landbrukshøgskole, Institutt for Skogfag, Ås. 21 pp. (In Norwegian.)

Map of the Nordic Countries

(Denmark's Greenland and Faroe Islands, and Norway's Svalbard and Jan Mayen islands are not shown on the map)

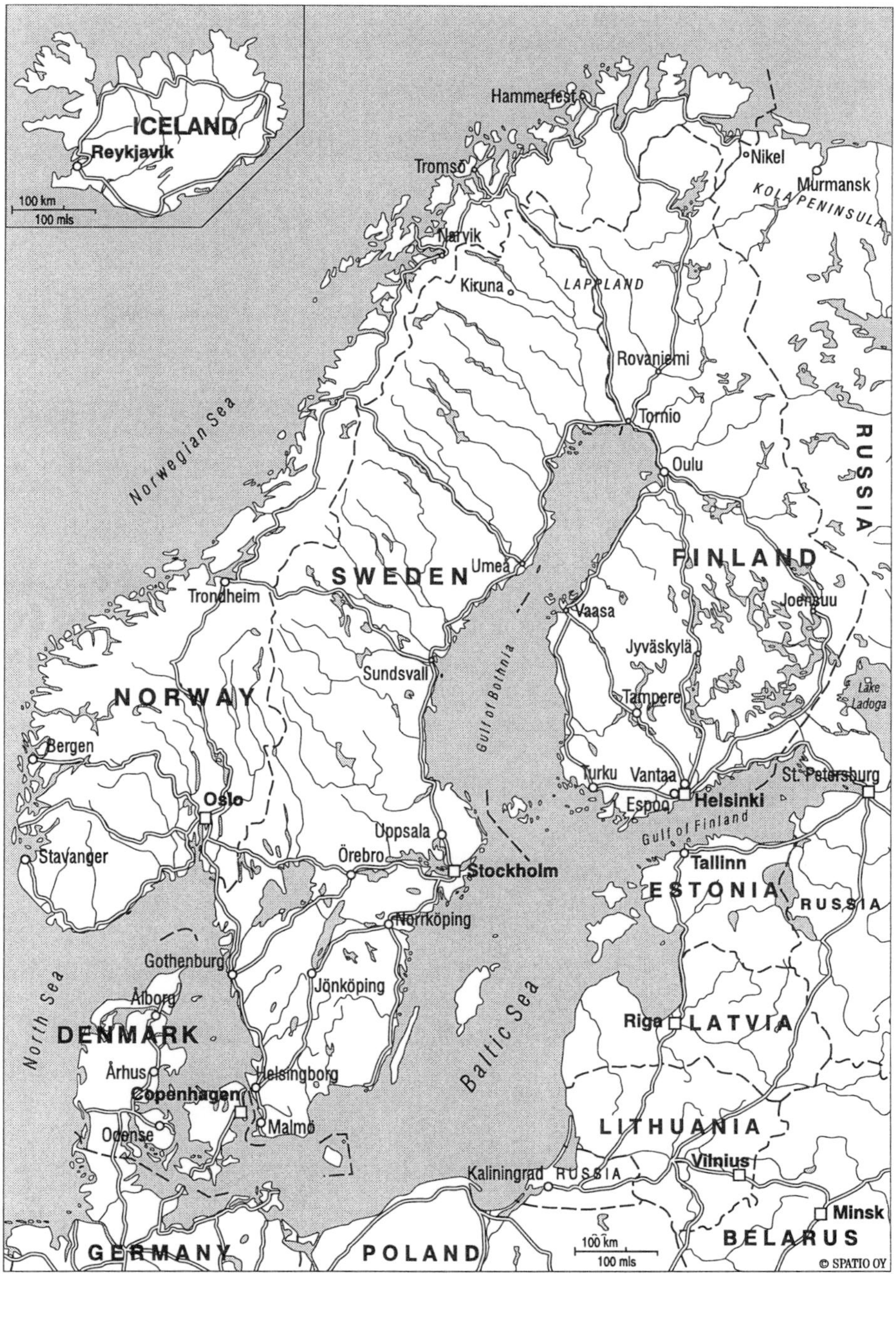

Statistics

A. Area and population

	Denmark	Finland	Iceland	Norway	Sweden
Total area, km^2	43 094	338 145	103 000	323 877	449 964
Population, 1000	5 162	4 998	262	4 299	8 585
Persons/km^2	120	16	3	14	21
Rural population, %	15	20	9	27	17

Source: Yearbook of Nordic statistics 1994. Nordic Council of Ministers, Nord 1994:1.

B. Structure of land use, %.

	Denmark	Finland	Iceland	Norway	Sweden
Forests	11	66	-	28	60
Other wooded land	-	11	1	3	9
Agricultural land	69	10	2	3	9
Other land	20	13	97	65	22

Source: The forest resources of the temperate zones: the UN–ECE/FAO 1990 forest resource assessment. Volume I. General forest resource information. 1992. United Nations, New York.

C. Forest resources

	Denmark	Finland	Iceland	Norway	Sweden
Forest, 1000 ha	466	20 112	11	8 697	24 437
Other wooded land, 1000 ha	-	3 261	123	868	3 578
Forest and other wooded land, ha/person	0.09	4.68	0.54	2.26	3.27

Source: The forest resources of the temperate zones: the UN–ECE/FAO 1990 forest resource assessment. Volume I. General forest resource information. 1992. United Nations, New York.

D. Tree species distribution, % of growing stock.

	Denmark	Finland	Norway	Sweden
Coniferous	52	82	80	84
Broadleaved	48	18	20	16

Source: The forest resources of the temperate zones: the UN–ECE/FAO 1990 forest resource assessment. Volume I. General forest resource information. 1992. United Nations, New York.

E. Productive forest area by owner groups, %.

	Denmark	Finland	Norway	Sweden
Public				
- state	26	24	9	18
- other public	12	5	6	8
Private				
- companies	16	9	4	24
- other	46	63	81	50

Source: Yearbook of Nordic statistics 1994. Nordic Council of Ministers, Nord 1994:1.

F. Forest products in the economy, million US $ in 1991.

	Denmark	Finland	Norway	Sweden
Production	584	8 959	2 400	10 756
% of GDP	-	7	2	4
Imports	1 766	512	708	1 168
Exports	417	8 238	1 517	9 873
% of trade	1	36	4	18

Source: Forestry statistics today and tomorrow 1961–1991...2010. 1993. FAO, Rome.

Subject index

F

G

H

Q

R

S